AF324298

BIOPHYSICS

Second Edition

BIOPHYSICS

Second Edition

Vasantha Pattabhi
N. Gautham

Alpha Science International Ltd.

Oxford, U.K.

Vasántha Pattabhi
N. Gautham
Centre of Advanced Study in Crystallography and Biophysics
University of Madras
Chennai, India

Copyright © 2001, 2009
First Edition 2001
Second Edition 2009
Reprint 2010

ALPHA SCIENCE INTERNATIONAL LTD.

7200 The Quorum, Oxford Business Park North
Garsington Road, Oxford OX4 2JZ, U.K.

www.alphasci.com

Printed from the camera-ready copy provided by the Authors.

ISBN 978-1-84265-517-7

Printed in India

This book is dedicated to
Late Professor G.N. Ramachandran, FRS,
whose life and work continues to inspire generations of biophysicists

எப்பொருள் யார்யார்வாய்க் கேட்பினும் அப்பொருள்
மெய்ப்பொருள் காண்பது அறிவு

குறள் 423

Wisdom lies in grasping the truth in whatever and by whomever said

Kural 423

Preface to the Second Edition

It is deeply gratifying that the response to the first edition of this book has been so good. Apart from indications to this effect from the raw sales figures, which, we are told, are impressive in themselves, the book has found its way into the reference lists of the syllabi of numerous courses in numerous universities. We have also come across a pdf copy of the book available for free download on the Internet, testifying to its popularity, albeit not contributing to the royalties!

In the eight years since the first edition was published, biophysics has developed rapidly, both in terms of techniques and instrumentation, as well in terms of new biophysical knowledge. Thus, for example, single-molecule technology is allowing the study of nano-scale biological materials at a level of detail hitherto only dreamt of. And computation now plays an increasingly large role in biological and hence also biophysical research. However, these advances have been at the forefront of the subject, while this book addresses itself to the fundamentals, which have remained largely unaltered. There have been no dramatic reversions of well-accepted theories, and the advances have been evolutionary, rather than revolutionary. Thus the subject matter in the second edition is not largely different from that in the first; indeed, the paragraphs on DNA nano-structures and on mass spectrometry are the only totally new material. Besides this, however, the entire text has been carefully reviewed to improve readability, and the formatting has been changed, hopefully making it more reader-friendly. Some figures have been redrawn, a few new ones introduced, and a few redundant ones deleted. All-in-all, we hope these changes will help the second edition to repeat the success of the first, if not surpass it.

As for the first edition, many people have contributed to the making of this one. Students are the main users of the book, and many of them have come forth to criticize aspects of it they thought could be better. We have included as many of such suggestions as we could. Dr. R. Malathi of our department spent a lot of time redoing some of the figures. We are grateful for her help. Our publishers have, as usual, been very patient with us, and for this, our thanks. Finally, and most importantly, our families have once again had to put up with our varying moods; and for this our deepest gratitude goes to Pattabhi, Selvi and Chitra.

Chennai, India

Vasantha Pattabhi
N. Gautham

Preface to the First Edition

The initiation of biophysics in India is synonymous with the starting of the Department of Crystallography and Biophysics at the University of Madras, in 1953, under the leadership of Professor G. N. Ramachandran. Even today, nearly half-century later, there are only a handful of Universities in the country which have full-fledged biophysics departments. However, most life science curricula in the country now comprise a course in biophysics, owing to its increasing importance and relevance to the study of any biological system. This book has been written to fill a long-standing lacuna of textbooks in biophysics, suitable for students at the senior undergraduate and postgraduate level in the various biological disciplines, such as biochemistry, molecular biology and even medicine.

Biophysics is an interdisciplinary subject, and its treatment varies with the background of the both the author and the reader. For example, some aspects of theoretical biophysics are indistinguishable from pure mathematics. On the other hand, modern biophysics has made profound contributions to subjects normally considered a part of classical biology, such as evolution. In this book we have not assumed any knowledge of higher mathematics on part of the reader as a prerequisite. We have emphasized an understanding of the underlying physical principles of the biological phenomena. Many of the explanations may therefore appear too elementary and too verbose to sophisticated readers. We believe however that the 'great divide' between students of the physical sciences who are mathematically fluent, and those of the biological sciences who are less so, is too deep to be bridged with one easy leap in a single textbook. Thus the present book does not set out to make biophysicists out of all biologists. However we believe it will cater to the needs of most biologists, who, especially in India, stop their acquaintance with mathematics at school. In addition, chemists and physicists may also find it useful as an introduction to the subject.

The choice of material covered in the book is based on an one semester course entitled 'Introductory Biophysics' offered mainly to students from the School of Biological Sciences, University of Madras. The teaching notes of the authors form the skeleton of the book and this was fleshed out on the basis of several years of classroom experience.

Chapter 1 gives an introduction to laws of physics and chemistry and is designed to reinforce and enhance slightly the school-level understanding of the subject, which students with a background of purely biological science would have. This chapter covers quantum mechanics, molecular orbital theory, essentials of weak and strong molecular interactions, stereochemistry, fundamentals of thermodynamics and radiation biophysics.

Chapters 2 and 3 deal with separation techniques and physico-chemical techniques used to study biomolecular structure. Strictly speaking this would be counted under physical biochemistry, rather than purely biophysics. The coverage is not exhaustive and is intended to give an overview of various techniques available to unravel the properties of biomolecules.

Chapters 4 to 9 describe some of the spectroscopy, microscopy, diffraction and computational techniques. These techniques are so powerful that every biophysicist must be armed with this knowledge so that it can be put to use when the need arises. These chapters aim at providing an understanding of the basic principles involved rather than thorough knowledge of the theory and practice of the techniques. The descriptions would enable most biologists to obtain an understanding of the relative merits and demerits of each technique, in order that the biochemical results, which one may come across in the general literature, become intelligible. X-ray crystallography and NMR spectroscopy have been dealt with in greater detail owing to their prominence in determining the biomolecular structures.

Results obtained from the application of the techniques of structural biology are reviewed in Chapter 10. This chapter is illustrated by the largest number of diagrams. Nevertheless, in a book of this nature, in which one is always trying to keep costs low, it is not possible to include every picture the authors may feel necessary. Also we would have dearly liked to include colour illustrations.

The later chapters try and fill the gaps between biophysicists and biochemists. The treatment of Energy Pathways in Chapter 11 involves a lot of physical chemistry. Biomechanics in Chapter 12 and Neurophysics in Chapter 13 involve a lot of physiology. Though the advancements in the application of physical principles in these fields has been phenomenal they are probably too advanced and too detailed for the intended audience of this book.

We have of course referred to several books and journal articles, review papers, etc, while writing this book. At the end of the book we have included a chapter-wise list for further reading, which would amplify and extend the contents of each chapter. This list is however neither exhaustive nor comprehensive. We have as far as possible selected books that would be available in most University libraries in the country.

This book was written as a project sponsored and funded by the University Grants Commission, India. We take this opportunity to thank the UGC and all its officers for the invitation and for support, financial and otherwise. We also thank the reviewers, whose corrections and suggestions we have tried to incorporate. Any errors that remain are, of course, ours. Our M.Sc., M.Phil and Ph.D. students of the past several years have taught us or have forced us to learn much of the material presented in this book. We thank them all most sincerely. For actual help in preparing the book, we thank Mr A. Johnson and Mr. Prem Raj Bernard. We would like to specially thank our respective family members for their enthusiastic support throughout. Writing this book has been a rewarding experience. But as we look back over it, we feel, with Thomas Hardy, 'the more written, the more it seems remains to be written'.

Chennai, India   **Vasantha Pattabhi**
                                                                       N. Gautham

Contents

3. Physico-Chemical Techniques to Study Biomolecules

4. Spectroscopy

5. Light Microscopy

9. Molecular Modelling   151

10. Macromolecular Structure   162

11. Energy Pathways in Biology

12. Biomechanics

13. Neurobiophysics

Chapter 1

Laws of Physics and Chemistry

1.1: Introduction

Physics is an exact science. Biology, on the other hand, has long been considered a descriptive science. By this we mean that in physics, numbers and their measurement play a crucial role. Physical theories may stand, or more precisely, fall on the basis of a single experimental observation. As a famous example, the accurate measurement of the speed of light along, and perpendicular to, the direction of motion of the earth by Michelson and Morley in 1887 showed that the speed did not depend on the relative motion of the observer, and thus laid to rest the theory that electromagnetic radiation was propagated through some material substance called 'ether'. In biology, a single contrary example does not usually spell the downfall of a theory. Likewise, a mass of data needs to be accumulated before a trend is guessed at and a theory formulated. However, in recent times, especially with the advent of molecular biology, molecular biophysics, computational biology and the like, biology also is being transformed into a more exact science. One aspect of this transformation originates from the applications of physics to physiology. Physical laws or concepts such as mechanics, hydrodynamics, optics, electrodynamics and thermodynamics are used to explain physiological observations like muscle contraction, neural communication, vision, etc. A second and more fundamental aspect of the transformation emerged from the search for universal principles governing the world around us. Almost from the time man was able to think he has been wondering as to what makes him and the living world distinct from the rest of the environment. By the beginning of twentieth century it was realized that the laws of physics and chemistry, which were applied to non-living things, could equally well explain forces controlling biology, and that no new fundamentally different principles were necessary to explain the organisms and interactions which make up the living world. In the hundred years that have passed since then, this concept has become stronger and today nobody thinks it necessary to invoke any special physical or chemical forces or laws in the study of biology.

Modern biology takes as its starting point this principle that biological systems are in no essential respect different from physical or chemical systems, though they are much more complex than the latter. All the laws of physics and chemistry that are used to describe, say, planetary systems or chemical reactions in a test tube, may be used to describe how a single living cell functions and how it could grow into a complex multicellular organism. Furthermore, and this is the crucial point, no *new* laws are required, laws that may be applicable to living systems alone and not to non-living matter. To restate this important point, biology is an extension of physics and chemistry, albeit far more complex. Before the advent of modern molecular biology, approximately half a century ago, biological systems were thought to be based upon an unknown principle that set them apart from non-living matter. This principle was known by the name *vis vitalis* in Latin (or *prana* in Sanskrit, or *chi* in Chinese). However, even at this time, questions regarding the exact nature of elementary life forms such as viruses blurred the boundaries between Life and Non-life. Darwinian principles of evolution, together with related theories on prebiotic systems showed it possible to think of Life as consisting of complex interactions among inert chemicals. This realisation led to the 'reductionist programme' in biology. According to this programme, biological systems are studied as component parts. These parts are subjected to analysis by physical and chemical techniques and theories. The understanding of the function of each separate portion so gained is put together, bit by bit, to build an understanding of the entire biological system.

The reductionist programme begins by reducing a living organism to its component tissues and organs. These are made of differentiated cells, and the next reductionist step is to understand cell function. A cell is a collection of a large number of molecules that are in intimate interaction with each other. To understand cell function therefore one needs to understand these molecules and their interactions. Molecular biology, as the name indicates, works out biological function on the basis of interactions between molecules. Some of these molecules are found also in non-living matter. But most are specific to biological systems. The macromolecules, in particular, belong to this category, namely proteins, DNA and polysaccharides. Another reductionist path also exists from organism to molecule. Here the emphasis is on function. A living system is characterised by many specific functions such as reproduction, metabolism, motion, etc. One may study each of these functions in molecular terms and put them together to obtain a molecular explanation of the entire biological system. In either case the essential idea is the same, and is expressed by the phrase 'the whole is the sum of its parts'. It may be noted that this epigram is not by any means a proven idea, and future studies may overturn this dogmatic representation of the living state. However, so far, the programme has been enormously successful, and there does seem to be any end to it in view. The development of the discipline of biophysics at the interface between biology and the physical sciences owes much to this programme.

In this chapter we will describe in an elementary way some of the basic concepts and laws of modern physics and chemistry, in particular those that are directly relevant to modern biology.

1.2: Quantum mechanics

By the end of the nineteenth century, physical science had arrived at a description of the external world in terms of two distinct types of entities. On the one hand were particles and particulate matter to which the laws of mechanics, as enunciated by Newton and developed by several others, were wholly applicable. Concepts such as mass, momentum, kinetic energy, etc. were descriptive of particles. On the other hand were radiation and other wave like phenomena to which the laws of wave propagation and other optical laws were applicable. Phenomena such as diffraction and interference could be explained in terms of the wave-like picture.

It was Max Planck who first suggested that there need not be a complete separation between the concept of a particle and the concept of a wave. In order to explain the observed results when the intensity of radiation from a perfectly radiating body (i.e. a so-called 'black body') was measured as a function of the wavelength, Planck proposed that electromagnetic radiation was emitted or absorbed not continuously but in small packets, or quanta. The energy contained within each quantum was given by the expression

$$E = h\nu$$

where ν is the frequency of the radiation and h is a universal constant called Planck's constant (h = 6.625×10^{-34} joule seconds). The black body could only absorb or emit integral multiples of this quantity, never any fraction of a quantum. Einstein extended this idea to say that not only is radiation absorbed or emitted in packets but it also propagates only as packets, called photons, each containing an amount of energy given by the Planck expression. Thus, radiation, which was thought to be wavelike, now could be treated as a particle and the same entity could both diffract as well as possess momentum. The French scientist De Broglie suggested that the same type of wave-particle duality could be applied to entities hitherto thought to be particles. Thus, he said, electrons and other sub-atomic particles could be thought of also as wave packets, with a wavelength given by

$$\lambda = h/p$$

where p is the momentum of the particle and h is again Planck's constant. In fact this equation is applicable not only to sub-atomic particles but to all matter. However Planck's constant is so small that the wave nature of matter becomes observable only when the mass of the particle, and hence its momentum, is also very small. When this happens, then according to De Broglie's theory, a particle such as an electron would not only possess mass and momentum but would also show diffraction and other wave-like effects. Hundreds of thousands of experiments performed all over the world since the proposals were made have firmly established the truth of the wave-particle duality of the physical world. Quantum mechanics is now accepted as the only correct description of physics. However for practical calculations on macroscopic objects like cricket balls and space rockets it is sufficient to use Newton's laws, which are now considered a special, macroscopic case the of quantum description. For calculations of atomic level phenomena we use quantum mechanics, chiefly as developed by Schrodinger and Heisenberg.

Schrodinger introduced a wave equation that replaces the classical equations of motion. The solution to this wave equation is a 'wave function', which is a function of space and time. The physical state of any system such as an electron (or an atom or a molecule or a system of molecules) can be described completely by the corresponding wave function. All physically meaningful and measurable quantities such as the energy, the momentum, the position and so on can be extracted from the wave function. The wave function is usually represented as Ψ and is obtained as a solution of the following 'eigenvalue' equation

$$H\Psi = E\Psi$$

where E is the characteristic energy of the system or the energy 'eigenvalue' and H is the operator called the Hamiltonian operator, which operates upon the characteristic wave function or the 'eigenfunction' Ψ to yield the energy values. A physical way of understanding Ψ is as a probability amplitude. In other words, since Ψ is a function of the coordinates x, y and z and the time t, $|\Psi|^2$ gives the probability of finding the system at the point x, y, z at the time t. Written explicitly, the

$$\frac{d^2\psi}{dx^2} - \frac{8\pi^2 m}{h}(E - V)\psi = 0$$

Schrodinger wave equation has the following form.

Here m is the mass of the particle and V is the expression for the potential energy. The other symbols

$$\frac{d^2\psi}{dx^2} + \frac{d^2\psi}{dy^2} + \frac{d^2\psi}{dz^2} - \frac{8\pi^2 m}{h}(E - V)\psi = 0$$

have been explained above. This is a one-dimensional equation, independent of time. To solve this equation one has to, first of all, define the appropriate expression for the potential energy V, which will depend on the problem studied. When this expression is inserted into the Schrodinger wave equation, the differential equation so obtained can be solved to find Ψ and E. In three dimensions the Schrodinger equation becomes

The time dependent form of the equation is as follows.

where i is the imaginary root.

This scheme of calculations involving the wave equation is called wave mechanics. It is

$$-\frac{h^2}{8\pi^2 m}\left(\frac{d^2\psi}{dx^2} + \frac{d^2\psi}{dy^2} + \frac{d^2\psi}{dz^2}\right) = \frac{ih}{2\pi}\frac{d\psi}{dt}$$

complemented by another, equivalent scheme developed by Heisenberg and Born, and called matrix mechanics. The starting point for these calculations was the discovery made by Heisenberg that there exists a physical limit to the accuracy with which one can simultaneously determine both the position and the velocity of a particle. This is due to the fact that any measurement necessarily implies that the particle under observation is being subjected to a disturbance, at the very least by a photon of light. If the particle is small, then the act of observation itself leads to uncertainty in the position. (If the particle were large, then since it is extended it cannot be said to possess a very precise location). Heisenberg stated these facts in the form of the Uncertainty Principle, which declares that no two canonically conjugate observables can be measured simultaneously with arbitrary precision. Position and Momentum are canonically conjugate variables. So are Energy and Time. Another such pair is Angular Momentum and Angular Position.

Both matrix mechanics and wave mechanics may be used to arrive at the physically meaningful and measurable variables that characterize any physical system. Which scheme of calculation is to be used depends on the actual information that is sought. The most spectacular and immediate success of quantum mechanics was in the theoretical description of the structure of atoms.

1.3: The electronic structure of atoms

Rutherford's model of the atom consisted of a central nucleus with a positive electric charge Ze, and Z electrons each with a negative electric charge of value e revolving around it. Z is the atomic number of the atom. In order to resolve certain inconsistencies in this picture, Bohr proposed an altered model, which was later shown by quantum mechanical calculations to be the correct one. Bohr postulated that the electrons, which revolve round the central nucleus, occupy only certain fixed orbits. These orbits

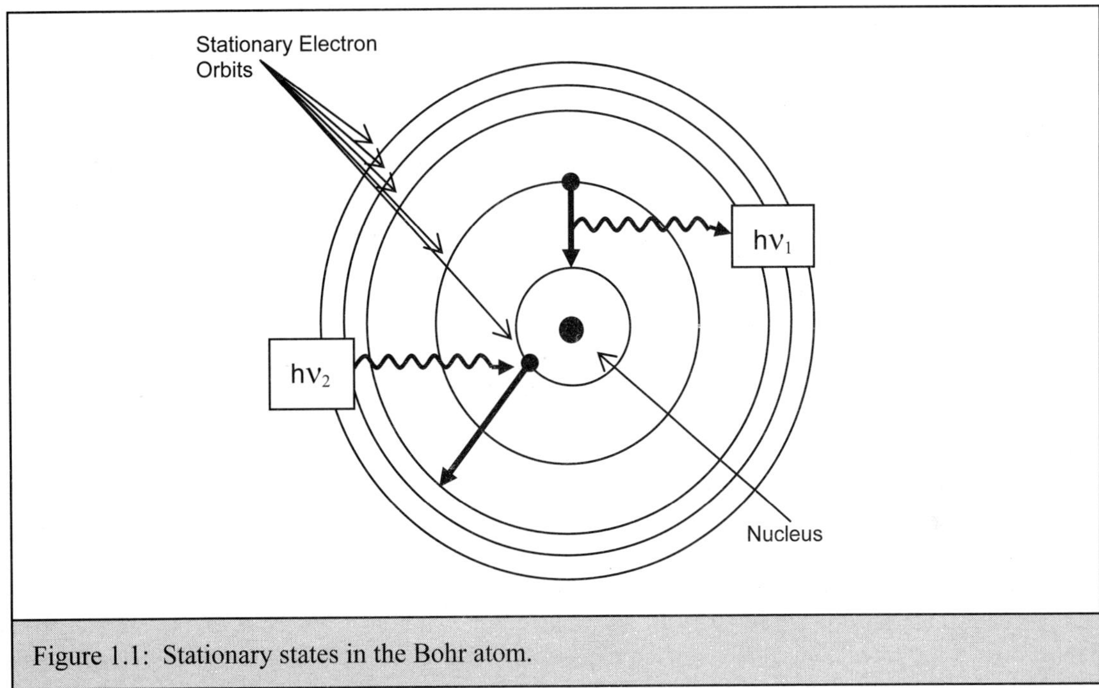

Figure 1.1: Stationary states in the Bohr atom.

are such that the associated angular momentum is always an integral multiple of h/2π where h is Planck's constant. They are called the stationary quantum states of the atom and can be calculated as the solutions of the Schrodinger wave equation (Figure 1.1).

When the electrons are in one of the stationary states, no radiation occurs, in spite of the fact that according to classical electrodynamics one would expect such radiation due to the accelerated motion of the charged electrons. Bohr also postulated that radiation is emitted or absorbed only when electrons jump from one stationary state (also called an orbital) to another. The difference in energy between the two states is given by the following expression.

$$\Delta E = h\nu = hc/\lambda$$

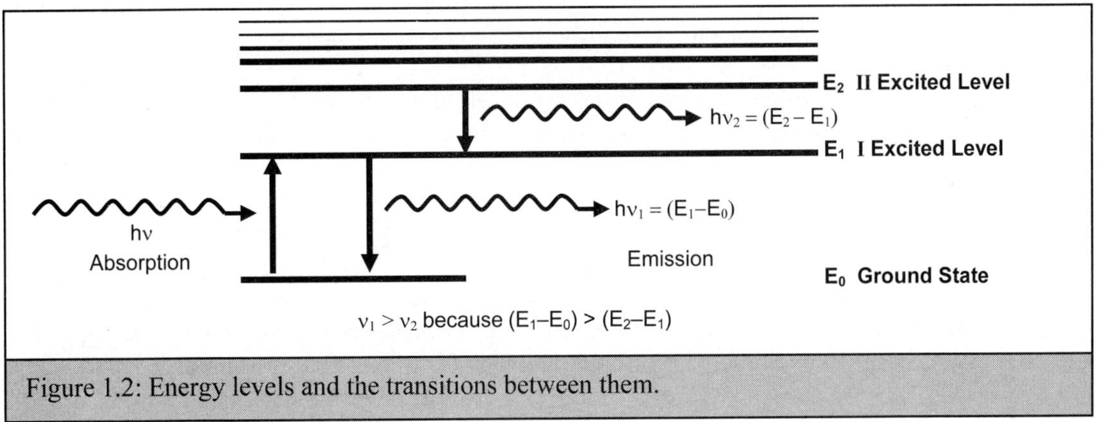

Figure 1.2: Energy levels and the transitions between them.

where ν and λ are the frequency and wavelength of the radiation, h is Planck's constant and c is the velocity of light. This equation determines the wavelength of the radiation emitted or absorbed (Figure 1.2). The motion of each of the electrons in the atom is described by a wave function ψ, known also as an atomic orbital. A combination of the wave functions of all the electrons leads to the wave function Ψ for the atom as a whole. Four quantum numbers are used to describe the orbital of an electron. These are the following.

a) The principal quantum number 'n'
b) The angular momentum quantum number 'l'
c) The magnetic quantum number 'm_l'
d) The spin quantum number 'm_s'.

The principal quantum number specifies the total energy of the electron. The expression for the energy of an orbital incorporates the square of this quantum number in the denominator – the larger the quantum number, the smaller the corresponding orbital energy. The angular momentum associated with the orbital motion of the electrons is described by the angular momentum quantum number. The principal quantum number can assume only integral values (n = 1,2,3,...). There can be more than one electron with the same principal quantum number, though for each electron at least one of the quantum numbers must be different. Electrons with the same value of the principal quantum number n constitute an electron shell. These shells are designated K, L, M,... etc., corresponding to n = 1,2,3,...etc. The angular momentum quantum number l can only take on the values 0,1,2,3,...n-1. In the presence of a magnetic field the orientation of the vector representing the angular momentum is described by the magnetic quantum numbers m_l, which can have values 0, ±1, ±2,...$\pm l$. Lastly the spin angular momentum is due to the spin of the electron about its own axis. The spin vector can assume one of two possible orientations in the presence of a magnetic field and the corresponding values of m_s are $+1/2$ and $-1/2$. Pauli's Exclusion Principle states that no two electrons can have all the four quantum numbers the same. Two electrons may occupy the same space orbital. Then their spins must be anti-parallel to each other, i.e. if one of them has $m_s = +1/2$, the other has the value $-1/2$. The orbital with the lowest energy is known as the ground state. For an atom with many electrons the ground state is when the electrons are arranged in the lowest energy orbitals consistent with the exclusion principle. Other states are called excited states.

When an electron (or the atom as a whole) makes a transition from one quantum state to another, energy is either absorbed or emitted. The radiated energy is equal to the difference in energy levels of the initial and final states. The frequency ν of the absorbed or emitted radiation can be written as

$$\nu = \Delta E/h$$

where ΔE is the difference in the energies of the two levels and h is Planck's constant. On atoms with more than one electron, the spin and orbital angular momenta of the individual electrons in an atom combine to produce a net angular momentum for the atom as a whole. Thus an atom with two electrons can have a total spin equal to zero when the electrons are paired (i.e. they have anti-parallel spins) or equal to one when they are unpaired (parallel spins). However, in the latter case the electrons must occupy different orbitals according to Pauli's Exclusion Principle. The orbitals are designated s, p, d, f, etc. according to the value of the angular momentum quantum number. The '1s' orbital is the lowest energy orbital with n = 1, l = 0 and $m_l = 0$. Two electrons with spins $+1/2$ and $-1/2$, respectively, can occupy this orbital. The orbital with the next higher energy value is the '2s' orbital, followed by the '2p' orbital and so on. Table 1.1 gives these details for the first three shells.

Table 1.1: Electron distribution in atomic orbitals of the first three shells.						
Atomic Shells	n	l	m_l	m_s	Orbital Designation	Number of Electrons Accommodated
K	1	0	0	+1/2, -1/2	1s	2
L	2	0	0	+1/2, -1/2	2s	2
		1	+1, 0, -1	+1/2, -1/2	2p	6
M	3	0	0	+1/2, -1/2	3s	2
		1	+1, 0, -1	+1/2, -1/2	3p	6
		2	+2, +1, 0 -1, -2	+1/2, -1/2	3d	10

1.4: Molecular orbitals and covalent bonds

An exact analytical solution of the Schrodinger equation is possible only for simple systems. In order to obtain the energy eigenvalues and eigenfunctions of systems that have more than one electron, a number of approximate methods are available. The molecular orbital method is one such approximate method wherein wave functions for the electrons in a molecule are constructed from atomic orbitals, i.e., the wave functions for single electrons in the atoms that form the molecule. When a linear combination of atomic orbitals is used to describe a molecular orbital then the method is called MO-LCAO approximation (Linear Combination of Atomic Orbitals).

Consider, for example a hydrogen molecule where two hydrogen atoms A and B come together to form a molecule. When the atoms are far apart their electrons each occupy the 1s orbital ($n = 1, l = 0$). But when they are brought together to form a molecule, the molecular orbital is given by

$$\Psi_{MO} = C_1\phi_A(1s) + C_2\phi_B(1s)$$

where ϕ_A and ϕ_B are atomic orbitals. C_1 and C_2 are constants such that $C_1 = \pm C_2$. Hence we get

$$\Psi_1 = C_1\phi_A(1s) + C_1\phi_B(1s)$$
$$\Psi_2 = C_1\phi_A(1s) - C_1\phi_B(1s)$$

Ψ_1 is called an even orbital and Ψ_2 is called an odd orbital. In the case of an even orbital the electrons are shared by the two nuclei while in an odd orbital there is no electron density between the nuclei. This orbital has higher energy (Figure 1.3). The even orbitals are bonding orbitals and the odd orbitals are anti-bonding orbitals. Note that if the hydrogen molecule is in the anti-bonding state, this does not mean that the electrons or the nuclei will fly apart. On the contrary, the bond between the two atoms remains, except that this is now unstable and of higher energy, and the molecule tends to make a transition to the lower energy bonding state. When the hydrogen molecule is formed, two electrons with anti-parallel spins occupy the bonding orbital. When the molecule is excited one of the electrons jumps to the anti-bonding orbital. Just as in the case of atoms, molecular orbitals also have associated

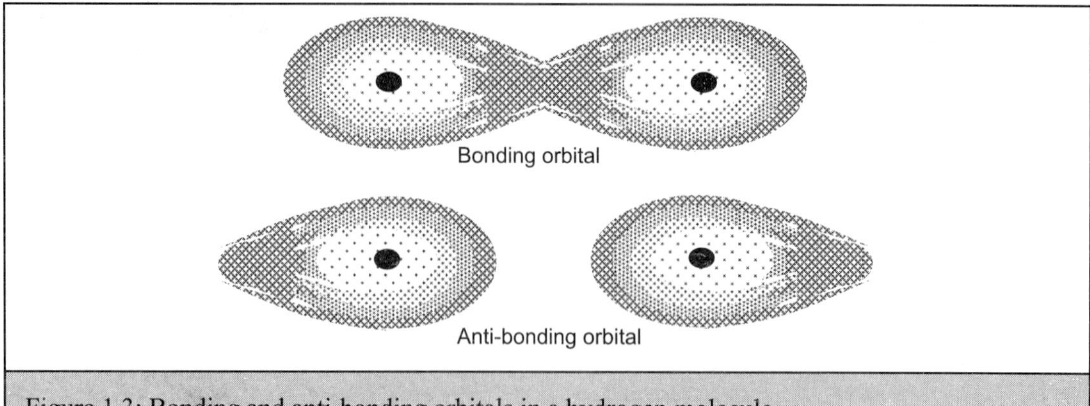

Figure 1.3: Bonding and anti-bonding orbitals in a hydrogen molecule.

quantum numbers. In a diatomic molecule like hydrogen, n is the principal quantum number and l is the quantum number that gives the angular momentum component along the inter-nuclear axis. l = 0, 1, 2.... are called σ, π, δ.... orbitals, analogous to the quantum number l in atoms. In diatomic molecules, σ orbitals are symmetric with respect to the inter-nuclear axis and π orbitals are anti-symmetric. The higher orbitals are more complex (Figure 1.4).

In multi-atomic molecules the atomic orbitals first combine to form a hybrid orbital before a molecular orbital is formed. Carbon, which is a pivotal atom in biology, is a classic example for the formation of hybrid orbitals. The electronic configuration of carbon is $1s^2 2s^2 2p^2$ (i.e. two electrons in the 1s orbital, 2 electrons in the 2s orbital and 2 electrons in the 2p orbital accounting for the six electrons in carbon). However a lower energy state is achieved if one of the electrons from the 2s level is promoted to a 2p state. Thus the atom has four unpaired electrons – one in the 2s orbital and three in the 2p orbital. These four orbitals combine to form hybrid orbitals. The stable way of doing this is to combine the 2s orbital with each component p_x, p_y and p_z of the p orbital in the following way.

$$\Psi_1 = 1/2 \ (\phi_s + \phi_{px} + \phi_{py} + \phi_{pz})$$
$$\Psi_2 = 1/2 \ (\phi_s + \phi_{px} - \phi_{py} - \phi_{pz})$$
$$\Psi_3 = 1/2 \ (\phi_s - \phi_{px} + \phi_{py} - \phi_{pz})$$
$$\Psi_4 = 1/2 \ (\phi_s - \phi_{px} - \phi_{py} + \phi_{pz})$$

(p orbitals have a butterfly-like appearance with their nodal point at the nucleus. The p_x, p_y and p_z orbitals are at right angles to each other, see Figure 1.4). The resultant hybrid is known as the sp^3 orbital and points towards the four corners of a tetrahedron (Figure 1.4). When these hybrid orbitals combine with the 1s orbitals of four hydrogens, a methane molecule is formed through four σ bonds (covalent bonds). The strongest bonds are formed when the overlap of the orbitals is a maximum.

Other schemes of hybridization are also possible. For example, sp^2 hybrids, also known as trigonal hybrids, are formed when 2s, $2p_x$, and $2p_y$ orbitals are mixed. This leads to three coplanar orbitals making an angle of 120° with each other (Figure 1.4). The p_z orbital is perpendicular to the plane. When a molecule such as ethylene (Figure 1.5) is formed, the sp^2 orbitals fuse with each other and with the 1s orbital of hydrogen to form localized σ orbitals. The p_z orbitals, which extend below and above the plane, also fuse together to form π orbitals. In benzene the sp^2 atomic orbitals of the six carbon

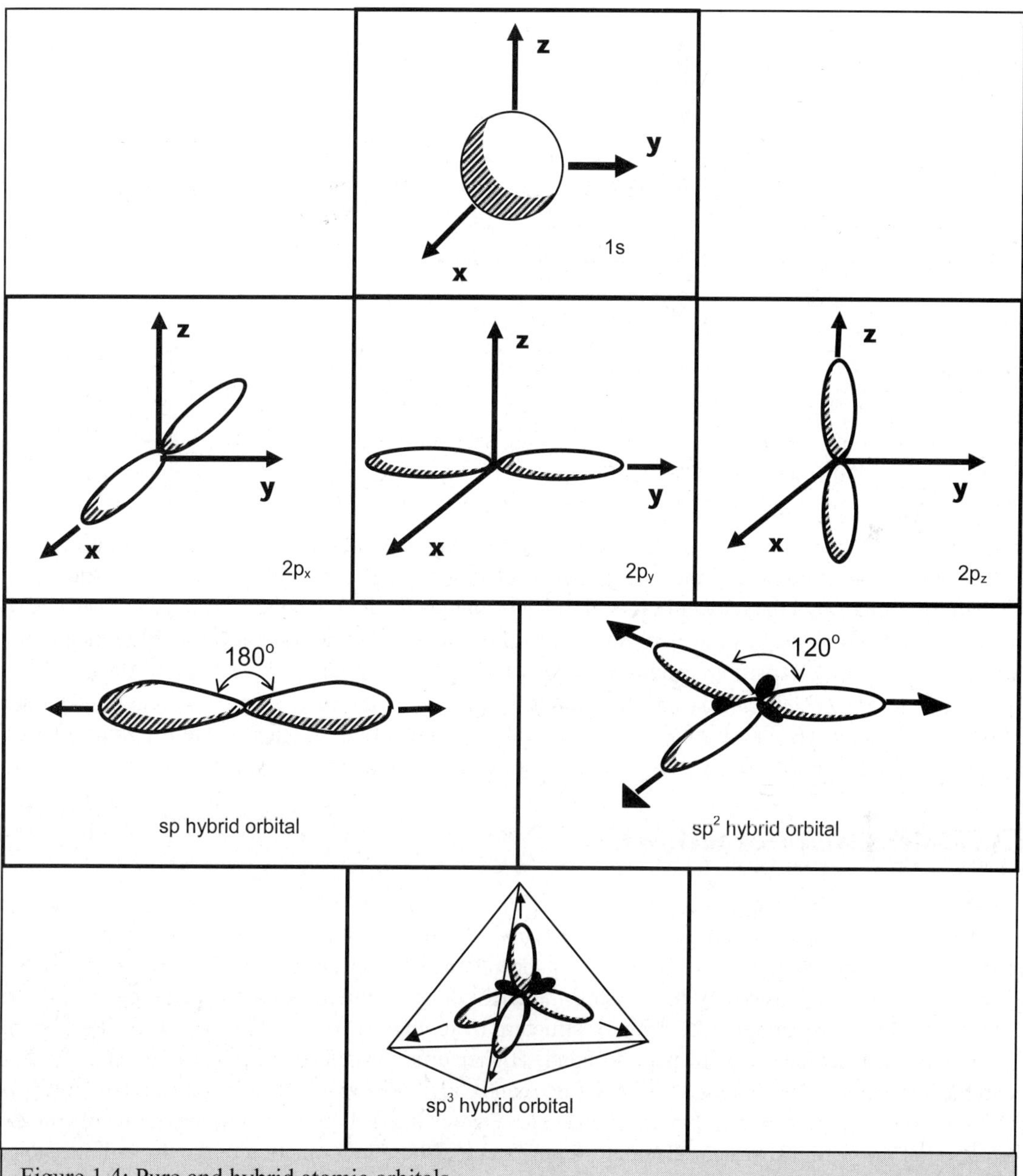

Figure 1.4: Pure and hybrid atomic orbitals.

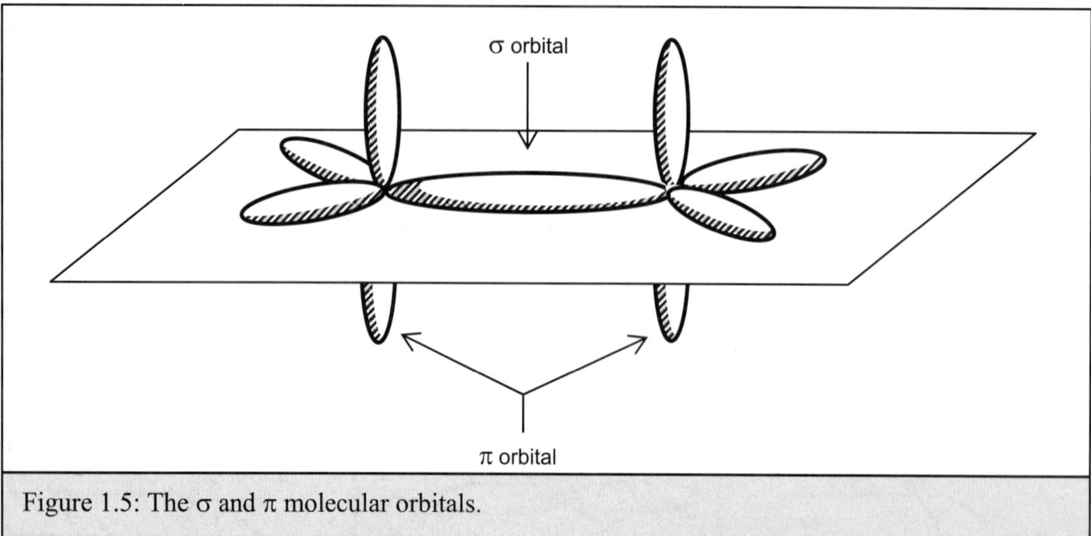

Figure 1.5: The σ and π molecular orbitals.

atoms, fuse together to form six σ orbitals in a closed ring. The six electrons in these orbitals are not localized to a particular atom but move freely in the ring. The orbitals are therefore known as delocalized orbitals and the molecule is said to have conjugated double bond or resonance structure. If two 2p orbitals combine to form a π molecular orbital it could be a bonding orbital (π) or an anti-bonding orbital (π^*). Transitions from one molecular orbital to another, for example from π to π^* or from n to π^* are called $\pi\pi^*$ or $n\pi^*$ transitions respectively. As in the case of atoms, electronic transitions in molecules also lead to emission or absorption of light. Owing to the differences in the energy levels, $\pi\pi^*$ transitions are about 100 times more intense than $n\pi^*$ transitions.

1.5: Molecular interactions

The structure of a biological molecule determines its function. In turn, the forces between the atoms in a molecule determine its structure. The interactions between the atoms in theolecule are classified as strong or weak, depending on whether or not the interaction can be disrupted by weak forces like thermal motion. (It is customary to measure the strength of an interaction by the energy required to disrupt it). Thus, for example, the primary structure of biological macromolecules is made of strong interactions such as the covalent peptide bond. Higher order structures like secondary, tertiary and quaternary structures are governed by weak forces and can, therefore, be disturbed by a relatively small increase in temperature, or a change of pH, etc. Strong interactions are implicated mainly in the formation of the chemical structure, and, to some extent, also in the formation of the three-dimensional molecular structure. Weak interactions, on the other hand, not only help to determine the three-dimensional structure, but also are involved in the interactions between different molecules. Any interaction within a molecule or between molecules can be understood as a sum of the interactions between pairs of atoms. When two atoms come close to each other a potential energy is associated with the system. This potential energy is a minimum (largest negative value) when the force of attraction between the atoms equals the force of repulsion, and is zero when the two atoms are separated by an

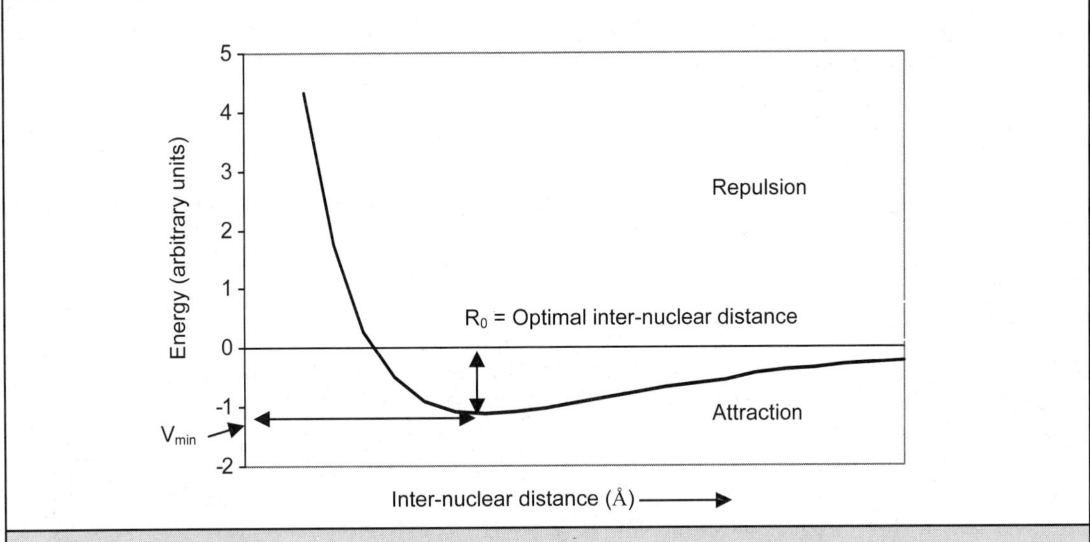

Figure 1.6: The potential energy between two atoms plotted as a function of the inter-nuclear distance.

infinitely large distance (Figure 1.6). The system as whole tries to attain an equilibrium minimum energy state and the final equilibrium distance between the two interacting atoms is a resultant of all the different forces, which act on each of them. Some of these forces are described below.

1.5.1: Strong interactions:

Covalent bonds, ionic bonds and metallic bonds are classified in this category. An ionic bond is formed due to electrostatic attraction between two oppositely charged ions. The structure of sodium chloride is an example of this type of bonding. In a crystal of common salt, every sodium ion is surrounded by six chlorine ions, and vice versa (Figure 1.7). Note that, strictly the entire crystal is one large 'molecule' of sodium chloride, consisting of sodium and chlorine atoms held together by ionic forces, and an isolated NaCl molecule has no independent existence[1]. In such ionic compounds an electron is transferred

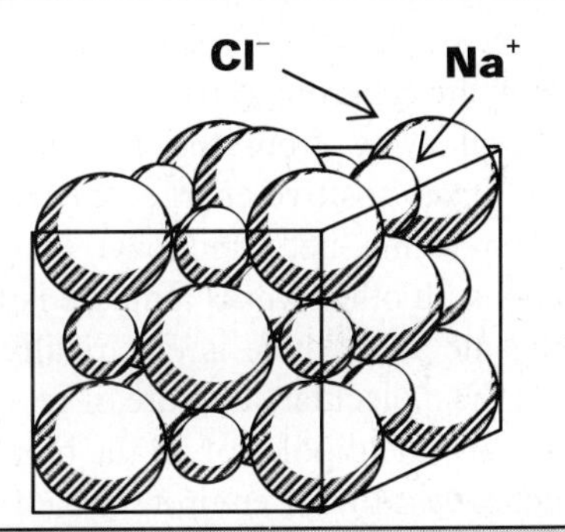

Figure 1.7: The structure of crystalline sodium chloride salt with Na^+ and Cl^- ions placed alternately in a face-centered cubic cell.

---

[1] When Lawrence Bragg determined the structure of NaCl by X-ray crystallography and suggested that NaCl did not exist as a single molecule by itself, he was subjected to much ridicule by chemists for daring to suggest that they give up their cherished prejudices. They said that as a physicist he should stick to subjects he understood, rather than speak what they thought was obvious nonsense. History of course proved Bragg right.

from an electropositive to an electronegative element in order to achieve a stable electron configuration. In NaCl, sodium looses one electron and becomes an ion, Na^+, with a single positive charge but a stable electronic configuration $1s^2 2s^2 2p^6$, with completely filled orbitals. In a similar fashion, chlorine acquires one electron to attain the same stable neon-like configuration and becomes Cl^-, again an ion, with a single negative charge. The positive ion and the negative ion come together to form a strong ionic bond and thence the crystal of sodium chloride.

Of the other strong interactions, covalent bonds have already been described. Metallic bonds are not relevant in biological systems and will not be further dealt with here.

1.5.2: Weak interactions:

Hydrogen bonds and van der Waals interactions are classified in this category. Van der Waals forces (also known as residual forces) are weak forces, which hold together inert elements and molecules. Most organic crystals, for example, are built of neutral molecules and this type of force plays an important role in their stability. Van der Waals forces act between all atoms and ions in all solids but the effect cannot be felt in the presence of strong interactions like covalent, ionic or metallic bonds. Van der Waals forces are basically electrostatic in nature in that they involve interactions between electric dipoles. When two bonded atoms have a difference in their ability to attract electrons (i.e. in their electronegativity), they form an electric dipole and hence possess a dipole moment. For example, in a hydroxyl group (OH) the electronegative oxygen atom draws the electron away from the hydrogen nucleus. There is thus a net (fractional) positive charge centered on the hydrogen nucleus, and a corresponding net (fractional) negative charge around the oxygen nucleus. In other words a dipole is formed (Figure 1.8). Note that the

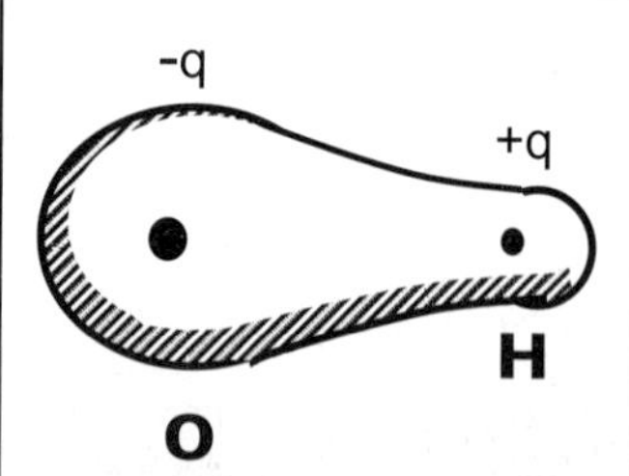

Figure 1.8: The dipole moment in a hydroxyl group arising out of the difference in electronic density surrounding the two nuclei. The charge q is less than one full electronic charge.

molecule as a whole is electrically neutral. The dipole moment may be a permanent feature arising from the molecular structure, if relatively electropositive atoms are bonded to relatively electronegative ones, and the dipoles of all the bonds do not cancel. In such a case a residual dipole moment may exist independent of the environment of the molecule. Temporary dipoles may also arise by induction due to, for example, the presence of an external electric field. Both permanent and temporary dipoles arise in solids in which the molecules are polar. But in the case of inert elements such as helium, neon, etc., and non-polar, symmetric molecules, such as methane, hydrogen, etc., London proposed that a dispersion effect would lead to momentary polarisation of the charge, and hence a dipole. Even in the case of a highly symmetrical system with no permanent dipole moment, the electronic distribution at any instant of time may not be perfectly symmetrical, due to the constant motion of the electrons. This lack of symmetry gives rise to an instantaneous dipole moment. This instantaneous dipole moment of an atom or a molecule will polarise the neighbouring atoms or molecules, and a momentary attractive force arises between the two. This adds up to give a net attractive van der Waals force due to London dispersion. There are thus three components of the van der Waals force, viz., the permanent dipoles in the molecule, the dipoles induced by an external electric field, and the ones induced by the London

dispersion effect. The relative contribution of each to the total force varies and depends on the type of molecule. At the level of the individual atoms, one
defines the term 'Van der Waals radius' as the distance of closest approach to the centre of the atom, in the presence of van der Waals forces alone. This radius can be experimentally determined by X-ray crystallography. In general, the van der Waals radius of an atom is larger than its covalent or ionic radius, showing that the effective radius of the atom within a molecule is different from (and less than) the distance of closest approach by other molecules or atoms.

Hydrogen bonds are weaker than covalent bonds but are stronger than van der Waals bonds. Hence the atoms connected by a hydrogen bond are separated by a distance (2.5 to 3.0 Å) greater than that of a covalent bond but shorter than if van der Waals forces alone were operative. Hydrogen bonds are generally found between two electronegative atoms. If X is an electronegative atom bonded to a hydrogen atom, it draws the electron of the hydrogen atom towards it, leaving the latter with a residual positive charge The magnitude of this residual charge is less than one, since the electron is not completely stripped away from the nucleus, but only displaced more towards the nucleus of the atom X. The H ion, with a fractional positive charge, is now attracted by the electrons surrounding the second electronegative atom Y. As the hydrogen ion is almost devoid of its only electron, it can penetrate the electron cloud of atom Y till the repulsive forces of the nucleus of atom Y stop it. Thus a bond is formed between atom X and atom Y, called the hydrogen bond. A hydrogen bond is generally written as X-H...Y since the H ion is asymmetrically disposed with respect to X and Y. It is particularly associated with X and loosely bonded to Y. Electronegative atoms are not all the same and they participate in hydrogen bonds of different strength according to their electronegativity value. Table 1.2 gives the electronegativity values of the atoms commonly encountered in biological systems. O-H...N bonds are the strongest with a bond energy of about 6.9 kcal/mol. Next in order come O-H...O (5.0 kcal/mol), N-H...N (3.1 kcal/mol) and N-H...O (1.9 kcal/mol) bonds. Carbon also participates in hydrogen bonds but these are usually weak and are difficult to distinguish from van der Waals interactions. Hydrogen bonds are among the most important of the interactions between biological molecules and confer specificity to the recognition of one molecule by the other. A prime example of this is formation of the highly specific A.T and G.C base pairs in DNA, which depends almost solely on the number and nature of the hydrogen bonds between the pairs.

1.6: Stereochemistry and chirality

The three-dimensional spatial arrangement of the atoms of a molecule is termed its stereochemistry. A molecule may be identified first by its chemical formula, then by its chemical structure, and finally by its molecular structure. For example, C_2H_6O is the chemical formula of ethyl alcohol. CH_3-CH_2-OH is the chemical structure of ethyl alcohol. The molecular structure is determined by the three dimensional arrangement of the atoms in space. The word 'configuration' is used to describe the distinctive arrangement of various atoms or groups attached to a chiral centre (as described below). The word 'conformation' describes the different arrangements of atoms that are obtained when parts of the molecule are rotated about one of the bonds. A change in the configuration necessarily involves breaking at least one covalent bond and reforming it after a rearrangement of the atoms. A change in the conformation necessarily does not involve any breaking of covalent bonds. Groups that are connected by a single bond can undergo rotation leading to different relative orientations with respect

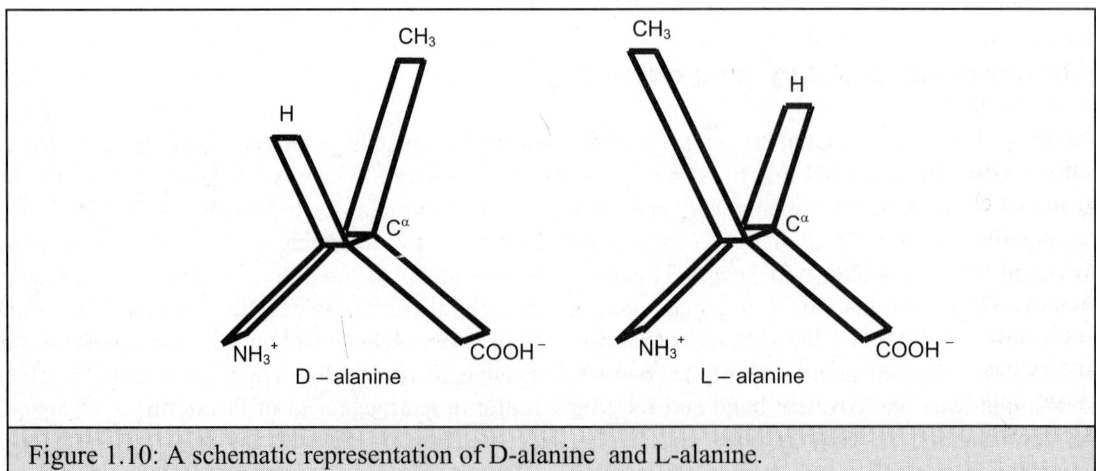

Figure 1.9: *Cis* and *trans* peptides.

to each other. For example, in ethyl alcohol, depending on the angle of rotation about the C—C bond, the relative orientations of hydrogen atoms and the hydoxyl group will vary. Energetically, the most favoured conformation about any bond is the 'staggered' one in which atoms attached at either end of the bond are at the largest possible distance from each other. An 'eclipsed' conformation increases steric contacts and is much less favourable. Arrangements of atoms or groups arrived at by rotation around a double bond are configurational isomers not conformational isomers. This is because the double bond does not allow free rotation. To produce the isomers, it becomes therefore necessary to break the double bond and then reform it after the rearrangement of the atoms. *Cis-trans* isomers fall under this category and are observed in the case of the amino acid proline and in polypeptides. Linear polypeptides consist almost exclusively of *trans* peptide units while cyclic peptides can have both *cis* and *trans* peptide units. (See Fig. 1.9).

In some molecules the differences in the arrangement of the atoms may not affect many of the physical properties of the molecule such as melting point, density, refractive index, etc., but may affect the interaction of the different molecules with polarised light. Such molecules are said to be optically active and are called optical isomers. Optically active molecules rotate in different directions the plane of polarisation of the plane polarised light incident on them. Levo-rotatory molecules rotate the plane in the clockwise direction and dextro-rotatory molecules rotate it in the anticlockwise direction. These

Figure 1.10: A schematic representation of D-alanine and L-alanine.

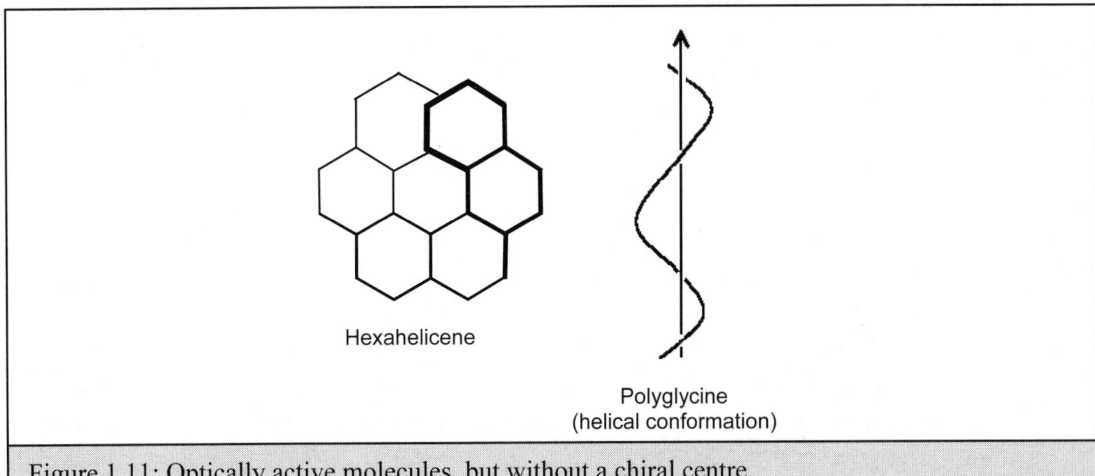

Figure 1.11: Optically active molecules, but without a chiral centre.

optically active molecules may differ in their biological behaviour. For example, only one of the optical isomers of aspargine is sweet to taste. Louis Pasteur showed that the difference in action between two optical isomers is due to the difference in molecular architecture, i.e. due to differences in the arrangement of the atoms in space. The two types of molecules are related by a mirror reflection and they cannot be super-imposed. Objects of everyday life that are enantiomers (i.e. related by mirror symmetry) are the right hand and the left hand, right-handed and left-handed screws, etc. The two alanine molecules shown in Figure 1.10 have the same number of atoms but differ in their stereochemistry and hence are known as stereoisomers. Compounds with four different substitutions at a carbon atom will exhibit optical activity. Such a carbon atom is generally known as an asymmetric carbon and a molecule containing such an atom is a chiral molecule. A mixture of two isomers, which have optical activity opposite to each other, is on the whole optically inactive since the effect of one is annulled by the other. Such a mixture is known as a racemic mixture.

A pair of enantiomeric molecules will behave in the same way with respect to symmetrical environments or achiral chemical reagents. However, their interaction with other chiral molecules will differ. An example of this is the very specific interaction between a protein and its substrate. In this case, the protein is made of L-amino acids. If the substrate is also optically active, then it could be either a levorotatory molecule or a dextrorotatory one, but not both. Stereoisomers not related by mirror symmetry are called diastereoisomers. Diastereoisomerism will occur in molecules with more than one chiral centre. Both configurational and conformational differences can lead to optical isomers – i.e. either enantiomers or diastereoisomers. Other than carbon, atoms like silicon with a valency of four can also act as chiral centres. Some molecules like hexahelicene and polyglycine show optical activity though they have no chiral centre. This is due to the helical structures assumed by these molecules, which can be either right handed or left handed (Figure 1.11). All the major biological molecules, like amino acids, proteins, DNA, sugars and lipids, are optically active. Out of the two or more optical isomers available for these molecules only one kind is commonly found in biological systems. Optical activity and Life seem to be inseparable. (One may speculate about the existence of an 'anti Life' governed by molecules of exactly opposite chirality to those on earth).

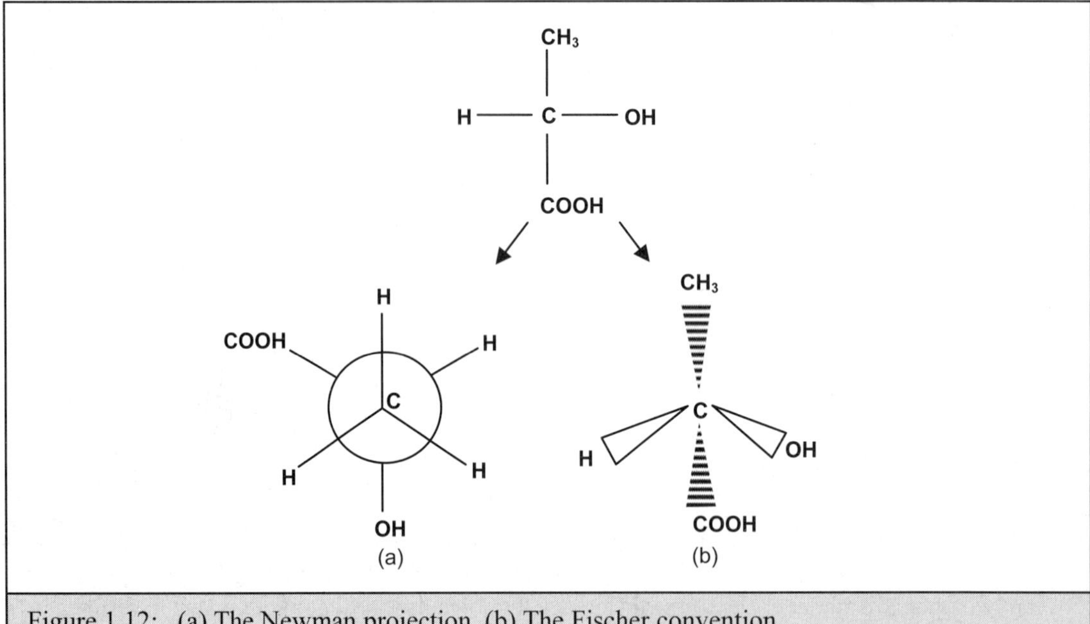

Figure 1.12:   (a) The Newman projection. (b) The Fischer convention.

1.6.1: Stereochemical nomenclature:

Molecules are three-dimensional entities. To represent them in two dimensions, i.e. on a piece of paper some standard methods are available. The most widely used are the Newman projection and the Fischer projection (Figure 1.12). In addition to these, there is another convention used to describe the arrangements of various groups around a chiral centre. This notation is known as the Cahn-Ingold-Prelog system and the enantiomers are called the [R] and the [S] configurations. The R-S convention is as follows: If the groups attached to the chiral centre are arranged according to the sequence rules (given below) as $a \rightarrow b \rightarrow c \rightarrow d$, then if the chiral centre is viewed from the side opposite to the lowest group (i.e. d in this case), and if $a \rightarrow b \rightarrow c$ occur in a clockwise direction, the molecule is in the [R] configuration. On the other hand if they occur in the anti-clockwise direction then it is in the [S] configuration (Figure 1.13). The sequence rules are as follows: 1) Atoms are arranged according to

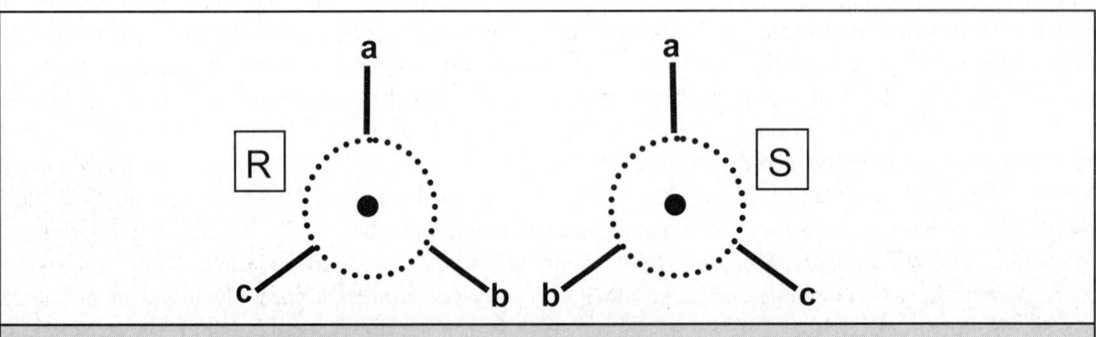

Figure 1.13: The Cahn-Ingold-Prelog nomenclature system to distinguish R and S enantiomers.

atomic number, e.g. $Cl > O > N > C > H$. 2) If the attached atoms are the same, then the next atom is used for ordering, e.g. $CH_3-CH_2-CH_2- > CH_3-CH_2- > CH_3-$. 3) If the substituents have atoms of the same rank then the one with more atoms of the highest rank gets preference, e.g. $(CH_3)_3-C- > (CH_3)_2-CH- > CH_3-CH_2-$ etc., and $CH_3-OH- > CH_3-CH_2-$ and so on.

Since at each chiral atom there are two ways in which the attached groups can be arranged, if there are n chiral centers then there will be 2^n stereoisomers. In the case of amino acids a different notation is used. In this molecule the C^α atom is the chiral centre. If we look down the $H - C^\alpha$ bond and if $CO \rightarrow R \rightarrow N$ is read in the clockwise direction, then it is the 'L' configuration. If it is read in the anti-clockwise direction then it is the 'D' configuration (Figure 1.10). All naturally occurring amino acids are in the 'L' configuration.

1.7: Thermodynamics

Thermodynamics deals with energy and the means of energy transfer, namely heat and work. Any system, physical, chemical or biological, can be considered as a thermodynamical system and the laws of thermodynamics can be applied to it, if it utilizes energy in any form. We note here that classical thermodynamics considered matter in bulk, before the atomic theory of matter was generally accepted, and derived laws and procedures to explain the observations. When concepts such as molecules were developed, these laws were reinterpreted to be consistent with the new concepts. This led to 'statistical thermodynamics', which has been one of the most successful theories in physics. Einstein made the following remark about thermodynamics (particularly classical thermodynamics) – 'It is the only physical theory of universal content concerning which I am convinced that, within the framework of the applicability of its basic concepts, it will never be overthrown'.

To introduce some of the thermodynamic concepts, consider the simple example of a closed container of hot water on a table. If the container is not disturbed in any way, the temperature of the water slowly decreases till it reaches the temperature of the room. This is achieved by transferring heat energy from the container to the room. Here the container of water is considered the system and the room is the environment. (In strict thermodynamic terms, the environment consists of not just the room, but the entire rest of the universe, i.e. everything except the defined system. In practice, however, it is sufficient to consider only the immediate surroundings.) A system is said to be *isolated* when there is no exchange of energy or matter with the environment. When there is an exchange of energy but not of matter then the system is said to be *closed*. When both matter and energy can be freely exchanged with the environment, then the system is an *open* one. If T_s and T_e represent the temperature of the system and the environment respectively, then, if $T_s > T_e$, heat will flow from the system to the environment. If $T_s < T_e$, the flow of heat will be in the other direction. Heat is one form of energy transfer, and is generally represented by the symbol Q. Q is positive when heat flows from the environment into the system and is negative when the opposite happens. Energy could also be transferred from the system to the environment (or vice versa) by means of work. To see this, consider a slightly different arrangement, in which the closed container of water is replaced by another that has a piston at one end. If the container is insulated from the surroundings, such that there can be no transfer of heat, the only way in which the two may attain equilibrium is for the system to do some work on the environment, in this case by raising the piston, which is counted part of the environment.

Depending on the weight of the piston, the gravitational force, and the difference in temperature (or the difference in the amount of energy) between the system and its immediate surroundings, the height to which the piston is raised will be large, or small. In either case, the effect of the transfer of energy is to raise the energy level of the surroundings, this time not as heat, but as work done on the piston. This leads to a change in the total potential energy in the environment. If the energy were transferred as heat, we would be able to measure a change in the temperatures before and after the transfer. Since the energy is now transferred as work, we are now able to measure a change in the position of the various objects, in this case the position of the piston, both with respect of the walls of the container, as well as with respect to the centre of the earth, which is part of the environment and is responsible for the gravitational force. Work is generally represented by the symbol W. W is positive when work is done on the environment and is negative when it is done by the environment.

In the previous example of the closed container of water, we consider that heat, i.e. energy, but not matter, can be exchanged between the container and the room. The system is a closed one and since initially the water is hotter than its surroundings, heat flows from the system to the environment, until $T_s = T_e$. At this point the system is said to be in equilibrium with the surroundings. The equilibrium in this case is a stable one and the temperatures remain equal, unless an external source (or sink) of heat disturbs one of them. The process by which the container of hot water has attained this equilibrium with its environment is called an irreversible process. On the other hand, in the second example, if the weight of the piston were to be changed slightly, or some other event tilted the equilibrium ever so slightly the other way, the piston could push upon the water, thus raising its temperature until the equilibrium state is again attained. In this case the process is reversible, and the equilibrium attained may not be a stable one. Since a very small change in the system is sufficient to tilt the equilibrium in one way or the other, a reversible process goes through infinitesimally small changes at an infinitely slow pace, and may be considered always at equilibrium. Reversible processes are therefore an idealisation of certain real processes, which strictly speaking, are always irreversible. A process can be considered reversible or irreversible depending on whether or not it can trace back the same path to reach the initial state. In macroscopic terms, processes in which heat is exchanged are usually irreversible, and those in which only work is exchanged are usually reversible. At the molecular level, it is not possible to decide if a process is reversible or not solely by this criterion. Natural or biological processes almost always take place spontaneously and transfer energy from one part of the system to another in a short time, and hence are not thermodynamically reversible processes.

Thermodynamics requires a measurement of energy. The units of energy are calories or joules, depending on whether it is expressed as heat or as work, respectively. One calorie is defined as the amount of heat required to raise the temperature of 1 gm of water from 14.5 to 15.5° C. Work is generally defined as

$$W = F \times d$$

where F is the force applied on the body and d is its displacement under the application of this force. One joule is defined as the work done in moving one kilogram through one meter in the absence of any external forces. The work done by a system is taken to be positive, while work done on a system is negative. The units of work and heat are inter-convertible, and one calorie equals 4.186 joules. This number is called the 'mechanical equivalent of heat'. Joule's experiment to measure this quantity consisted of using a falling weight connected to miniature turbine to churn a sealed beaker of water, thereby raising its temperature.

1.7.1: The three laws of thermodynamics:

Though work and heat are related to energy, they are not intrinsic properties of a system, and cannot be used to characterise the state of a system at any given moment. In thermodynamic terms, they are not state functions. Energy, like pressure, temperature, etc., is a state function, and can be used to characterise the system. A state function is property that depends only upon the state of the system, and not upon how the system was brought into that state, or upon the history of the system. A system can go from one state to another through an infinite number of different processes or paths. The amount of work done or the amount of heat transferred in following these paths depends on the path and is therefore different for each one of the paths. In every case however, the difference in energy depends only on the initial and final states, and not on the path followed. This may also be expressed as follows. The change in Q and W, i.e. δQ and δW, for each one of these paths is different. However, the change in energy, which is the sum of these two quantities, remains the same and is independent of the path taken. Thus

$$dU = \delta Q + \delta W = U_f - U_i \tag{1.1}$$

U_f and U_i are the final and initial values of the energy, respectively. Note that there is no meaning in writing δQ as Q_f-Q_i, or δW as W_f-W_i, since these depend on the actual path taken. Equation (1.1) is the first law of thermodynamics. It can be paraphrased as 'energy can neither be created nor destroyed', and is essentially a statement of the law of conservation of energy. The law may also be understood as stating that though work and heat are not state functions, their sum, i.e. energy, is a state function.

The first law tells us about the energy balance but not about the direction in which the process proceeds. The fact that every thermodynamic process always goes in a particular direction may be illustrated as follows. Consider an iron rod lying on a table. It will never happen, spontaneously, that all the heat energy in the rod gathers together at one end making that end of the rod hotter than the other. On the other hand, if there is an imbalance in the heat at the two ends of the rod to begin with, then it always happens that, as time proceeds, the difference in temperature between the two ends becomes smaller and smaller and the heat at both ends tends to equalize. The above process therefore has a particular direction. This intuitive concept is codified as the second law of thermodynamics, which makes it possible to predict the direction of the natural process. This law is stated in many different ways. One of the conventional statements is "It is not possible to convert heat completely to work with no other change taking place". It is also stated in terms of heat flow, as follows: "It is impossible for heat to flow from one body to another which is at a higher temperature, with no other change taking place". Another statement is the following – 'a spontaneous thermodynamic process will always result in an increase in the entropy of an isolated system'

The third law of thermodynamics states that the entropy of a pure solid or liquid is zero at the absolute zero temperature. In order to explain this further, here we may define the entropy of a system as a quantity related to the number of different states that it could take. We now recall that the quantum mechanical picture of a system, in particular of a molecule or atom, postulates that it has several discrete energy levels including one 'ground state' energy level. At any given moment the total energy of a collection of such objects may be distributed over its various levels. However, when the temperature is at the very lowest level, i.e. 0 K or the absolute zero of temperature, the ground state is the only possible state. The value of the entropy, which is proportional to the logarithm of the number of accessible states, is thus zero at 0 K. The third law is also sometimes stated as follows - "It is impossible to attain absolute zero temperature through a finite number of operations."

There are several thermodynamic variables that are used to describe a thermodynamic system. Some of them are described in detail below.

1.7.2: Entropy:

Entropy is a variable used in a precise statement of the second law of thermodynamics. Classically, it is defined as a thermodynamic function whose change is independent of the path of transformation of the system and is given by

$$dS = \delta Q/T \tag{1.2}$$

i.e. the change in entropy, dS, is measured as the ratio of the change in the heat energy, δQ, to the temperature at which the change takes place in a reversible process. The difference in entropy between any two states can be obtained by integrating the above equation.

$$dS = S_f - S_i = dQ/T$$

Note that since the heat Q depends on the specific path followed by the process, the integration is carried out over the reversible path of the process. Though Q is not a state function, it can be shown that by the above definition S, the entropy, is state function, independent of the path of the process. Therefore the change in entropy during a process may be written as the difference between S_f and S_i, the final and initial states of the system respectively. This difference is independent of the path taken during the process. Thermodynamic equations most often involve the difference in entropy between two states of the system, rather than the absolute entropy of a system and it is this quantity which is of interest in almost all the cases. In the case of a reversible process, the change in entropy is calculated in a straightforward way, using the above equations. In the case of an irreversible process, the change in entropy between two equilibrium states is calculated by finding a reversible path between the two states and calculating the entropy change for that path. As mentioned earlier, the second law of thermodynamics can be restated to say that the entropy of an irreversible process in an isolated system always increases,

$$\Delta S = S_f - S_i > 0$$

Thus naturally occurring spontaneous irreversible processes, such as biological processes, always result in an increase in entropy. Boltzmann showed that in molecular terms, entropy could be defined in such a way that it will be a measure of order in a system

$$S = k \log(w)$$

where k is Boltzmann's constant and w is the number of configurations a system can take. A larger number of configurations imply a less ordered system. Thus to state that a spontaneous process moves toward the maximum entropy is the same as saying that the system will move toward maximum disorder or randomness. A scale to measure entropy is given by the third law of thermodynamics. The zero of the scale is the value at absolute zero temperature. Entropy is measured in units of Joules/K.

1.7.3: Enthalpy:

Consider again the example of the cylinder of water with a piston on top. We have given the expression for the change in energy for such a system as

$$dU = \delta Q + \delta W$$

If we assume that the piston is kept fixed, then we have a system that is at constant volume. There is therefore no change in the position of the object, and thus no work is done. All the change in energy of the system comes therefore from the change in the heat energy. The measurement of such energy experimentally is simply and conveniently made by enclosing the system in a rigid container (a 'bomb') and measuring the change in temperature inside. Most chemical and biochemical reactions however occur in conditions where the volume is allowed to change freely and only the pressure is kept constant. That is, these reactions are 'open'. With respect to the example above, this is equivalent to having a weightless piston and an infinitely long cylinder. Any change in the thermodynamic state of the system could therefore be reflected as a rise in temperature, or a change in the position of the piston, or both. In such a system the measurement of the energy change, as defined above, is no longer as convenient. It therefore becomes necessary to define another state function that would serve the purpose better than U. Such a function is the enthalpy H.

The enthalpy of a system is defined as

$$H = U + PV$$

where U is the internal energy, P is the pressure and V is the volume. P and V are intrinsic properties and hence thermodynamic parameters. Their product is expressed in units of energy. Therefore enthalpy is also expressed in units of energy and is known as the heat content of a system. Since H is a combination of state functions and parameters that are independent of the path of transformation of the system, enthalpy too is independent of the path. The change in enthalpy is obtained by considering the differential form of the equation, which is

$$dH = dU + PdV + VdP$$

At constant pressure VdP = 0. Therefore,

$$dH = dU + PdV \qquad (1.3)$$

But from the first law of thermodynamics we know that

$$dQ = dU + dW$$

where W, the work done, equals PdV (at constant pressure). Therefore

$$dQ = dU + PdV = dH \qquad (1.4)$$

Thus we see that due to our definition of enthalpy, we may measure it as the change in heat energy in a constant-pressure reaction. Enthalpy is an intrinsic property of the system, or a state function. Its change is equal to the amount of heat exchanges with the environment at constant pressure. A reaction is exothermic when heat is let out to the surroundings and the enthalpy change is then negative. On the other hand, in an endothermic reaction heat is absorbed from the surroundings, and the enthalpy change is positive. Enthalpy plays the same role in a constant-pressure process as the energy U in a constant volume process. In chemistry and biochemistry enthalpy is a useful and important thermodynamic function. Enthalpy is measured in the same units as the internal energy U, i.e. joules or calories.

1.7.4: The free energy of a system:

A linear combination of thermodynamic state functions such as internal energy (U) and entropy (S) is also a state function. Such combinations are known as thermodynamic potentials. These potentials are often required to study thermodynamic systems under different conditions. Enthalpy, defined above, is one such potential, and as mentioned above, is required to characterise a constant-pressure process. The Helmholtz free energy is another thermodynamic potential or state function. It is used to characterise processes that occur at constant temperature and volume, and is used widely in biology. It is defined as

$$F = U - TS$$

The change in the Helmholtz free energy is thus

$$dF = dU - TdS - SdT$$

The physical meaning of this differential equation is that a change in free energy will involve a change in internal energy, a change in temperature and a change in entropy. At constant temperature, i.e. for an isothermal process, $dT = 0$. In that case

$$dF = dU - TdS \tag{1.5}$$

Now the first law of thermodynamics states that

$$dU = \delta Q + \delta W$$

But, from equation (1.2)

$$\delta Q = TdS$$

Therefore

$$dU = TdS - \delta W \tag{1.6}$$

Now combining (1.5) and (1.6) we have

$$-dF = \delta W \tag{1.7}$$

Therefore work done under isothermal conditions always leads to a decrease in the Helmholtz free energy (as inferred from the negative sign of dF). This may also be stated as follows – a process for which the change in the Helmholtz free energy dF is positive cannot proceed spontaneously at constant temperature and volume. It can only proceed if some work (positive δW) is done on the system. If dF is negative, then the reaction will proceed spontaneously at constant volume and temperature, and work will be done by the system. In either case the amount of work done by or on the system is given by dF. Note that this represents only the reversible work done. If any irreversible component such as friction exists in the system, this will be appropriately added to or subtracted from the total work. It is also seen from the above relations and definitions that when the system is in equilibrium state, the entropy is at a maximum and the Helmholtz free energy is at a minimum. This is because the direction of any spontaneous thermodynamic reaction (i.e. one not limited by any imposed constraints) is such as to always increase the entropy and to decrease the Helmholtz free energy.

Another thermodynamic potential function is used to describe processes that occur at constant temperature and pressure. This is Gibbs free energy. Since biological processes are almost always constant-pressure processes, this state function is particularly useful in describing these. Gibbs free energy is defined as

$$G = U - TS + PV$$

Under constant pressure and temperature (i.e., $dT = 0$, $dP = 0$) the change in this potential is given by the appropriate differential form of the equation,

$$dG = dU - TdS + PdV = dH - TdS \tag{1.8}$$

Note that the definition of Gibbs free energy involves the enthalpy. Thus the change in Gibbs free energy is a function of the change in enthalpy as well as entropy. We can show that this definition of Gibbs free energy implies that in a system at constant temperature and pressure, this energy will decrease as a result of any spontaneous process until the system attains equilibrium. Thus, in common with the other definitions of energy, the direction of a spontaneous reaction is again one that reduces the Gibbs free energy, i.e. $dG < 0$ for reactions at constant temperature and pressure. Unlike Helmholtz free energy, it is a little more complicated to relate this principle to the work done by or on the system. Since the volume is not a constant, reversible work due to changes in volume are represented as P-V work. Apart from this there is other reversible work that might occur, such as electrical work, which is represented by δW_{rev}. Analogous to the case of Helmholtz free energy, we can show that

$$-dG = \delta W_{rev} + (P\text{-}V \text{ work})$$

Using this relation we can state that if dG is negative, then δW_{rev} is the work exclusive of the P-V work that can be carried out by the system, if the change is carried out reversibly, and is thus the maximum work that can be obtained from the system. If dG is positive, then the process will not be spontaneous and δW_{rev} is the minimum amount of work required, apart from P-V work, that is required in order that the reaction takes place. We emphasize once again that dG refers to the change, and not the total quantity, of Gibbs energy that takes place during the reaction. Likewise dS and dF are the changes in entropy and Helmholtz free energy respectively. At equilibrium dG, dS and dF are all zero. And spontaneous thermodynamic reactions proceed in a direction that minimizes the free energy and maximizes the entropy.

1.7.4: Chemical potential:

In a closed system the mass and chemical composition of the constituents remain constant. But in an open system, such as a biological process, the mass and chemical composition may change. Hence, thermodynamic quantities like U, S, F, etc. will also depend on the quantities of chemicals present in the system. Let n_1, n_2, n_3... be the quantities of the 1st, 2nd, 3rd, etc. components. Then the change in

$$\mu_i = \left(-\frac{\partial G}{\partial n_i}\right)_{T,P,n_j,j \neq i}$$

free energy by the addition of 1 mole of component i to the system at constant pressure and temperature, when the quantities of the other components remain unaltered, is given by

μ_i is known as the chemical potential of the component i. Generally n_1, n_2, n_3, etc. are expressed in gram-moles. For a one-component system, the chemical potential is nothing but the Gibbs free energy measured per mole of the component. For a multi-component system the total Gibbs free energy is given as

$$G = \Sigma_i \, \mu_i \, n_i$$

The free energy of a system cannot be defined as an absolute number, since what is measured is only the change in the free energy during a process. Therefore a relative scale is defined for every element in its stable state, such that the free energy is zero at a temperature of 25° C and a pressure of 1 atmosphere. The free energy of a compound is therefore just the change in free energy due to its formation from its elements. The change in the standard free energy in the course of a reaction (denoted ΔG^O) can be computed from the standard free energy of formation of reactants and products.

$$\Delta G^O = -RT \log_e K$$

where K is an equilibrium constant, R is the gas constant and T is the temperature. For the reaction

$$A + B \text{ (reactants)} = P + Q \text{ (products)}$$

if C_A, C_B, C_P, C_Q, are the concentrations of the components,

$$K = (C_A C_B)/(C_P C_Q)$$

When ΔG^O is negative, i.e. if the concentration of the products is higher than that of the reactants at equilibrium, then the reaction will proceed spontaneously in the direction of the products and is exothermic. If it is positive, it requires energy for the reaction to go in the direction of the products i.e. the reaction is endothermic. Thus standard free energy change is an indicator of the direction of the reaction. However, it does not quantify the rate of reaction.

1.7.5: Oxidation-reduction potential:
Oxidation-reduction reactions, which are very common in biochemical processes, involve the transport of electrons. A substance is said to be 'oxidized' when it looses electrons and said to be 'reduced' when it gains electrons. A biological reaction takes place in an aqueous medium and hence the reactant can pick up a proton from the surrounding medium and become reduced. Or it can bind to an oxygen atom and become oxidized. Consider the following reaction.

$$A_{oxid} + B_{red} \leftrightarrow B_{oxid} + A_{red}$$

Since in an oxidation-reduction reaction there is an exchange of electrons, the free energy of of the reaction is expressed in units of electric potential. The above equation can be split as follows:

$$A_{oxid} \leftrightarrow A_{red}$$

where A_{oxid} and A_{red} are called a redox couple. Similarly,

$$B_{oxid} \leftrightarrow B_{red}$$

The change in free energy for one half of the reaction is

$$\Delta G^O = \Delta G^O + RT \log_e (A_{oxid}/A_{red}) = \Delta G^O + RT \log_e (\alpha / (1-\alpha))$$

where α is the fraction of A that is oxidised. Converting this to electrochemical potential, we obtain

$$E_A = E^O_A - (RT/ZF) \log_e (\alpha / (1-\alpha)) \tag{1.9}$$

where E is the oxidation-reduction (redox) potential, F is the Faraday constant, and Z is the number of transported electrons in the reaction. The redox potential is measured against a standard, which is generally a hydrogen electrode. A standard hydrogen electrode consists of an electrode immersed in 1 M H^+ solution that is in equilibrium with H_2 gas at 1 atmospheric pressure (1 bar). The potential E measured while inserting the hydrogen electrode in a 1 N acid solution (i.e., pH = 0) is taken to be zero. Against this standard, the redox potential of other substances can be measured. As biological processes take place at pH = 7 (i.e. neutral pH) it is more convenient to use the redox potential measured at this value as the standard. For such a measurement equation (1.9) is generally written as

$$E_R = E_m - (RT/ZF) \log_e (\alpha / (1-\alpha))$$

where E_R is the redox potential measured against a standardized electrode and E_m is the midpoint potential, measured when half of the substance is in the oxidized state and the other half is in the reduced state. A strong reducing agent (such as NADH) has a negative redox potential while a strong oxidizing agent (such as O_2) has a positive redox potential.

1.8: Radioactivity

An atom consists of a nucleus surrounded by electrons. The nucleus in turn consists of protons and neutrons. The atomic number Z of the atom is the number of electrons present, which is also equal to the number of protons in the nucleus. Since each electron carries a single negative charge, each proton carries a single positive charge, and each neutron is electrically neutral, this makes the atom as a whole electrically neutral. The mass number A of the atom is given by the sum of neutrons and protons, i.e. A = Z + N, where Z = atomic number and N = number of neutrons. Atoms with the same atomic number but different mass numbers are known as isotopes. The number of isotopes for a given element may vary from two to as many as 20. For example, hydrogen has three isotopes, $_1^1H$ (hydrogen), $_1^2H$ (deuterium) and $_1^3H$ (tritium). The superscript stands for the mass number. In general an element is represented $_Z^A X$ where A is the mass number and Z is the atomic number. In an atomic nucleus when the protons and neutrons are equal in number, the isotope is generally a stable one. However, elements with higher atomic numbers frequently have a neutron to proton ratio of more than one. These isotopes are unstable and are known as radioisotopes. Radioisotopes are found in nature (e.g. uranium, plutonium, thorium and radium) though they are more often made artificially. Radioactive isotopes emit particles and/or electromagnetic radiation to become a stable isotope. This process is known as radioactive decay. A neutron in the nucleus can emit an electron (also known as a negatron to indicate is nuclear origin) to become a proton

$$\text{neutron} \rightarrow \text{proton} + \text{negatron}$$

Similarly a proton can become a neutron by emitting a positron. A positron has the same mass as an electron but is positively charged.

$$\text{proton} \rightarrow \text{neutron} + \text{positron}$$

Both the above processes alter the N/Z ratio. For example

$$_{6}^{14}C \rightarrow {}_{7}^{14}N + \beta(\text{-ve})$$
$$_{11}^{22}Na \rightarrow {}_{10}^{22}Ne + \beta(\text{+ve})$$

Sometimes the term 'β particle' is used to denote either a positron or a negatron. Radioactivity resulting in the emission of β particles is termed 'beta decay'. Radioactive decay can also result in the emission of α particles and γ-rays. An α particle is a helium nucleus (two protons and two neutrons). The emission of an α particle leads to a reduction in mass number by 4 and atomic number by 2, e.g.

$$_{88}^{226}Ra \rightarrow {}_{86}^{222}Rn + {}_{2}^{4}He\ (\alpha\text{-particle}) \rightarrow ... \rightarrow {}_{82}^{206}Pb$$

Since the product of the first decay, radon, is also radioactive, it initiates a further decay series, which finally leads to the formation of lead. γ-ray emission does not lead to any change in atomic number or mass. γ-rays are electromagnetic waves at the extreme short wavelength, high frequency end of the spectrum. γ-ray emission is frequently accompanied by α and β particle emission. Like the atom as a whole, the nucleus too has a ground state energy level and excited states. γ-ray emission arises when an excited nucleus returns to its ground state.

1.8.1: Rate of radioactive decay:
Radioactive decay always occurs at a specific rate that is characteristic of the source. The rate of decay follows an exponential law. If we consider a sample containing N nuclei, then the quantity of nuclei that undergo decay in a given time period is proportional to the number of nuclei available for decay. The rate of change in the number N of radioactive atoms is given by the differential equation

$$-\delta N/\delta t = \lambda N$$

where λ is the proportionality constant or the decay constant.

$$\lambda = -(\delta N/N)/\delta t$$

λ may also be considered the fractional rate of decay during a very short period of time, during which N may be treated as a constant. Integrating this equation over the time t we get

$$\log_e (N_t/N_o) = -\lambda t$$

or

$$N_t = N_o \exp(-\lambda t)$$

where N_t = the number of radioactive atoms present at time t, N_o = the number present at time t = 0 (i.e. the initial number). From the last equation we see that there will always be some parent nuclei present after any finite time, however long. It is therefore usual to specify the half-life of a radioactive material rather than its lifetime. The half-life of a radioactive nuclide is defined as the time taken for

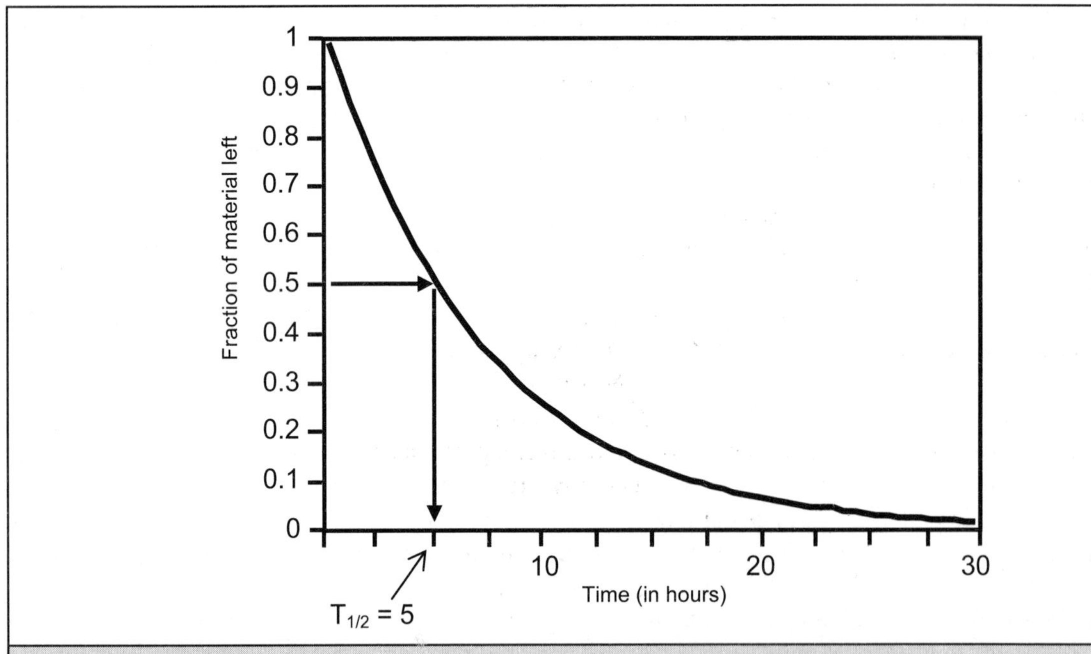

Figure 1.14: The radioactive decay curve with the decay constant $\lambda = 0.1386$, corresponding to a half-life of 5 hours.

one half quantity of the sample to decay from the parent nuclide to the daughter nuclide (Figure 1.14). Substituting this definition into the above equation we have $N_t = 0.5N_o$. Thus half-life is given by

$$T^{1/2} = \log_e (2/\lambda)$$

The half-lives of radioactive materials vary from 10^{19} years to 10^{-7} secs. The half-lives of some of the isotopes are:

^{3}H --   12.26 years
^{14}C --   5760 years
^{32}P --   14.20 days
^{42}K --   12.40 hours

1.8.2: Measurement of radioactivity:

Apart from the half-life, radioactive decay can also be expressed as disintegrations per minute (dpm) or per second (dps). The unit of radioactivity used is curie (C_i) and is defined as that quantity of radioactive substance in which the number of nuclear disintegrations is equal to that in 1 gram of radium. Thus 1 curie = 3.7×10^{10} dps. Millicurie (1 mC_i = 1/1000 of a curie) and microcurie (1 μC_i = 1/1000000 of a curie) are the units generally used in biological studies. Specific activity is the amount of radioactivity per unit mass of the sample. Hence, the unit of specific activity is mC_i/mg or μC_i/mg. The amount of radiation that falls on a sample can be measured in roentgens or in rads. Roentgen is generally used in cases where electromagnetic radiation is involved, while rad can be used both for radiation and particles. One rad is defined as the dose required to produce 100 ergs of absorbed energy per gram of target. A dose of one rad can increase the temperature of one gram of material by 2×10^{-6} K.

Radioactivity can be measured by counting devices or by photographic methods. In both cases advantage is taken of the ability of the radioactive emission to interact with matter. Geiger-Muller counters are based upon the ability of the radioactive emission to ionize matter. Scintillation counters are based on its ability to excite an atom leading to emission of visible light as the atom returns to the ground state. The photographic method is advantageous when we require a permanent record of the experiment conducted. In this method the ionizing radiation from a radioactive source interacts with a photographic emulsion (silver halides) and leaves an image. Autoradiography, which uses photographic method, is a sensitive technique by which the distribution of radioactivity in a biological specimen can be determined, for example, the sites of localization of a radiolabeled drug in the cell.

1.8.3: Effects of radioactivity on matter:

α particles ($^4\text{He}^{2+}$), due to their size, are slow and interact with atoms causing ionization and excitation. Ionization takes place when an outer electron of an atom is completely removed. Excitation takes place when the electrons are raised to a higher excited orbital. Though the initial energy of alpha particles is reasonably high (3 to 8 Mev) their energy is dissipated very quickly and hence they are not very penetrating. β particles with a negative charge (negatrons) are small compared to α-particles and hence their interaction with matter is very much less. They are thus more penetrating. They are rapidly moving particles causing ionization and excitation, though less than like α-particles. γ-rays are electromagnetic radiation and carry no charge or mass. They travel very large distances without appreciable loss of energy, and are thus highly penetrating. γ-rays also interact with matter and produce secondary electrons. A low energy γ-ray can transfer all its energy to an orbital electron, which is then ejected out of the atom as a photoelectron. This phenomenon is known as photoelectric absorption of the radiation. A medium energy γ-ray transfers only a part of its energy to an orbital electron (which is then ejected) and proceeds further with reduced energy. The change in energy between the impinging photon and the emitted photon can be measured as a change in the wavelength and is known as the Compton effect. A high- energy γ-ray, when it interacts with a nucleus, transfers all its energy to it and excites it. The excited nucleus in turn emits another photon of the same or lesser energy. The photon may then spontaneously decay to produce a positron - negatron pair in a process known as 'pair-production'. Note that the photon is not usually considered a particle of matter, more as a 'particle of radiation'. Nevertheless, it can also decay.

1.8.4: Biological effects of radiation:

Ionizing radiation can affect both somatic and sex cells. The somatic effects depend on the amount of radiation and it can cause reddening of the skin, loss of hair, etc. It is also known that radiation can cause carcinogenesis. Further, biological specimens consist of a large amount of water, and irradiation of these can lead to hydrolysis, further leading to the production of reactive ions and free radicals that may damage tissue or induce cancer.

$$H_2O \rightarrow \ H + OH \quad \text{or} \quad H_2O \rightarrow H_2O^+ + e^-$$

Unlike genetic damage, which is caused by even very small amounts of radioactivity, serious somatic effects occur only with exposure to higher doses. Even so, a dose of about 500 rads can immediately kill a human being. Further, irradiation can release free electrons which in turn can produce secondary electrons by collision. These free electrons can cause a great deal of damage in the cell.

Genetic effects occur when the reproductive cells are irradiated, leading to mutations in DNA The damage to DNA due to irradiation can be in the form of breaks in the strands or alterations in the base sequence. For a given dose, mRNA is the most vulnerable, while DNA synthesis itself is the least affected. Thus the biological effects of radioactivity could be disastrous to life.

1.8.5: Applications of radio isotopes:

^{3}H, ^{14}C and ^{32}P isotopes are used in autoradiography of biological samples. Radioactive isotopes used in biological experiments are known as radiotracers. The greatest advantage of the radiotracer method is its extreme sensitivity. Metabolites that occur in very low concentrations in the system can be identified and therefore, radiotracer experiments are frequently performed *in vivo*. Apart from their use in tracing metabolic pathways, in pharmacology radioisotopes are used to study drug accumulation, the rate of metabolism of the drug, and also the metabolic products of the drug under consideration. On the analytical side, enzyme substrate binding studies can be carried out using radio labeled substrates. Radioimmunoassay is yet another area where radioisotopes are used. Radioisotopes have other biological applications like sterilization of food, clinical diagnosis, ecological studies, etc. Isotopes like ^{60}Co, ^{226}Ra, ^{222}Rn are used in radiation therapy of cancer.

Summary

- Physics is an exact science. Biology, on the other hand, is considered a descriptive science.

- With the advent of molecular biology, molecular biophysics, computational biology and the like, biology also is being transformed into a more exact science.

- Modern biology assumes that Biological systems are in no essential respect different from physical or chemical systems, though they are much more complex than the latter

- According to the reductionist programme in biology, biological systems are studied as component parts. The understanding of the function of each separate part so gained is put together to build an understanding of the entire biological system.

- Max Planck proposed that electromagnetic radiation was emitted or absorbed not continuously but in small packets, or quanta.

- De Broglie suggested that the same type of wave-particle duality could be applied to entities hitherto thought to be particles.

- Schrodinger introduced a wave equation that replaces the classical equations of motion.

- Heisenberg's Uncertainty Principle states that there exists a physical limit to the accuracy with which one can simultaneously determine both the position and the velocity of a particle.

- In the Bohr model of the atom, the electrons that revolve around the nucleus occupy only fixed orbits.

- When an electron (or the atom as a whole) makes a transition from one quantum state to another, energy is either absorbed or emitted.

- An exact analytical solution of the Schrodinger equation is possible only for simple systems.
- Wave functions for the electrons in a molecule, i.e. the molecular orbitals, are constructed from atomic orbitals, i.e. the wave functions of the elctrons in an atom.
- Covalent bonds, ionic bonds and metallic bonds are classified as strong molecular interactions.
- An ionic bond is formed due to electrostatic attraction between two oppositely charged ions.
- Hydrogen bonds and van der Waals interactions are classified as weak molecular interactions.
- The van der Waals force is made up of three dipole-dipole interactions, viz., between the permanent dipoles in the molecules, between the dipoles induced by an external electric field, and between the ones induced by the London dispersion effect.
- Hydrogen bonds are weaker than covalent bonds but are stronger than van der Waals bonds.
- Hydrogen bonds are generally formed between two electronegative atoms, one of which is bonded to a hydrogen atom.
- The three-dimensional spatial arrangement of the atoms of a molecule is termed its stereochemistry.
- The word 'configuration' is used to describe the distinctive arrangement of various atoms or groups attached to a chiral centre (as described below). The word 'conformation' describes the different arrangements of atoms that are obtained when parts of the molecule are rotated about one of the bonds.
- Optically active molecules rotate in different directions the plane of polarisation of the plane polarised light incident on them. The two types of molecules are related by a mirror reflection and they cannot be superimposed.
- Thermodynamics deals with energy and the means of energy transfer, namely heat and work.
- The first law of thermodynamics may be paraphrased as 'energy can nether be created nor destroyed'.
- The second law of thermodynamics states "It is not possible to convert heat completely to work with no other change taking place".
- The third law of thermodynamics states that the entropy of a pure solid or liquid is zero at the absolute zero temperature.
- The change in entropy is measured as the ratio of the change in the heat energy to the temperature at which the change takes place in a reversible process.
- Enthalpy may be measured as the change in heat energy in a constant-pressure reaction.
- Elements with high atomic numbers frequently have a neutron to proton ratio of more than one. These isotopes are unstable. They tend to decay or break apart, or undergo fission, and are known as radioisotopes.
- The rate of decay follows an exponential law.
- The half-life of a radioactive nuclide is defined as the time taken for one half quantity of the sample to decay from the parent nuclide to the daughter nuclide.

Chapter 2

Separation Techniques

2.1: Introduction

The identification, purification and separation of the components of a mixture of molecules are of considerable importance in molecular biology and biophysics. There are several techniques available for this purpose and in this chapter we shall describe two of the most frequently used ones, namely Chromatography and Electrophoresis.

2.2: Chromatography

Chromatographic techniques can be classified into four main categories based on the type of molecular interactions used for separation, viz. ion-exchange chromatography, adsorption chromatography, partition chromatography and molecular exclusion chromatography. The techniques can also be classified in another way, on the basis of the mode of separation and the support system used, as column chromatography, thin layer chromatography and paper chromatography. The particular technique that is used for a given sample depends on its characteristics. Sometimes, several techniques may be used one after the other to obtain extra-pure sample.

The basic principle behind all the different types of chromatography is however the same. Chromatographic separation takes advantage of the fact that the sample distributes or partitions itself to different extents in two different, immiscible phases. This property is described by the partition or distribution coefficient K_d. If we consider two immiscible phases A and B,

$$K_d = \text{(Concentration of the sample in phase A)/(Concentration in phase B)}.$$

The effective distribution is then defined as the total amount of the substance present in phase A divided by the total amount of substance present in phase B. The two immiscible phases could be a solid and a liquid, or a gas and a liquid, or a liquid and another liquid. Generally, one of the two phases is a stationary phase and does not move. The other is a mobile phase and moves with respect to the first. To understand the use of chromatography to separate and purify molecules, one may consider the

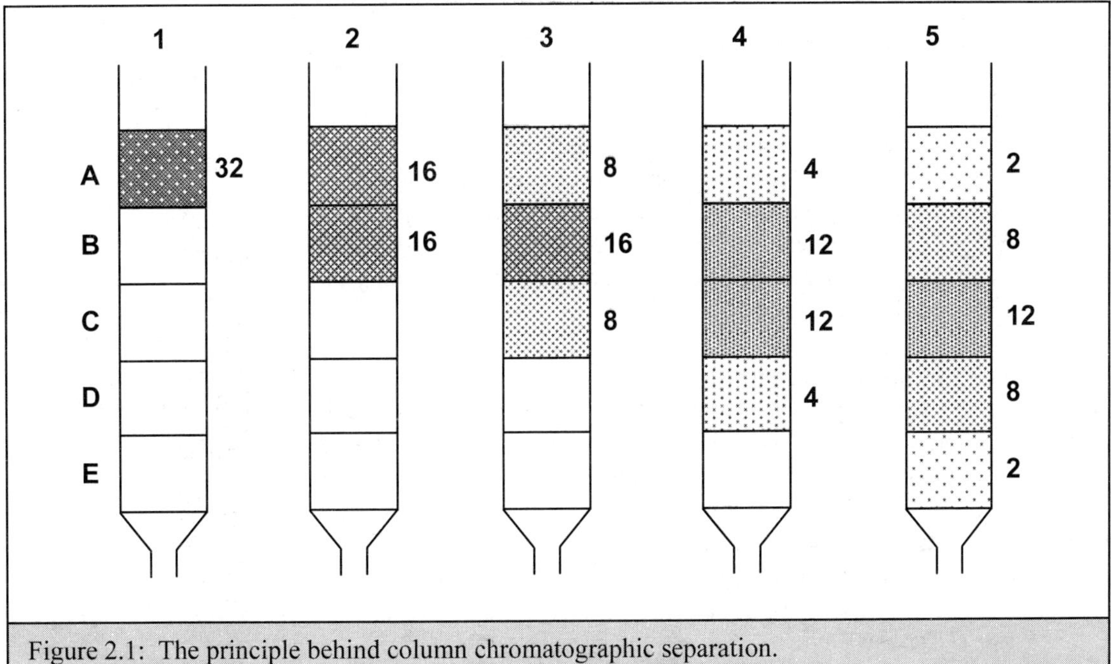

Figure 2.1: The principle behind column chromatographic separation.

column mode, with a stationary solid phase and a mobile liquid phase. To separate a mixture of molecules, a column is packed with the granular stationary phase to a height of a few centimetres and is then surrounded by the liquid phase. Assume that the column diameter is such that 1 cm^3 of liquid will surround 1 cm length of the column material. First consider the behaviour of a solvent (i.e. the liquid phase) containing a single solute. Let 32 μg of the sample in 1 cm^3 of solvent be added to the column. It would occupy position A in Figure 2.1. If the distribution coefficient as defined above were 1.0, the solute would distribute itself equally between the liquid phase (the solvent) and the solid phase (the support material in the column). Now, if another 1 cm^3 of the solvent (this is also sometimes more properly called the 'elutant', especially when it is particularly used to elute or bring out the sample) is added to the column, half the solute will move from position A to position B, since the distribution coefficient is 1.0. There will now be 16 μg of sample each at positions A and B. If yet another 1 cm^3 of the solvent is added to the column, the situation will be as shown in the third panel of Figure 2.1, with 8 μg in position A, 16 μg in position B and 8 μg in position C. When this process is repeated, the pattern shown in the other panels in Figure 2.1 will be observed. At every step of the equilibration, 50% of the sample will be left bound to the solid phase at each position. After five equilibration steps the sample is present all through the column but has a maximum concentration at the centre of the column.

If the solvent contains two solutes with different partition coefficients, at each step the amount in the solid phase will be different for the two. After a number of such steps the peaks of concentration for the two components will be at different positions along the column, thus achieving a separation of the mixture. If the process is allowed to continue sufficiently long, the different components of the mixture will flow out of the column at different times and can be collected separately. This is the basic principle of chromatography. We now consider the different applications of this principle.

2.2.1: Column chromatography:

Adsorption chromatography, partition chromatography, ion exchange chromatography, exclusion chromatography and affinity chromatography all use the column mode (Figure 2.2), where the stationary phase is packed into glass or metal columns. The nature of the stationary phase depends upon the particular form of chromatography. It is generally an adsorbent, in other words, a resin or a gel.

2.2.2: Thin layer chromatography:

In this case, the stationary phase is applied to a glass, plastic or metal foil plate as a uniform, thin layer and the sample is applied at the top edge of the layer using a micropipette or syringe. This technique can be used for partition chromatography, adsorption chromatography and exclusion chromatography. The advantage of this method is that a large number of samples can be studied simultaneously. Also, it is quick and simple.

2.2.3: Paper chromatography:

In this technique, the cellulose fibres of a sheet of special chromatographic paper act as a support or the stationary phase (Figure 2.3). There are generally two methods that are employed in paper chromatography – the ascending method, and the descending method. In both of them the solvent is placed at the bottom of a jar or tank so that the chamber is saturated with solvent vapour. The chromatographic paper is held vertically inside the tank. In the ascending method, the lower end of the paper dips into the solvent and the sample spot is applied just above the surface of the solvent. As the solvent moves up the sheet of paper by capillary action, the sample constituents move along with it and get separated. In the descending method, the upper end of the paper, where sample is applied, is held in a trough of solvent at the top of the tank, and the rest of the

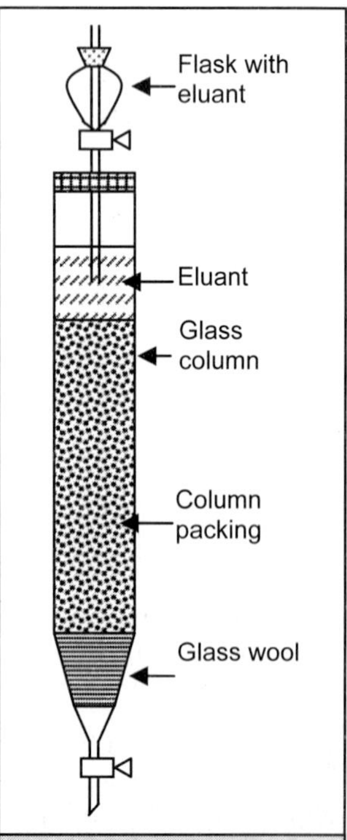

Figure 2.2: The column mode of chromatography.

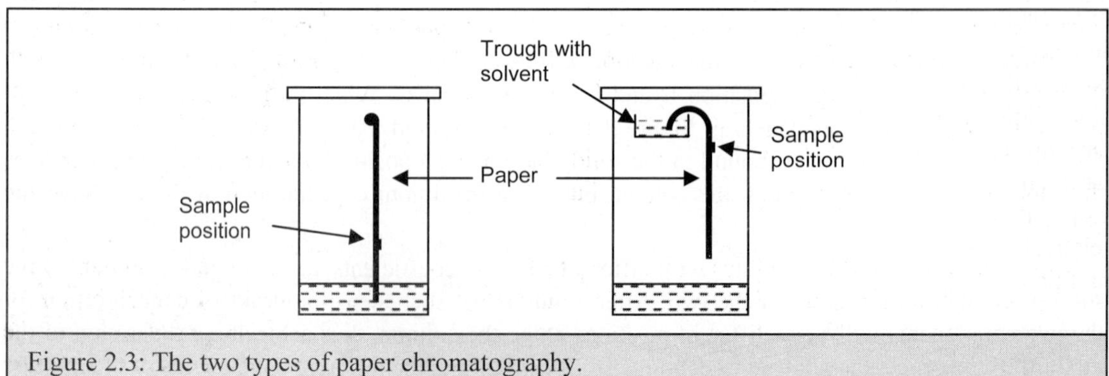

Figure 2.3: The two types of paper chromatography.

paper is allowed to hang down, without touching the solvent at the bottom of the tank. A solvent reservoir is required at the bottom of the tank to maintain the vapour pressure of the solvent in the enclosed volume of the tank. In other words it is required to keep the paper 'wet'. Again by capillary action (and gravity), the solvent moves down the sheet of paper, taking the sample along with it and separating the components. In both the ascending and descending methods, the components are detected by spraying the paper with specific colouring reagents, for example, ninhydrin, which reacts only with amino acids and proteins to produce a blue coloured spot. Sometimes fluorescent dyes are used. For example, ethidium bromide is used with DNA samples. When the paper is subsequently examined under ultraviolet light, the position of the fluorescent or ultraviolet absorbing spots can be seen. The corresponding compounds are identified on the basis of their R_f values where

$$R_f = (\text{distance moved by solute from the origin/distance moved by solvent})$$

The R_f value is the relative displacement of solute with respect to the solvent front and is a constant for a given compound under standard conditions. Thus by looking up a table of R_f values, it is possible to identify the compound after separation.

As mentioned earlier, chromatographic techniques are also classified on the basis of the molecular property of the support system or sample that is used in the separation. Some of these are considered below.

2.2.4: Adsorption chromatography:

This technique separates the sample into its components based on the degree of their adsorption[2] by the adsorbent and on their solubility in the solvent used. Silicic acid (silica gel), aluminium oxide, calcium carbonate and cellulose may be used as the stationary phase in this method. The optimum combination of adsorbent and eluting solvent will depend upon the type of sample separation under consideration. For example, hydroxyapatite (calcium phosphate) is used for DNA separation as this adsorbent can bind double stranded DNA but not single stranded DNA or RNA. Normally any organic solvent can be chosen as the mobile phase, e.g. hexane, heptane, proponal, butanol etc.

2.2.5: Partition chromatography:

There are two types of partition chromatography.

(a) Liquid-liquid chromatography: In this type of partition chromatography, the sample separation is achieved based on its differing solubilities in two (immiscible) liquid phases. The sample is dissolved in a relatively non-polar solvent and is passed through a column that contains immobilised water or some other polar solvent in a supporting matrix such as silica gel. When the sample moves through the column it gets partitioned between the mobile phase and the immobile, polar phase. In the reverse phase liquid-liquid chromatographic method, the stationary phase is a non-polar substance like liquid paraffin supported by the matrix. High Performance Liquid Chromatography (HPLC) is a liquid-liquid partition chromatographic technique. It is widely used for the separation of biological compounds either as a normal phase method, or as a reverse phase method. In the normal phase method, a relatively non-polar compound is dissolved in a non-polar liquid phase. Normal phase liquid-liquid partition chromatography is thus useful to separate non-polar compounds from relatively more polar

---

[2] Adsorption is the process by which a substance is absorbed on to the surface of another. There is almost no penetration beyond a few molecular layers.

compounds. In the reverse phase, a polar compound is dissolved in a polar liquid phase. This form of chromatography is therefore is useful for separating polar compounds from relatively less polar compounds.

(b) Counter current chromatography: This is also partition chromatography with this major difference, that there is no supporting matrix. The sample is equilibrated between two immiscible solvents (e.g. ethyl acetate and water) in a test tube. The two solvents will separate into two phases in the test tube, the upper phase and the lower phase, and the sample will mix in different concentrations in the two phases. The partition coefficient is defined as

$$K = (\text{concentration in the upper phase/concentration in the lower phase})$$

After equilibration the upper phase is withdrawn and transferred to a fresh volume of lower phase solvent in another tube. The original lower phase is added to a fresh volume of upper phase solvent, and so on. The process is repeated several times, yielding several test tubes, half of them finally containing the upper phase solvent, and half containing the lower phase solvent. A solute that is more soluble in upper phase ($K > 1$) will concentrate in tubes containing the upper phase solvent. If its solubility is greater in the lower phase ($K < 1$), then it will concentrate in tubes of the lower phase solvent. Counter current chromatography is successfully used for cell organelle fractionation. It is especially useful in the separation of natural products

2.2.6: Gas liquid chromatography (GLC):
The two phases used are a liquid and a gas. The method is very sensitive and the results obtained are highly reproducible. The partition coefficient of a volatile sample between the liquid and the gas phases, as it is carried through the column by the carrier gas, may be largely different from unity. This feature is exploited to separate or purify the compound. The stationary or immobile phase is generally non-volatile and thermally stable at the temperature of the experiment. The base is generally made up of celite (diatomaceous silica), and is coated on the inside of a narrow, very long (up to 100 meters), coiled, steel or glass tube. Usually organic compounds with high boiling point are coated on a base to form the stationary liquid phase. At the temperatures used for chromatography, the organic compound forms a thin liquid layer on the surface of this solid support. The sample to be separated is packed into the tube at one end, and placed in an oven at an elevated temperature. An inert gas like argon, nitrogen or helium is passed through the tube. The sample is volatilised and is carried away by the mobile gas phase. As the sample passes over the support, it distributes itself between the gas and liquid phases.

2.2.7: Ion exchange chromatography:
Many biological samples like amino acids and proteins have ionisable groups. The samples can therefore be separated based on the net charge acquired by them at a certain pH. The ion exchanger is a resin and can be either a cationic exchanger or an anionic exchanger. The cationic exchanger has negatively charged groups and attracts positively charged molecules. The anionic exchanger has positively charged groups and attracts negatively charged molecules. Commercially available ion exchangers are made by co-polymerising styrene with divinyl benzene. The two materials become cross-linked, with the permeability of the resin varying with the degree of cross-linking. When this insoluble, cross-linked resin is placed in water, a network of charged groups is formed. When the sample, suspended in the liquid phase, comes in contact with the resin, the charged groups in the sample replace the charged ions on the surface of the resin and the sample becomes strongly bound to

the resin. The usual mode of ion exchange chromatography is column chromatography or HPLC. The resin is usually packed into a column to act as the stationary phase, and the sample dissolved in an appropriate, aqueous buffer is the liquid phase injected into the column.

As an example, suppose that the sodium salt of a strongly acidic resin, which contains SO_3^- groups, is placed in a suspension. Normally the Na^+ ions would be bound to the negative sulphate groups in the resin. However, on exposure to the sample, the positively charged groups present in the sample may displace some of the Na^+ ions from the surface. This exchange of ions at the exchange site leads to the distribution of the sample between the solid and liquid phases. The higher the charge on the molecule to be exchanged, the tighter it binds to the resin. At low pH and salt concentration most amino acids will firmly bind to the resin by the following reaction

$$RSO_3^-... Na^+ + N^+H_3R' \leftrightarrow RSO_3^-...N^+H_3R' + Na^+$$

The next and last step in the separation procedure is to elute the bound molecule from the resin by reducing the attractive force between the resin and the charged molecule. This is achieved by changing the pH or the ionic concentration, or by affinity elution. In affinity elution, an ion with greater affinity than the bound molecule is introduced into the system. If more than one type of molecule is bound to the column, a stepwise elution can also be done by using a second solvent, a third solvent and so on. A widely used matrix medium for ion exchange chromatography is DEAE-cellulose column in which the Diethyl aminoethyl group acts as an exchanger. Cellulose fibres, polyacrylamide gel, cross-linked dextran (also known as sephadex) are materials commonly used as the matrix.

2.2.8: HPLC

HPLC or High Performance Liquid Chromatography is a form of column chromatography that may, in principle, be used with various column materials, but is in practice used for ion exchange or adsorption chromatography. One method of increasing the resolution in any column chromatography is to increase the length of the column. Beyond a certain length, however, this procedure is subject to the law of diminishing returns, chiefly for two reasons: the time it takes for the liquid phase to flow through the column increases enormously; concomitantly, the diffusion of the different solutes into other zones increases, thus reducing the resolving power. HPLC is a way of working around this problem. The columns are small, but very densely packed, with very fine particles of size approximately 10 microns. By applying a pressure gradient between the two ends of the column, the transit time of the mobile phase through the column is drastically reduced, thereby cutting down the diffusional spreading. HPLC thus leads to quick separation of the molecules at very high resolution. The columns are about 0.5 cm in diameter and about 15 cm in length. The pressure difference between the two ends of the column is about 10 kg/cm^2. This means that very small quantities of the sample may be used to obtain high resolution very quickly.

2.2.9: Molecular exclusion chromatography:

The separation of samples based on their molecular weights is known as molecular exclusion chromatography (MEC), gel filtration or molecular sieve chromatography. The matrix, i.e. the solid support, for MEC is made up of polyacrylamide, sephadex or agarose gel. Polyacrylamide and sephadex are supplied in the dehydrated form. When soaked in the aqueous solvent, they take it up rapidly and swell up forming a slurry. When this slurry is used for packing a column, pores of uniform size are formed by the gel and are used for separating the sample according to their molecular size and

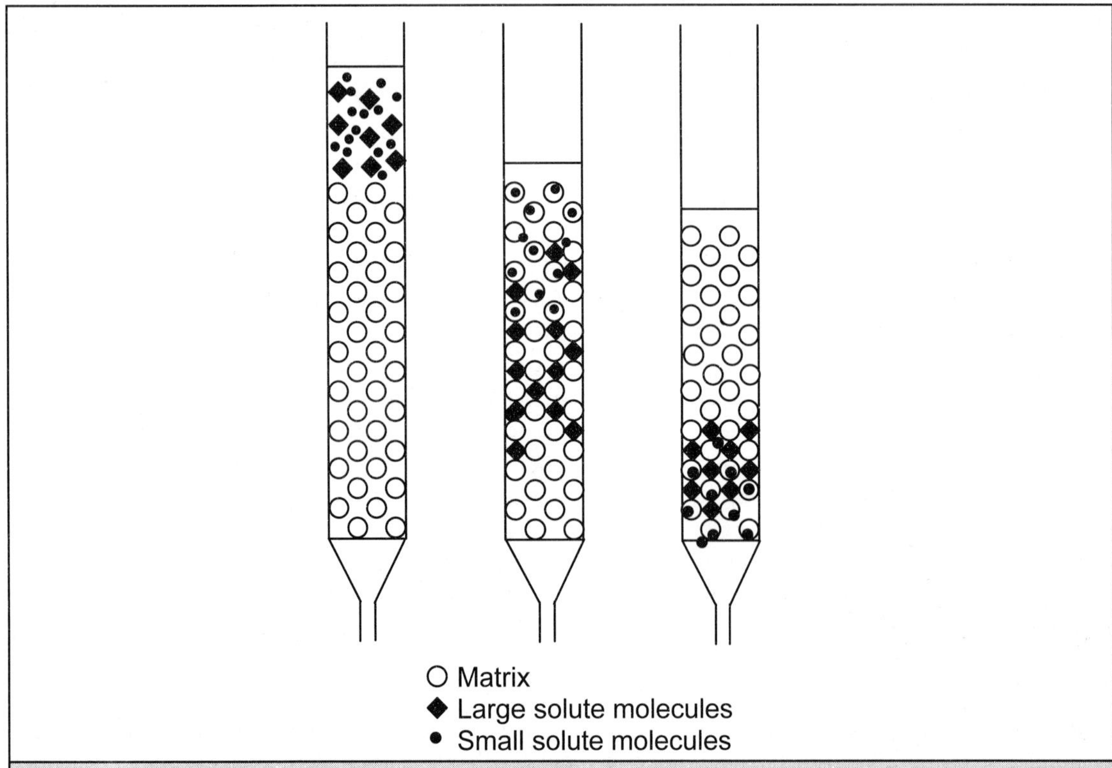

Figure 2.4: Molecular exclusion chromatography. The column is filled with solvent containing large and small solute particles.

shape (Figure 2.4). A column of the gel (the porous material) is kept in equilibrium with a suitable solvent in which the molecule to be separated is dissolved. Large molecules in this solution, are excluded from the pores, and pass through the interstitial space reaching the bottom of the column quickly. Smaller particles penetrate through the pores of the gel and are distributed in the solvent inside as well as outside the molecular sieve. Smaller molecules are thus retarded and reach the bottom at slower rate. For a given sample the distribution coefficient (K_d) is dependent upon its size. If $K_d = 0$ the molecule is completely excluded by the sieve and is found only in the external bulk solvent; if $K_d < 1$ the molecule is concentrated more in the bulk solution than in the pores; if $K_d = 1$ the molecule distributes itself equally between the pores of the sieve and the external bulk solvent; if $K_d > 1$ the molecule is prefers to attach itself to the inner surface of the pores by an adsorption process. The chromatography column consists of loosely packed resin particles. V_T is the total volume of the column. It is made up of three terms: the void volume (V_0) or the volume of the exterior solvent; V_g or the solid volume of the gel particles; and V_i or the internal volume of the pores that are accessible to solvent. Therefore

$$V_T = V_0 + V_g + V_i$$

The eluting volume is the volume of solvent that must flow through the column before a particular solute emerges. If the solute is completely excluded from the pores (i.e. if $K_d = 0$) the eluting volume is

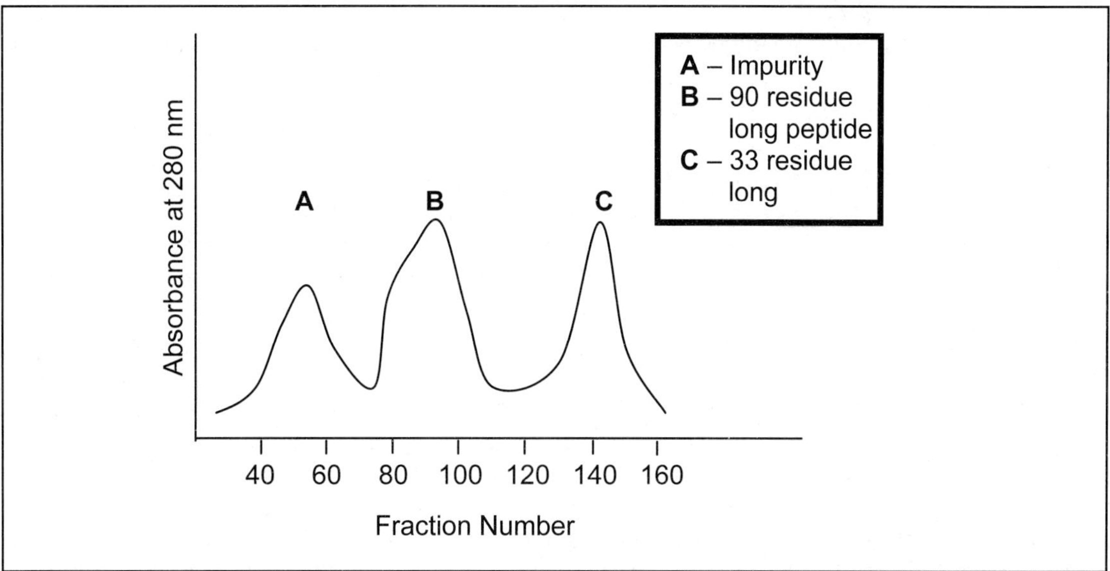

Figure 2.5: Separation of a mixture of peptides by a Sephadex column – schematic sketch of the absorbance v/s fraction number indicating the difference in movement of the peptides through the column corresponding to the different molecular weights.

the same as the void volume, and $V_e = V_0$ for such a sample. However, to elute a solute that can enter the pores, the solvent flowing through the column will have to displace an additional volume given by $K_d V_i$.

Thus the total eluting volume is given by the equation

$$V_e = V_0 + K_d V_i \qquad (2.1)$$

The partition coefficient K_d can be determined from this equation if the elution volume V_e is measured. V_0 and V_i are determined prior to the start of the experiment and are constant for a given column. V_0 is determined by measuring the elution volume of a particle larger than gel pores. V_i is determined by the weight of water intake by dry resin. Two substances with different K_d values (K_d and K_d') and different sizes can be separated if the sample volume is less than or equal to V_s, where

$$V_s = (K_d - K_d') \, V_i$$

Figure 2.5 shows the relationship between the elution volume and the molecular weight of various proteins when they pass through a Sephadex G-200 column. Since the elution volume is related to the partition function by equation (2.1) above, it is possible to correlate K_d to the molecular weight. From the best-fit line in Figure 2.5, it is possible to deduce this relationship to be given by the function

$$K_d = -A \log (M) + B$$

where A and B are constants and M is molecular weight. Thus by measuring the elution volume, it is possible to determine the partition coefficient and thence the molecular weight. Separation is thus based on the molecular weight, and molecules of different molecular weights will be present in different elution volumes that flow through the column.

2.2.10: Affinity chromatography:

Separation by affinity chromatography is based on a biological property of macromolecules rather than on a physical property. It is a highly sensitive technique and, in principle, can give absolutely pure sample in a single step. The technique requires that the macromolecule be able to bind to a specific ligand attached to an insoluble matrix under certain conditions, and then detach itself under certain other conditions. In other words the binding should be specific and reversible.

$$\text{Macromolecule} + \text{ligand} \leftrightarrow \text{Macromolecule . ligand}$$

When a complex mixture of proteins is passed through a chromatography column to which a ligand is attached, then the protein with specificity for that ligand alone will bind to the column and the rest of the material will simply flow through. The bound molecule can then be recovered by a change in the conditions such as pH of the solution, salts or temperature, or by adding a compound that interacts even more strongly to the ligand on the matrix, thus displacing the molecule of interest. The method therefore requires a preliminary knowledge of the biological specificity of the compound to be separated. In practice, agarose, cellulose, and polyacrylamide gels are used as the insoluble matrix. It is generally possible to select a ligand that has absolute specificity for the compound of interest or to select a ligand that has group selectivity. Group selectivity means that the ligand will selectively bind to a closely related group of compounds. For example, 5'AMP can reversibly bind to many NAD dependent dehydrogenases. The other end of the ligand, which is bound to the matrix, should be such

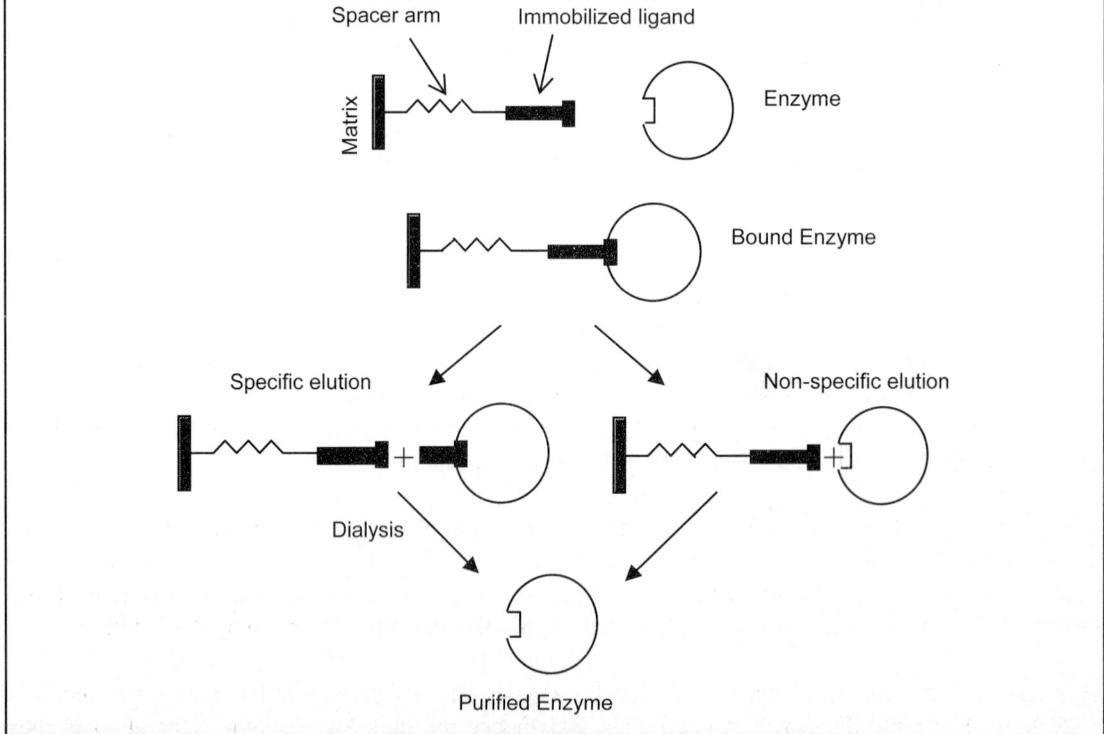

Figure 2.6: Affinity chromatography. The enzyme attaches itself to a specific ligand that is bound to the matrix. It is then eluted out, either specifically or non-specifically.

that it will not interfere with the reversible binding of the macromolecule. –NH, –COOH and –SH groups are the most common groups used for binding the ligand to the matrix. In addition, there is generally a spacer molecule between the ligand and the matrix, to minimize the interference of the groups in the matrix support material (Figure 2.6). The spacer arm usually has about six to ten carbon atoms. 1,6-diamino hexane and 6-amino hexanoic acid are examples of commonly used spacers. Some of the ligands used in affinity chromatography are: 5'AMP (with a specificity for NAD dependent dehydrogenases), lectins (different lectins are specific to different carbohydrates found on cell surfaces), lysine (specific to RNA) and concanavalin A (specific to certain polysaccharides and glycoproteins). A highly specific ligand for any macromolecule may be produced by raising antibodies against it and then binding the antibodies to the matrix.

2.3: Electrophoresis

Important biological molecules like peptides, proteins, carbohydrates and nucleic acids have ionisable chemical groups. Hence, under suitable conditions, they exist in solution as charged species. Even non-polar molecules can be made to exist as weakly charged species by preparing phosphate, borate or similar derivatives. Molecules with similar charge may have different charge/mass ratios since their molecular weights could be different. Likewise, molecules of the same mass, but different charge also may be differentiated on the basis of the charge/mass ratio. In an electric field, molecules with different charges or different charge/mass ratios would migrate at different rates. The separation of molecules by utilizing these differences in rates of migration is known as electrophoresis.

An elementary theory the rate of migration (i.e. the electrophoretic mobility) of different molecular species may be developed as follows. If a molecule has a net charge q, the application of an electric field E will result in a force

$$F = qE$$

This force will accelerate the particle in a fluid till a steady state is reached when the frictional force is equal and opposite to the applied force. If v is the steady state velocity then

$$fv = qE$$

where f is the frictional force. Therefore

$$v = qE/f$$

For a spherical molecule with radius r and charge Ze (where e is the charge on electron and Z is the atomic number),

$$v = ZeE /(6\pi\eta r)$$

since $q = Ze$ and $f = 6\pi\eta r$, where η is the viscosity of the solution. The electrophoretic mobility u is defined as the velocity per unit electric field.

$$u = v/E = Ze/(6\pi\eta r)$$

However, due to the presence of counter ions in any aqueous solution, this equation does not completely account for the electrostatic mobility. Such ions can neutralize some of the charges on the macromolecule if they bind tightly to it. Alternatively unbound or loosely bound ions can modify the ionic strength of the solution and thus alter the electric field that is felt by the polymer. In such a situation the electrophoretic mobility for a spherical particle is modified by a factor K, where K is a screening parameter, with dimensions of L^{-1} (inverse of length). Therefore

$$u = (ZeK)/(6\pi\eta r)$$

Thus mobility will increase with increasing charge on the macromolecule. This gives a way of calculating the charge on a molecule and its molecular weight based on its electrophoretic mobility. However the above discussion holds true only for a strictly spherical molecule. Proteins in general are not spherical, and the above discussion applies only loosely. A rigorous derivation of the mobility for a real protein is complicated and beyond the scope of this book.

Electrophoresis procedures can be generally divided into three categories: (1) Moving boundary electrophoresis. (2) Zone electrophoresis. (3) Continuous flow electrophoresis.

2.3.1: Moving boundary electrophoresis:
In this method, the sample is dissolved in a buffer and placed in a cell. An additional quantity of the buffer is layered on top of it. When the electric field is applied, closely related molecules will tend to move through the solvent as a band since their charges are similar. During this motion, boundaries are formed between groups of molecules with different electrophoretic mobility, and hence the name of the technique. The separations can be seen by direct optical observations. This method suffers from many experimental difficulties such as errors due to convection, and is therefore not widely used.

2.3.2: Zone electrophoresis:
In this method the mixture to be analysed is applied as a small spot or very thin band to a supporting medium, such as a gel. The use of the supporting medium prevents problems related to convection that are encountered in the moving boundary technique. When electric field is applied the molecules move as distinct bands or zones. The molecules can be identified by staining, UV absorption etc. There are several different types of zone electrophoresis. We treat some of these below.

2.3.3: Low voltage electrophoresis:
The electrophoresis unit (Figure 2.7) consists of a power pack, which supplies DC current across two electrodes, buffer reservoirs, a support, and a transparent insulating cover. The low voltage power pack can either give a constant voltage or a constant current and has an output of 0 to 500 V and 0 to 150 mA. Either stainless steel or platinum electrodes are used. The supporting medium, which can be paper or cellulose acetate, must first be saturated with buffer and held on a sheet of insulating material like

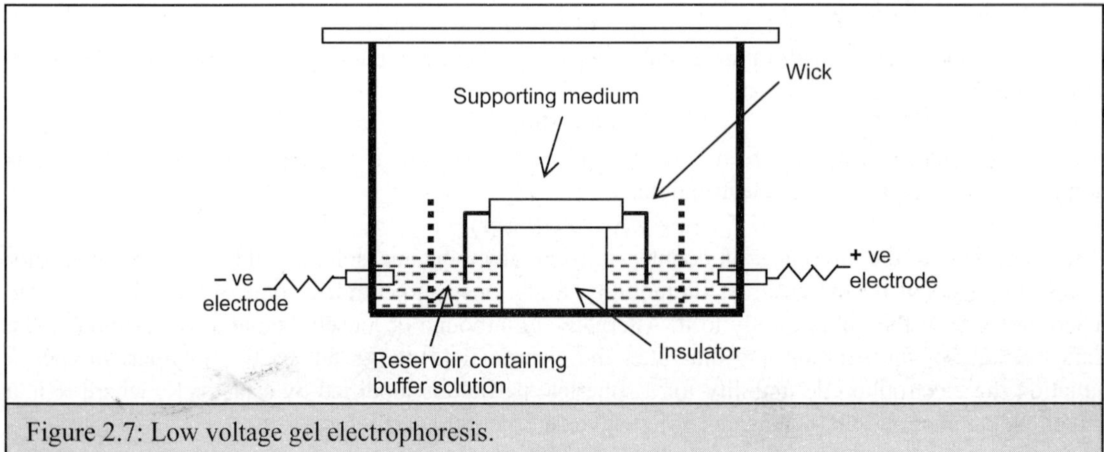

Figure 2.7: Low voltage gel electrophoresis.

Perspex. The sample is then applied as a small spot using a micropipette. The location of the spot depends on the nature of the molecular mixture present in the sample. For example, if there are molecules with opposite charges, then they will separate and move towards opposite electrodes and hence the sample must be applied at the centre. The two reservoirs on either side of the medium are isolated from the electrodes so that any pH change there does not affect the buffer. After application of the sample, power is switched on for the required amount of time. At the end of the experiment, the migration of the different constituents of the mixture may be analysed. The experiment may be performed in different ways. For example, instead of paper or cellulose acetate, thin layers of silica, alumina or cellulose may be applied on to a glass plate and used as the supporting medium. This technique is similar to TLC (thin layer chromatography) and is known as thin layer electrophoresis (TLE).

2.3.4: High voltage electrophoresis:

A disadvantage with low voltage paper electrophoresis is that diffusion processes may influence the migration of the molecules so that it is not strictly according to the charges on them. One way to overcome this is to resort to high voltages, of the order of 10,000 volts. Such a voltage can produce potential gradients up to 200 volts/cm. If the separation between the electrodes is d metres and if V is the potential difference in volts then the potential gradient is V/d volts/metre. The force on the charged molecule is then Vq/d newtons, where q is the charge on the molecule in coulombs. The migration of the molecules is caused by this force and is proportional to it, and an increase in the potential gradient leads to an increased rate of migration. The distance migrated by the ions will depend on the voltage as well as the amount of time for which it is applied. This technique, which uses the same apparatus as the low voltage set up, results in higher resolution and faster separation. However, because of the high voltage, it generates considerable heat and the apparatus requires cooling.

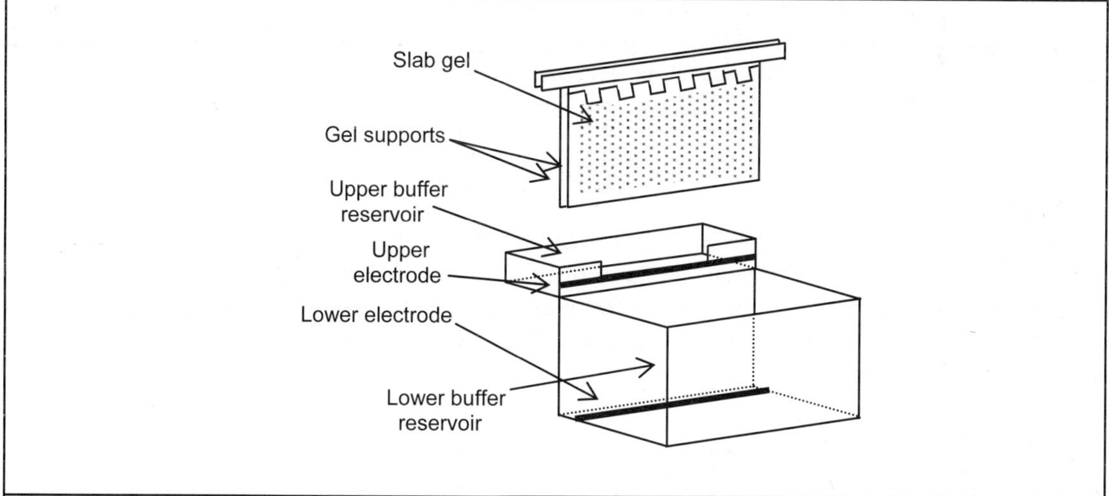

Figure 2.8: Vertical slab gel electrophoresis apparatus. Most of the equipment (except obvious parts such as electrodes and clamps) is made of Perspex.

2.3.5: Gel electrophoresis:

When applied to the separation of a mixture of proteins, simple zone electrophoresis has limited resolving power, due to the similarity in mobility of the components. When zone electrophoresis is combined with molecular exclusion effects using gels, much higher resolutions can be achieved. The molecular sieving property of the gel helps in separating molecules that have similar charge properties but differ in size and shape. Starch, agar, and polyacrylamide gels are used for this purpose. The gels can be run as horizontal slabs or as vertical slabs. Vertical slabs of the type shown in Figure 2.8 are most frequently used. Gels are prepared in glass or Perspex containers. In the case of slab gels, the gel is cast between two clean glass plates that are clamped together but held apart by spacers. The samples to be separated are applied using a micro syringe into the wells set in the gel. In the vertical slab method the two ends of the plate containing the gel are kept in contact with the lower and upper reservoirs of buffer respectively, thus completing the electrical circuit between lower and upper electrodes. The components of the mixture migrate with different mobility on application of an electric field, and can be identified at the end of the experiment by staining techniques. Normally, a known sample is run along with unknown samples in order to compare the migration characteristics. A major advantage of this method is that even very small quantities of proteins can be separated and identified.

2.3.6: Sodium dodecyl sulphate poly acrylamide gel electrophoresis (SDS-PAGE):

In this technique, all the proteins in a mixture are converted into structures having similar charge characteristics, though they may differ in molecular weight. Sodium dodecyl sulphate is an anionic detergent that binds tightly to proteins and denatures them. When an excess of SDS is added to the solution (about 1.4 gm per gm of amino acid), then the protein molecules in the mixture are completely surrounded by SDS. The total negative charge acquired due to SDS binding overwhelms the variations in charge in the different protein molecules. Hence, when a drop of the solution is placed on the polyacrylamaide electrophoresis gel, all the protein molecules move towards the anode, with their mobility varying according to their molecular weight, size and shape. Reducing agents like β-Mercaptoethanol may also be added to break any intrachain or interchain disulphide bonds. The apparent mobility $\mu(c)$ of the molecule at the gel concentration c is given by

$$\log(\mu(c)) = -k_x c + \log(\mu(0))$$

where $\mu(0)$ is a constant for a set of similar proteins, and k_x is a constant depending on the extent of crosslinking of the gel. It has been shown that the mobility μ and molecular weight M are related by the following equation.

$$\mu = b - a \log (M)$$

where b is a constant which contains the term $\mu(0)$, and a is a function of gel concentration c. This equation fits the experimental observations extremely well and is used to determine the molecular weight of an unknown sample by running it on a gel side-by-side with another mixture containing standard proteins of known molecular weights.

2.3.7: Iso electric focussing:

Isoelectric focussing is an electrophoretic method where a pH gradient is used along with the voltage gradient. It is extremely useful in separating proteins according to their isoelectric points. The isoelectric point of a protein is defined as the pH at which the average net charge on the molecule is zero. In the electrophoretic gel the region around the anode is kept at a lower pH as compared to the region near the cathode. A pH gradient is established such that the samples to be separated have their isoelectric points within this pH range. At a pH below their isoelectric point, the particles will move towards the cathode, and at a pH above the isoelectric point they move towards the anode. At the isoelectric pH, the particles will remain immobile. Thus, when placed in the combined pH and voltage gradient on the gel, the particles move towards their isoelectric points no matter where they are initially applied. This method gives very high resolution and is especially suitable for separating isoenzymes. Even differences in pH of the order of 0.01 are enough to separate the molecules. Substances known as ampholytes generate the pH gradient between the electrodes. Ampholytes are molecules with both positive and negative charges, for e.g. polymers containing several amino and carboxyl groups. Ampholine, Biolyte and Pharmalyte are some of the commercially available ampholytes. When an electric field is applied to a mixture of ampholytes with a wide range of isoelectric points in a column or gel, they distribute themselves according to their charge and thus establish a pH gradient. A small amount of sample may then be added to the medium. The different proteins in it migrate along the column or gel till they reach the pH corresponding to their respective isoelectric points. Each protein then remains at that position as a sharp band and may be viewed and analysed by staining, absorbance etc.

2.3.8: Continuous flow electrophoresis:

In continuous flow electrophoresis the experiment is conducted in free solution without a supporting medium. This reduces frictional effects between the sample ions and the solution and causes rapid migration of the components of the mixture. This is a form of the moving boundary method. The sample, in the carrier buffer, is applied through the central tube as a continuous flow. The electric field is applied between the outer and inner concentric cylinders and the outer cylinder is rotated to maintain a stable flow of the buffer solution. Electrophoresis takes place continuously as the sample is carried upwards by the flow of carrier buffer from the bottom of the inner tube. As it moves upwards the components are separated radially. At the top a series of radial slits separates the buffer stream into thirty or more fractions, each containing a separated portion of the original mixture. Large-scale separations can be achieved by this method. This technique is therefore used for enzyme purification on an industrial scale, blood plasma fractionation. In the laboratory it is used for collection of large amounts of proteins on a preparative scale.

Summary

- Chromatography and electrophoresis are the two commonly used techniques for the separation of the components of a mixture of molecules.

- Column chromatography, thin layer chromatography, paper chromatography, adsorption and partition chromatography, gas liquid chromatography (GLC), ion exchange, molecular exclusion and affinity chromatography are the techniques which are commonly employed.

- Adsorption, partition, ion exchange, exclusion and affinity chromatography all use the column mode.

- Ion exchange chromatography is useful in separating the components based on the net charge acquired by them at a certain pH.

- In the molecular exclusion chromatography the separation is based on their molecular weight.

- Separation by affinity chromatography is based on the biological property of macromolecules rather than the physical property.

- When a complex mixture of macromolecules is passed through the affinity chromatography column to which a ligand is attached, then the molecules with specificity for that ligand will bind the column and the rest will flow through the column.

- Electrophoresis takes advantage of the fact that molecules with differing charge/mass ratio will migrate at differing rates in an electric field.

- Moving boundary electrophoresis, zone electrophoresis, gel electrophoresis, and isoelectric focussing are some of the frequently used techniques.

- Isoelectric focussing is an electrophoretic method where pH gradient is used along with the voltage gradient. This method is useful in separating proteins according to their isoelectric points.

- In the SDS-PAGE (Sodium Dodecyl Sulphate – Poly Acrylide Gel Electrophoresis) all the proteins are converted into structures with similar charge characteristics and hence their mobility varies according to their size and shape.

- This method is used to determine the molecular weight of an unknown sample by running it on a gel side by side with another mixture containing standard proteins of known molecular weights.

Chapter 3

Physico-Chemical Techniques to Study Biomolecules

3.1: Introduction

Biological macromolecules have molecular weights of the order of 10,000 daltons. Their functions depend on their three dimensional structures. It is therefore useful to know the structure, size and shape of these molecules in order to understand the mechanism of function. There are several techniques that can give information on these aspects. Table 3.1 lists some of the physical techniques that are used in studying large biomolecules.

In this chapter we will focus our attention on some of the hydrodynamical methods listed above and discuss in detail the principles underlying them.

3.2: Hydration of macromolecules

The molecular volume V of any isolated molecule is given by

$$V = (M<v>_0)/N_0,$$

where M is the molecular weight, and $<v>_0$ is the specific volume, equal to $1/\rho$, where ρ is the density. N_0 is Avagadro's number. However, in the case of proteins and nucleic acids, there may be counter ions and water molecules tightly bound to the molecule and therefore it cannot be treated in isolation. It is normal laboratory practice to store and use proteins and nucleic acids in ~ 0.1 M buffered salt solutions, thereby simulating approximately the physiological environment. Hence, when we study the thermodynamic properties of a macromolecule solution we must treat it as a three-component system consisting of salt, water and the macromolecule. Such a multi-component system is very difficult to deal with. However, since we are mainly interested in the interaction of the macromolecules with water, we may approximate the system as consisting of two components– water and macromolecule. The total volume of such an ideal solution is given by the following equation.

Table 3.1: Physical techniques used to study biological molecules and the information they yield.

Technique	Information obtained
1. Osmotic pressure	Molecular weight
2. Diffusion	Molecular volume
3. Sedimentation	Molecular weight and volume
4. Viscosity	Molecular shape
5. Birefringence	Molecular shape
6. Optical Rotatory Dispersion	Helical content
7. Light scattering	Molecular size and weight
8. X-ray scattering	Molecular size and shape
9. X-ray crystallography	Three dimensional structure (atomic positions)
10. Electron Microscopy	Size and shape
11. Electromagnetic absorption, nuclear magnetic resonance, electron spin resonance	Intramolecular forces, Orientation of groups, Three dimensional structure

$$V_{tot} = g_1 <v>_1 + g_2 <v>_2$$

In this equation, g_1 and g_2 are the weights of water and macromolecule respectively, measured in grams, and $<v>_1$ and $<v>_2$ are the specific volumes of the water and the macromolecule, measured in cubic centimetres per gram. In addition to the bound water molecules, we may have bulk water, which may occupy holes or cavities in the macromolecule, thus changing its shape and size. (Bulk water is the the term used to describe the water molecules not specifically bound to the macromolecule. That is, water molecules not in one of the hydration shells are considered part of the 'bulk' water.) Taking all this into consideration, the volume occupied by the hydrated macromolecule is given by

$$V_h = (M/N_0)(<v>_2 + \delta_1 <v>_1)$$

where the ratio (M/N_0) (or the molecular weight divided by Avogadro's number) gives the weight of a single macromolecule. $<v>_2$ is now the partial specific volume of the solute, and is defined as the change in volume of the solution when a small amount of solute is added at infinite dilution (i.e. when the addition is made in such a manner that the concentration of the solute in the solution remains zero). $<v>_1$ is the partial specific volume of water $= 1/\rho$, where ρ is the density of pure water, and δ_1 is the hydration, measured in units of grams of bound water per gram of macromolecule. The hydration value of a protein gives an estimate of the amount of water that is bound to a protein. This value may be calculated from the above equation, in which the quantities $<v>_2$ and δ_1 are experimentally measurable quantities. Such measurements have yielded hydration values of the order of 0.3 to 0.4 gm of water per gram of protein. By comparing this value with the shapes and sizes of most proteins, we may deduce that the thickness of the hydration shell is about 2.8 A. In other words, one layer of water molecules, closest to the surface of the proteins remains tightly bound, and forms the hydration shell.

3.3: Role of friction

The frictional forces between protein molecules and the solvent affect the hydrodynamic properties like diffusion, viscosity and sedimentation of macromolecules. Since this frictional force opposes the motion, we can include this in the equation of motion as follows.

$$F - fv = m(dv/dt)$$

where f is the frictional coefficient, (dv/dt) is the acceleration and m is the mass of the molecule[3]. In the case of spherical particles, the translational frictional force f is proportional to the fluid viscosity η and radius r of the particle. Thus the frictional force for spherical particles is given by

$$f_{sph} = 6\pi\eta r$$

This relation, known as Stokes' law, is true when the interaction of the particle with the fluid is strong. On the other hand, when there is virtually no interaction at all between the fluid and the particles, the frictional force is given by

$$f_{sph} = 4\pi\eta r$$

The frictional force can also act as a damping force for the rotational motion of the particle and in that it is given by

$$f_{rot} = 6\eta V$$

where V is the volume. Since most biological molecules are not spheres, the previous equations will have to be modified for the shape factor. As a significant fraction of macromolecules are globular, compact irregular bodies or an ellipsoid may be a better approximation than a sphere. Ellipsoids are classified into prolate and oblate ellipsoids. An oblate ellipsoid is disc shaped and is obtained by rotating an ellipse about its minor axis (Figure 3.1). A prolate ellipsoid can be taken as an approximation for a rod and is obtained by rotating an ellipse about its long or major axis. For equal volumes, ellipsoids have greater surface area compared to spheres and hence their frictional coefficients will be larger. The frictional forces or coefficients may be measured by diffusion

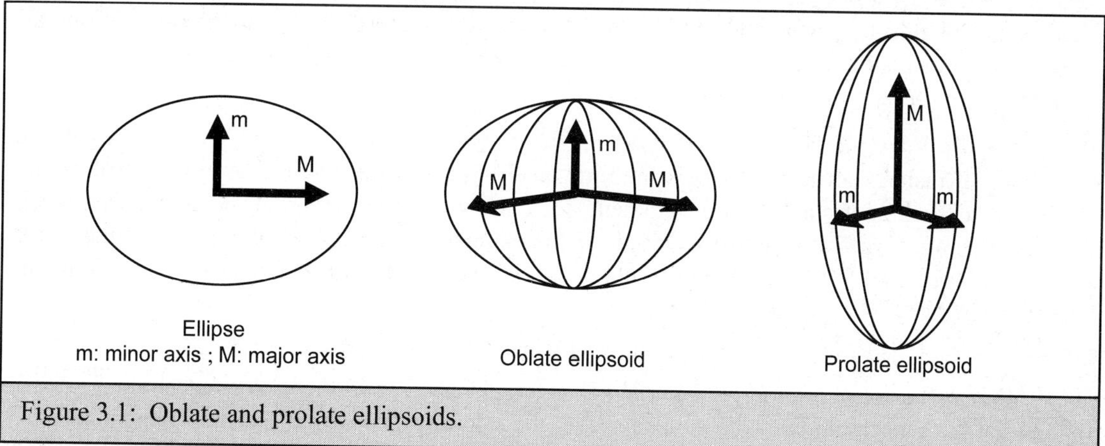

Figure 3.1: Oblate and prolate ellipsoids.

---

[3] This is the general equation of motion F = m(dv/dt) or (mass x acceleration) modified to take into account the opposing frictional force which is taken to be proportional to the velocity v.

experiments, and can be used to determine the shapes of molecules.

3.4: Diffusion and Osmosis

3.4.1: Diffusion:

Molecules in liquids and gases are in constant motion. The movement of the molecule could be described as either diffusion or sedimentation, and the driving force can be the concentration gradient, the force of gravity or a centrifugal force. Diffusion can be considered as movement of molecules from a higher concentration to a lower concentration (Figure 3.2). Several biological processes take place due to diffusion. A few examples follow. The exchange of O_2 and CO_2 in lungs and tissues are diffusion-driven processes. Similarly nutrients like water-soluble vitamins; minerals etc. are absorbed by diffusion in gastro-intestinal tract. In biochemical experiments, dialysis is based on diffusion properties of the solvents, the solutes and the membranes. Small organisms such as protozoa depend on diffusion to acquire nutrients and dispose of waste. However, since diffusion is effective only when the ratio of surface area /internal volume is large, large organisms have evolved circulatory systems (such as blood flow) to transport substances from the site of its production to the site where they are utilised or disposed off.

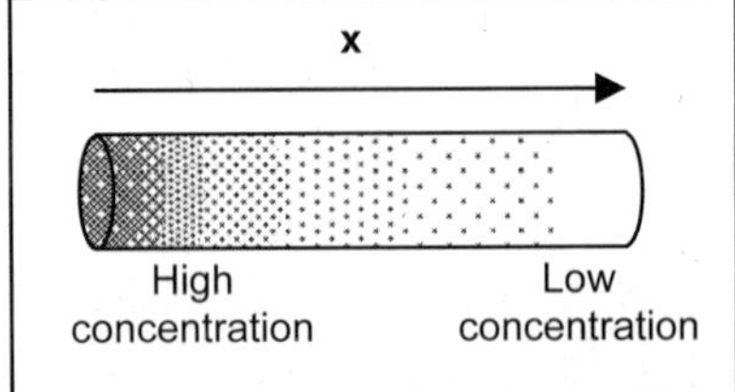

Figure 3.2: Diffusion of molecules from higher concentration to lower concentration.

   Diffusion is quantitatively characterised by a parameter called the diffusion coefficient, D. This coefficient is measured as the net flow of particles per unit time across unit area of an imaginary plane perpendicular to the concentration gradient. The diffusion coefficient D is related to the frictional coefficient f by the following equation.

$$D = kT/f \qquad (3.1)$$

where k is Boltzmann's constant and T is the absolute temperature. For a stochastic (i.e. random) process like Brownian motion, the root mean square distance (r_{rms}) travelled by a diffusing particle is given by

$$(r_{rms})^2 = Dt$$

where D is diffusion coefficient and t is the time taken. For a diffusion process in the presence of a driving force, such as a concentration gradient, Fick's law applies. According to Fick's first law of diffusion, the rate of diffusion across a boundary (the number of molecules which pass through a cross section A in unit time) for a single solute component diffusing in a system at constant temperature and pressure is given by

$$dn/dt = -DA(\delta C/\delta r)$$

where $\delta C/\delta r$ is the concentration gradient and D is the diffusion coefficient. A concentration gradient implies that the concentration of the molecules (i.e. the solute) varies with distance r in one dimension. The concentration of the solute or the concentration gradient can be measured as a function of position, by the absorbance at various points. Alternatively it can be measured as the refractive index gradient using a light scattering technique known as Schlieren optics. The concentration gradient across a boundary is given by Fick's second law of diffusion as

$$(\delta C/\delta r)_t = C_0/(4\pi Dt)^{1/2}\exp(-r^2/4Dt) \qquad (3.2)$$

where C_0 is the total solute concentration difference across the boundary. In the graph of concentration gradient versus distance, plotted at different times (Figure 3.3), the downward change in the concentration gradient as time passes is clearly observed. The x co-ordinate of this graph is defined in such a way that the maximum value H_m of the gradient is at x = 0. The rate of change of concentration gradient with time (d/dt ($\delta C/\delta r$)) is zero at the maximum value of the gradient. Differentiating equation 3.2 with respect to time and setting it equal to zero, we obtain

$$H_m = C_0/(4\pi Dt)^{1/2}$$

When H_m^2 is plotted versus 1/t, the slope of the straight line so obtained will give the value of D, which in turn can be used to determine the molecular weight using equation (3.3) below, provided the partial specific volume <v> is known. <v> is defined as the volume occupied by 1 gram of the solute in the solution. For non-hydrated spheres the volume V is given by

$$V = (M\text{<v>})/N_0$$

where M is the molecular weight and N_0 is Avagadro's number. But $V = (4/3)\pi r^3$. Therefore

$$M = (4\pi r^3 N_0)/(3\text{<v>})$$
$$= (4\pi N_0/3\text{<v>}) \times (f/6\pi\eta)^3$$
$$= (4\pi N_0/3\text{<v>}) \times (kT/6D\pi\eta)^3 \qquad (3.3)$$

where is η the viscosity. The last step follows from Stokes' law. Thus an experimental measurement of the diffusion constant of a molecule enables the calculation of its molecular weight. Diffusion experiments are not easy to perform, due to experimental difficulties such as the slow rate of diffusion of macromolecules (diffusion constants are only of the order of 10^{-7} to 10^{-6} cm^2/sec), gravitational instabilities, convective mixing, etc.

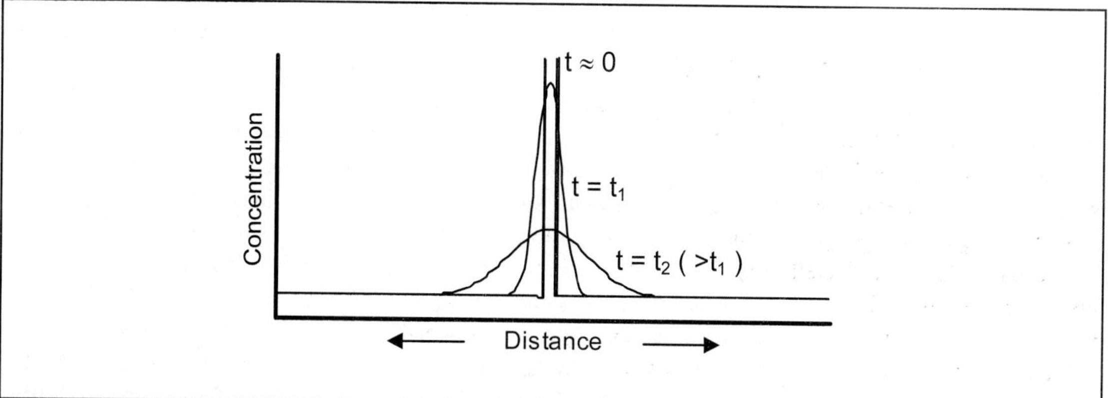

Figure 3.3: A plot of the concentration as a function of distance at three different times. At t = 0, the sample is applied at the centre and is concentrated only here. It starts diffusing until at t = t_2 it is spread out over a wide area.

3.4.2: Osmosis:

The movement of solvent molecules across a semi-permeable membrane that separates two solutions of different solute concentrations is known as Osmosis. The flow of solvent occurs from a solution of low concentration of solute to a solution of high concentration. For example, when an aqueous solution is separated from pure water by a semi-permeable membrane that allows only water molecules to pass through, then the pure water present outside the membrane moves into the solution in order to equalise the concentration of the solute. If the membrane is permeable both to water and solute, then the solute will diffuse through the membrane from the solution to the pure water, while water molecules will move in other direction across the membrane, till equilibrium is reached. An osmotic pressure difference between the liquid on both sides of the membrane is then said to exist. Osmostic pressure will depend on the nature of the solvent as well as the nature and concentrations of the solute. Solutions that exert the same osmotic pressure are iso-osmotic. If two iso-osmotic solutions are placed on either side of the membrane, there will be no change in volume of the solutions, i.e. the two solutions are already in osmotic equilibrium, and no net transport of solute or solvent molecules occurs across the membrane. Two such solutions are also called isotonic solutions. For example, if a cell is in direct contact with a solution of 0.9 % NaCl, then there is no change in the cell volume, and the 'cell tone' is maintained. Thus 0.9% NaCl is said to be an isotonic solution with respect to the contents of a normal cell. Isotonic solutions are often used in the treatment of dehydration. A solution with higher osmotic pressure is said to be hypertonic, and one with lower osmotic pressure is called hypotonic.

In multi-cellular organisms the fluid inside the cells, and that surrounding the cells contain dissolved substances ranging from small inorganic ions to huge molecular aggregates Osmotic pressure can destroy the cell wall and in order to prevent this, most plants and bacteria use a rigid cell wall, which can withstand the pressure. In some cases the cell is surrounded by a solution of similar osmotic pressure, i.e. an isotonic solution.

3.5: Sedimentation

The sedimentation or separation of particles from a suspension in a liquid may be achieved either just by the force of gravity or by applying a centrifugal force. For example, if a solution of mud (composed mainly of silica) in water is left to stand undisturbed, it will separate out after some time as a relatively clear layer of water at the top of the container, and a thick layer of mud at the bottom. This sedimentation of mud particles is due to the force of gravity. However, when we perform a sedimentation experiment with macromolecules, the force of gravity is insufficient to make the molecules to undergo sedimentation. Under these circumstances the solution containing the macromolecules is subjected to a strong inertial force such as centrifugation in an ultracentrifuge. An ultracentrifuge is an instrument that can spin the sample at speeds of several thousand revolutions per minute. Thus the suspended particles in the sample are subjected to centrifugal forces of the order of 3,00,000 times the acceleration due to earth's gravity, or 3,00,000 g. This force is countered by the force of buoyancy of the suspended particle in the solution. Since the centrifugal forces act also on the solvent molecules, the net velocity of the suspended particles in the solution subject to the centrifugal force depends on the molecular weight, size and shape of the particles. In general, the larger the mass of the particle, and the more compact it is, the faster its movement through the solution.

As the particles are spun in an ultra-centrifuge, the radial or centrifugal force F_C is given by

$$F_C = m\omega^2 r$$

where ω is the angular velocity, and r is the radial distance of the particle from the axis of rotation. As the particles sediment, diffusion effects also take place simultaneously, and the motion of macromolecules in an ultracentrifuge is the resultant of both the effects. Therefore the term for mass in the above equation is replaced by one for the effective mass, or the difference between the weight of the particle and the buoyant force due to displacement of the medium. The effective mass is given by $(m-V\rho)$, V is the volume and ρ is the density of the solvent. ($V\rho$ represents the mass of solvent displaced by the particle of mass m.) Making this substitution, the centrifugal or radial force is given by

$$F_C = (m-V\rho)\omega^2 r$$

This force is opposed by the frictional force F_r that is proportional to the velocity of the particle

$$F_r = f \times (radial\ velocity) = f(dr/dt)$$

where f is the frictional coefficient. When the centrifugal force is balanced by the frictional force then $F_C = F_r$ and the particles will sediment with uniform velocity, since there is no net force on them and thus no acceleration. Hence

$$(m-V\rho)\omega^2 r = f(dr/dt) \tag{3.4}$$

in which ω is expressed in radians per second. Multiplying both sides by Avogadro's number N_0 we get

$$N_0(m-V\rho) = N_0 f(dr/dt)/(\omega^2 r)$$

The velocity of sedimentation depends on the size and shape of the sedimenting particle and is proportional to the applied force. Therefore

$$v = s\omega^2 r$$

where s is a proportionality constant depending on shape and size of the molecule and the nature of the liquid. s is called the sedimentation coefficient and, according to the above equation, is defined as the velocity per unit centrifugal force applied.

$$s = velocity/\omega^2 r = (dr/dt)/(\omega^2 r)$$

Therefore,

$$N_0(m-V\rho) = N_0 fs$$

or

$$N_0 m(1-V\rho/m) = N_0 fs$$

But $N_0 m$ is the molecular weight M. Therefore

$$s = [M(1-V\rho/m)]/N_0 f$$

Again, $V/m = \langle v \rangle$, the partial specific volume. Substituting,

$$s = [M(1-\langle v \rangle \rho)/N_0 f]$$

or

$$M = N_0 fs/(1-\langle v \rangle \rho)$$

From equation (3.1)

$$f = kT/D.$$

where f is the frictional coefficient, k is Boltzmann's constant, T is the absolute temperature and D is the diffusion coefficient. Substituting,

$$M = N_0kTs/D(1-<v>\rho) = RTs/D(1-<v>\rho) \qquad (3.5)$$

where R is the gas constant and is equal to N_0K. This equation is called the Svedberg equation and is arrived at by combining the sedimentation and diffusion equations. If M and D are known, s, the sedimentation coefficient, or f, the coefficient of friction, can be calculated. Else, molecular weight M can be determined from knowledge of D and s. The sedimentation coefficient for any molecule at 20^oC and in pure water is taken as the standard for comparison. The expression for this value in terms of the coefficient s at any temperature and in any liquid is given by

$$s_{20,W} = [s\eta(1-<v>\rho)_{20,W}]/\eta_{20,W}(1-<v>\rho)$$

In this equation, η is the viscosity of solution at the experimental temperature, and $\eta_{20,w}$ is the viscosity of water at 20° C, a constant. The sedimentation coefficient is measured in Svedberg units ($1 S = 10^{-13}$ sec). For example the sedimentation coefficient of one of the subunits of ribosome is 30S and the other is 50S.

3.6: The ultracentrifuge

The ultracentrifuge is an instrument used to measure sedimentation coefficients by rotating the sample at speeds as high as 100,000 rpm (revolutions per minute). It was developed around 1923 by T. Svedberg, a Swedish biochemist. After the invention of this instrument it was possible to demonstrate that proteins are macromolecules with homogeneous compositions. Macromolecules in cells can be solubilised by disrupting the cells either by mechanical means or chemically using detergents. After such lysis, ultracentrifugation can help to purify the molecules, at least partially. In ultracentrifugation, the molecules are subjected to centrifugal force of several thousand g. Molecules can be separated as they sediment at different rates. Their molecular weights can be determined by estimating the rates of sedimentation. Alternately, molecules can be separated by exploiting their different buoyant densities in a solution with a gradient of density. Such a gradient may be prepared using a dense, fast diffusing substance like CsCl. Figure 3.4 shows an ultracentrifuge with its component parts. The sample is held in a centrifugal cell placed in rotor made of titanium or aluminum. This assembly is rotated by an electric motor at speeds up to 100,000 rpm. The entire rotor assembly, including the sample cell, is kept inside an armoured chamber to protect the users against any possible explosive accident. In order to avoid excessive heating that may be caused by friction with air, the rotor chamber is evacuated and refrigerated. In order to avoid collision of the particles in the sample with the cell wall, the sample cell is shaped like the sectors of a circle drawn around the axis of rotation. Due to the high speeds, the balance of weight on the rotor is critical.

A centrifuge can be either a preparative instrument or an analytical one. In a preparative centrifuge the sample is spun for a fixed length of time to separate mixtures or purify samples. In an analytical centrifuge, the movement of the boundary during sedimentation can be monitored using optical devices like Schlieren optics, Rayleigh interference or absorbance, and the period and rate of centrifugation can

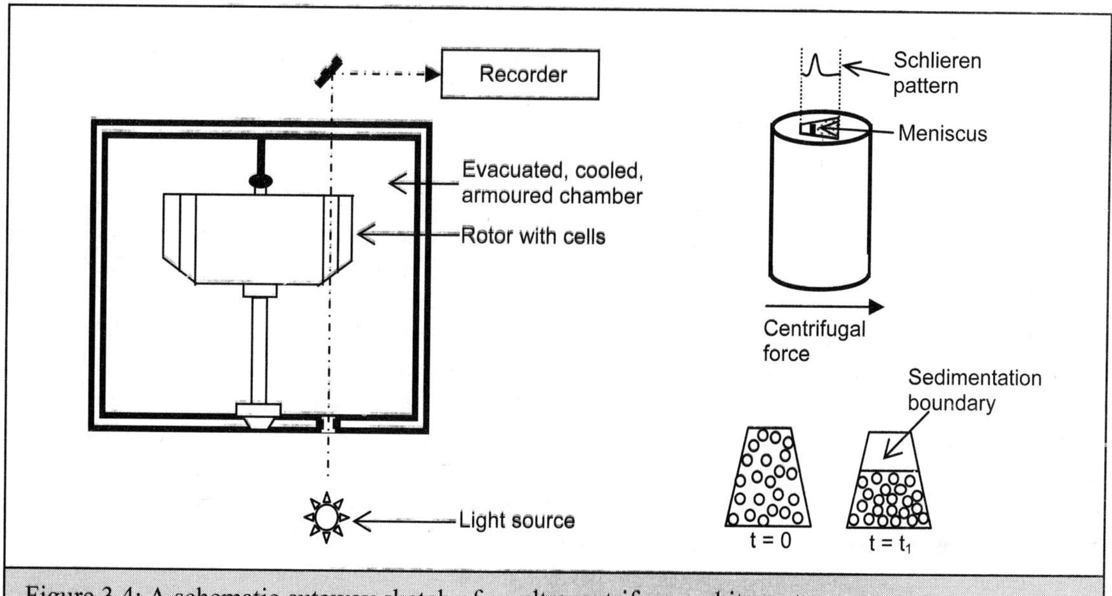

Figure 3.4: A schematic cutaway sketch of an ultracentrifuge and its parts.

be accordingly adjusted. The Schlieren optical system is based on the fact that light does not deviate as long as it moves through a solution of uniform concentration, but undergoes refraction in a solution with varying densities. The change in refractive index with change in concentration is recorded and is used in the determination of the sedimentation coefficient.

Molecular weight determination can be carried out by the sedimentation velocity technique or the sedimentation equilibrium method. In the sedimentation velocity method, the ultracentrifuge is operated at high speeds and the movement of the sedimentation boundary is recorded as a function of time. This is a measure of the rate of sedimentation. Rayleigh's interference method is used to make this record, since the Schlieren optical system is not very sensitive to small changes in concentration. The Rayleigh interference system uses a double sector cell, one for the solvent and the other for the solution. By measuring the displacement of interference fringes, the system measures the difference in refractive index between the reference solvent and the solution.

In the equilibrium method the sample is centrifuged till equilibrium is reached between sedimentation and the diffusive movement of the particles, i.e. there is no net movement of solute particles in the cell and the concentration gradient remains stable. In other words, the net flux J is zero. This means that

$$J = s\omega^2 rC - D(\delta C/\delta r) = 0$$

or

$$s\omega^2 rC = D(\delta C/\delta r)$$

$$s/D = (\delta C/\delta r)/\omega^2 rC$$

Substituting in the Svedberg equation (3.5), we obtain the expression for the molecular weight.

$$M = [RT/(1-<v>\rho)\omega^2 rC](\delta C/\delta r)$$

Thus the molecular weight can be determined from a measurement of the concentration gradient. Expressing the above equation in terms of concentration at two different distances from the rotor axis, r_a and r_b we get

$$M = 2RT(C_b - C_a)/[(1 - \langle v \rangle \rho)\omega^2(r_b^2 - r_a^2)C_0]$$

where C_a and C_b are concentrations corresponding to distances r_a and r_b, and C_0 is the initial concentration. The advantage in using this equation instead of the previous one is that r_a and r_b can be measured at the meniscus and at the bottom of the cell respectively. In addition, the equation is independent of the size and shape of the molecule. Moreover, at the meniscus and at the bottom the equilibrium condition is surely satisfied, i.e. the flux is zero. This equation is, however, valid only for homogeneous samples and therefore the molecule to be analysed will have to be pure. The rotor should be spun at low speeds as otherwise there will be no solute molecules at the meniscus and a pellet will form at the bottom. Molecular weights ranging from a few hundred daltons to several thousand daltons can be determined using the equilibrium method and the molecular weight decides the rotor speed. For example, a molecular weight of 50,000 daltons will require a speed of 10,000 rpm. The main disadvantage of this method is that long spinning times may be required before equilibrium is reached. Nevertheless, this technique is desirable when the diffusion coefficient D is not known.

Another method suggested by Yphantis uses a high-speed equilibrium. The centrifuge is operated at such a high speed that the concentration of the solute at the meniscus is zero. Now the difference in refractive index between any point in the cell and meniscus is proportional to the concentration of the solute at that point.

In another technique known as density gradient ultracentrifugation, a solution of low molecular weight (e.g. CsCl) is centrifuged in a cell, so that a density gradient along the length of the cell is formed at equilibrium. If we add a substance of high molecular weight to this solution then it will be suspended at a position where its buoyant density is equal to the density of the low molecular weight solution. The position where it is suspended is measured and is used to calculate its molecular weight. This method is widely used in biochemical and biophysical research. If we have a multicomponent system with varying densities then the components will be suspended at different positions in the cell. In practice, a thin layer of the macromolecular sample is layered on a larger volume of the solvent and centrifuged. The components of the system will sediment according to their sedimentation coefficients and separate out as bands (Figure 3.5). This is called zonal sedimentation. The advantage of this technique is that very little ($\sim$ µg) sample is used.

Apart from molecular weights, sedimentation methods can also yield useful information regarding the shape and size of the particles. An expression relating the shape and size of the particles to the sedimentation velocity may be obtained by manipulating equation (3.4)

$$v_0 = \text{velocity} = [2r_p^2(\rho_p - \rho_m)\omega^2 r]/9\eta$$

where ρ_p and ρ_m are the densities of the particle and the medium

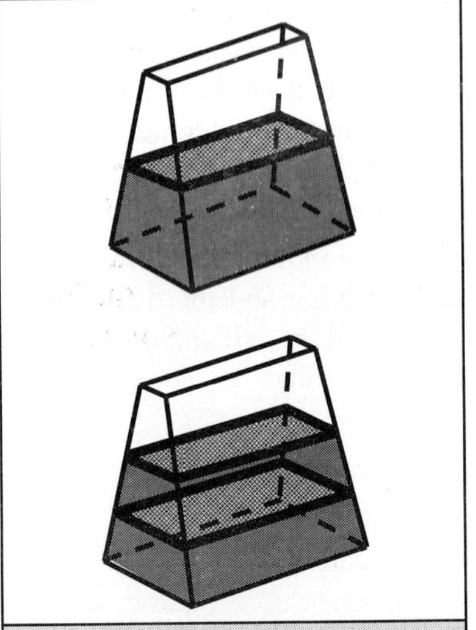

Figure 3.5: Zone sedimentation.

respectively, (ρ_m is a correction for the buoyant force), r is the distance of the particle from the axis of rotation, η is the viscosity and ω is the angular velocity. Thus the sedimentation velocity rate is proportional to the radius r_p of the particle and to its density. The equation is true for perfectly spherical particles. In the case of particles that are not spherical, a correction factor is applied. This is called the shape factor F and is equal to (f_1/f_2), where f_1 is the frictional coefficient of the molecule, and f_2 is the frictional coefficient of an equivalent sphere. Thus r_p, which indicates the size of the particle, and the shape factor, which gives information about the deviation of the particle from sphericity, both may be obtained from sedimentation experiments.

3.7: Viscosity

Viscosity is the resistance to liquid flow. It is increased by the presence of a solute. The increase depends on the concentration and other structural properties of the solute, and therefore information regarding these properties may be obtained from measurements of

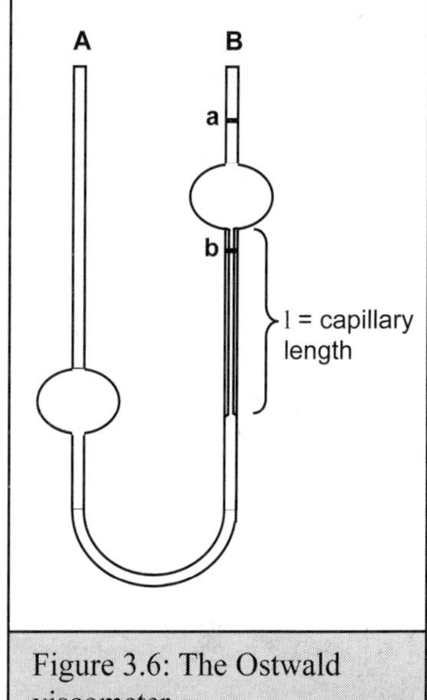

Figure 3.6: The Ostwald viscometer.

viscosity. Viscosity measurement is simple and inexpensive, and may be made using a capillary viscometer. A typical example is the Ostwald viscometer shown in Figure 3.6. The solution whose viscosity is to be determined is poured into the lower bulb through arm A. It is then sucked up through arm B to fill the upper bulb and rise past the mark **a**. Now it is allowed to fall back through the capillary under gravity. The time taken for the solution to fall from point **a** to point **b** is noted. The viscosity η may then be calculated using the following expression, derived from Poiseuille's equation for the viscosity of a liquid flowing through a capillary

$$\eta = (\rho g r^4 t \pi)/(8lV')$$

where ρ is the density of the liquid, g is the acceleration due to the earth's gravity, l and r are the length and radius respectively of the capillary, and t is the time taken by the liquid to fall back from **a** to **b**. V' is a measure of the total hydrostatic pressure to which the liquid flow through the capillary is subjected, and is calculated as the integral of dV/h between a and b, where h measures the height of the liquid column and dV is the volume element at the point h. It is difficult to measure the absolute viscosity of a liquid by this method, especially since V' is difficult to estimate. However, a comparison with a reference standard can be made. Thus if η_0 is the viscosity of the reference, and ρ_0 its density,

$$\eta/\eta_0 = (\rho/\rho_0)(t/t_0)$$

where t_0 is the outflow time for the standard. The reference liquid is usually chosen to be the solvent, which is then used to dissolve the particles of interest in order to measure the viscosity of the solution.

A theoretical treatment of the viscosity of a solution of particles is required before we can obtain shape and size information from viscosity measurements. The viscosity η of a solution, relative to the viscosity η_0 of the pure solvent, can be expressed as a power series

$$\eta = \eta_0(1 + k_1C_2 + k_2C_2^2 + \dots)$$

Axial Ratio	Prolate Ellipsoid		Oblate Ellipsoid	
	v	$\beta \times 10^{-6}$	v	$\beta \times 10^{-6}$
1	2.5	2.12	2.5	2.12
10	13.63	2.41	8.04	2.14
30	74.51	2.78	21.58	2.15
100	593.70	3.22	69.10	2.0

Table 3.2: Simha and Scheraga-Mandelkern factors for ellipsoids.

where C_2 is the solute concentration in gm/cm^3. Einstein showed that, ignoring the higher order terms, the above equation may be simplified to obtain the relative viscosity for a dilute solution of large rigid spheres as

$$\eta/\eta_0 = 1 + 2.5\phi$$

where

$$\phi = (V_h N_0 C_2)/M = C_2 <v>$$

in which V_h is the hydrated volume, (M/N_0) is the mass of the molecule in grams, C_2 is the solute concentration in gm/cm^3 and $<v>$ is the partial specific volume. Thus the change in viscosity on addition of the solute particles does not depend on their size but depends on the volume they occupy in solution.

For particles that are shaped as rigid ellipsoids, an intrinsic viscosity $[\eta]$ may be obtained from the expression

$$[\eta] = (v V_h N_0/M)$$

v is called the Simha factor and contains information on the shape of the molecule. It varies as the ratio of the major and minor axes of the ellipsoids. The above equation can be more generally written as

$$[\eta] = K M^{\alpha}$$

where K and α are constants. Measurements of the intrinsic viscosity together with sedimentation or diffusion measurements can lead to a good estimate of molecular weight by eliminating shape effects. Table 3.2 lists the Simha factors for prolate and oblate ellipsoids with different axial ratios. Thus, for example, rod shaped particles always have a Simha factor greater than 15. Table 3.3 lists some experimentally determined molecular weights using the Simha shape factor.

Sample	$[\eta]$	M (cm^3/gm)
Near spherical particles		
Ribonuclease	3.3	13,700
Hemoglobin	3.6	68,000
Rod like particles		
Myosin	217	4,93,000
DNA	5,000	60,00,000

Table 3.3: Experimentally determined intrinsic viscosities and molecular weights.

Table 3.4: Experimentally determined molecular weights compared with theoretical values from chemical structure.

Molecule	From chemical structure	From Sheraga-Mandelkern equation	From sedimentation equilibrium
Ribonuclease A (bovine)	13,683	15,200	13,700
Lysozyme	14,211	12,400	14,500
Serum albumin (bovine)	66,296	59,000	68,000

Another factor, which is also related to shape of the particles, is given by the Scheraga-Mandelkern equation

$$\beta' = \{N_0^{1/3} v^{1/3}\} / \{(162\pi^2)^{1/3} F\} = \{N_0 s [\eta]^{1/3} \eta\} / \{M^{2/3}(1-<v>\rho)\}$$

where $F = f/f_{sph}$ is the shape factor, f is the frictional coefficient of the particle, f_{sph} the frictional coefficient of an equivalent sphere, N_0 is Avogadro's number, s is the sedimentation coefficient of the solute particles, $[\eta]$ and η are the intrinsic and relative viscosity of the solution respectively, M is the molecular weight of the solute particles, $<v>$ is the partial specific volume of the solute and ρ is the density. The values of the Sheraga-Mandelkern factor are tabulated in Table 3.2 as $\beta = 100^{1/3}\beta'$. β is less sensitive to the axial ratio as compared both to the Simha factor v, as well as to another factor in common use called the Perrin factor. For most globular proteins the axial ratio is less than 10 and the β value is close to 2.20×10^6. Using these values, and the Sheraga-Mandelkern equation, molecular weights can be determined with accuracy better than 10%. Table 3.4 shows some experimentally determined molecular weights using the Scheraga-Mandelkern equation.

3.8: Rotational diffusion

Molecules in solution undergo both translational and rotational displacements due to the Brownian motion of the solvent. Each solute molecule undergoes rotation about its centre of mass. This movement is characterized by a coefficient called the rotational diffusion coefficient $[\theta]$, which represents the average angle of rotation per unit time and is dependent on the rotational friction experienced by the particle. For non-spherical particles, a single frictional coefficient is insufficient to describe the rotational friction. In such cases, the frictional forces acting on the particle are described by a tensor[4]. Therefore $[\theta]$ is also a tensor and is given by the equation

$$[\theta] = kT/f_{rot}$$

where f_{rot}, the coefficient of rotational friction, is a tensor. The determination of $[\theta]$ is especially useful in the case of molecules with high axial ratios (length/width) since it gives a measure of the length of

---

[4] A tensor is matrix of numbers that describes a given property in different directions. In the present instance, the frictional tensor will consist of a 3 x 3 matrix with the diagonal elements giving the frictional coefficients along the three principal axes of rotation, and the off-diagonal elements giving the coefficients for rotation about the other axes.

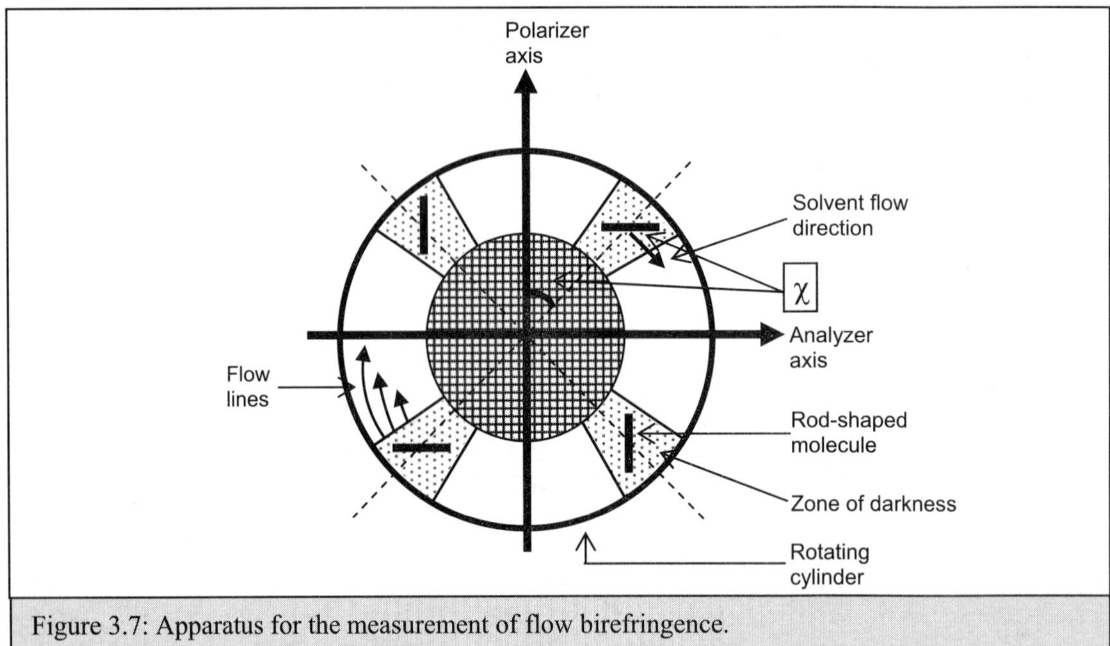

Figure 3.7: Apparatus for the measurement of flow birefringence.

the molecule in solution. $[\theta]$ may be measured by a variety of techniques like flow birefringence, electric birefringence and the polarisation of fluorescence. In order to observe rotational diffusion, the equilibrium orientations of the molecule in normal solution must first be perturbed, and then some property of the molecule, which is dependent on the orientation of the molecule, must be measured.

3.8.1: Flow birefringence measurements:

Flow and electric birefringence methods are based on the differences between the refractive index measured in different directions that are produced by oriented molecules in solution. In flow birefringence measurements, the solution to be studied is placed between two concentric cylinders (Figure 3.7), of which one (generally the outer one) is rotated with respect to the other. This establishes a velocity gradient along the radius of the cylinder, which in turn produces a torque on the molecule. Under the influence of this torque, the molecules align their long axis along the flow lines, perpendicular to the radius (Figure 3.8). The higher the axial ratio of the molecules, the better is the alignment. Now a parallel beam of light is passed through a polariser[5], and then through the solution, and then through an analyser. To start with, as the molecules are randomly oriented, the solution is optically isotropic and the analyser blocks the entire incident light. When the outer cylinder is rotated, the molecules become oriented as described above and the solution becomes birefringent, i.e. the refractive index of the solution is not the same in all directions. The light transmitted through the solute now becomes elliptically polarised and hence some of it will pass through the analyser. An image of the analyser appears bright everywhere excepting in four perpendicular regions where the optic axis of

---

[5] See chapter 5 for a discussion of the polarisation of light.

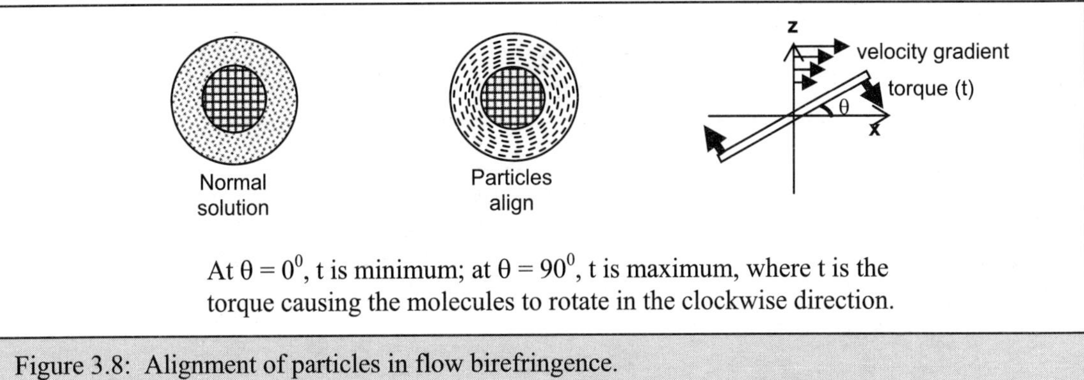

At $\theta = 0^0$, t is minimum; at $\theta = 90^0$, t is maximum, where t is the torque causing the molecules to rotate in the clockwise direction.

Figure 3.8: Alignment of particles in flow birefringence.

the molecule is parallel to the plane of polarisation of the incident light (Figure 3.7). These four regions form a dark cross, corresponding to regions of zero birefringence. The cross is seen initially at 45° to the polariser axis when there is only a slight degree of orientation of the molecules. As the degree of orientation increases, the dark cross rotates to align itself with the cross formed by the polariser and analyser.

The position of the cross is measured by the extinction angle χ (Figure 3.7). A plot of χ versus shear gradient G is made to determine $[\theta]$. The shear gradient depends on the speed of the outer cylinder, and the distance between the cylinders. Therefore χ, which is the angle between the macroscopic optic axis of the solution of oriented molecules and the direction of the solvent flow, is a measure of the degree of orientation. Rod-shaped molecules tend to orient as shown in Figure 3.7, with their long axis parallel or perpendicular to the polariser axis. The measurement $[\theta]$ then enables the calculation of the frictional tensor using the equation above.

3.8.2: Electric birefringence:

Instead of a velocity gradient, an electric field can be used to orient the sample. Electric birefringence is then an optical measurement of the orientation caused by the electric field. An electric pulse, generally a square wave, is applied to the solution for a sufficiently short duration such that there is no electrophoretic mobility, only rotational reorientation of the molecules (Figure 3.9). The macromolecules in the solution get preferentially aligned in a particular direction in the electric field. The degree of alignment depends on the electric properties of the molecule and is measured by the difference between the refractive index of the solution parallel to the applied field and the one perpendicular to it. This is known as form birefringence. (Intrinsic birefringence, as opposed to form birefringrence refers to the difference in the refractive index between two perpendicular directions within a rod-like molecule such as DNA.) The experiment to measure the electric (form) birefringence consists of passing light through crossed polarisers, with the sample in the electric field placed in between. The light is finally received by a photomultiplier tube connected to an oscilloscope. Since the molecules are oriented by the electric field, a static measurement of the birefringence at the equilibrium state will yield little information about the particle's shape. However, the decay in the birefringence, when the electric field is removed will depend on the molecular shape. This is measured

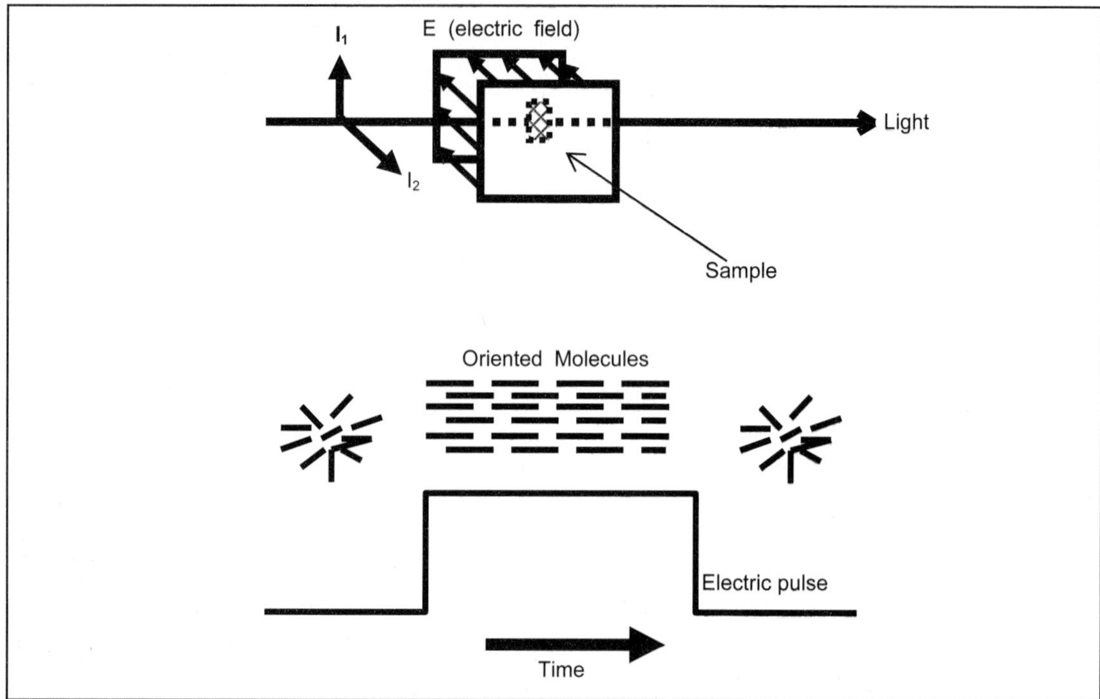

Figure 3.9: Electric dichroism. The sample is placed in an electric field. On application of an electric pulse, the molecules in the sample become oriented along one direction. Removal of the electric field results in the molecules reverting to a disoriented state.

as the decay of the birefringence signal. The rotational diffusion coefficient of the molecule may be calculated from a measurement of the decay.

3.9: Light scattering

Light scattering is a technique widely used to determine the molecular weight of macromolecules in solution. A beam of light of a particular wavelength illuminates the sample. Radiation scattered by the sample, at the same wavelength, is measured as a function of the angle between the incident beam and the detector. (The experiment is extremely sensitive to impurities hence great care must be taken to exclude dust particles from the solution.) The chief difference between light scattering and visible range spectroscopy is that, in light scattering, only elastic scattering by the particles in the solution is considered. This is the reason the measurement of scattering is made at the same wavelength as the incident beam. In this respect, the technique is similar to diffraction. However, in diffraction, we measure the interference patterns formed by radiation scattered from different particles. In light scattering, the solution is dilute enough to consider that the molecules scatter independently and are randomly oriented. Moreover they do not absorb the incident radiation, and therefore the scattered radiation has the same wavelength. Hence the term elastic scattering.

In general, there are two situations – (a) when the particles are smaller than the wavelength of the incident and scattered radiation, and (b) when they are comparable in size to wavelength. Consider first the situation when the molecules are smaller than the wavelength. Since light is electromagnetic radiation, it has electrical as well as magnetic components. But for the purposes of the present discussion we concentrate only on the electric field component E, and ignore the magnetic component. If the radiation is of frequency v and the maximum amplitude is E_0, then the electric field component at a given spatial point, measured at the time t is given by

$$E(t) = E_0\cos(2\pi vt)$$

According to classical electromagnetic theory, when light strikes matter, electrons are displaced from their equilibrium positions with respect to the nucleus. This results in an induced dipole. The dipole moment p so induced is proportional to the electric field and is given by

$$p = \alpha E = \alpha E_0\cos(2\pi vt)$$

where α is a constant of proportionality called polarisability. (The polarisability may be described as a measure of the ease with which an electron can be displaced from its equilibrium position.) Since the incident electric field is oscillating, the dipole also oscillates in magnitude. The oscillating dipole gives rise to scattered radiation, the electric field component of which is given by

$$E_s = [4\pi^2 v^2 \alpha^2 E_0\cos(2\pi vt)]/c^2 r$$

in which c is the velocity of light, and r is the distance of the dipole from the detector. The ratio of the scattered radiation I_s to the incident intensity I_0 is given by

$$I_s/I_0 = |E_s|^2/|E_0|^2 = (16\pi^4\alpha^2)/\lambda^4 r^2$$

in which we have substituted $v = c/\lambda$. (In the above equation we consider only the magnitudes, and hence the term $\cos(2\pi vt)$ cancels out.) Thus the ratio is inversely proportional to the fourth power of the wavelength λ. This is called Rayleigh scattering. If, instead of unpolarised or circularly polarised light, we use light polarised along a particular axis then the induced dipole oscillates only in that direction. As a consequence the scattered radiation will also be polarised and the above equation has to be modified for such effects. Thus

$$I_s/I_0 = [(8\pi^4\alpha^2)/\lambda^4 r^2](1 + \cos^2\theta)$$

for a single scattering particle. If there are N particles, then

$$I_s/I_0 = N[(8\pi^4\alpha^2)/\lambda^4 r^2](1 + \cos^2\theta)$$

The Rayleigh ratio, R_θ, is the relative intensity of the scattered light and is given by

$$R_\theta = [r^2/(1+\cos^2\theta)V][I_s/I_0]$$

Substituting for $[I_s/I_0]$ from the previous equation yields

$$R_\theta = (8\pi^4\alpha^2 N')/\lambda^4$$

where $N' = N/V$ is called the number density of particles. N' may also be written as CN_0/M, where C is the concentration, N_0 is Avogadro's number and M is the molecular weight. In all the above equations, V is the volume of the solution. When we substitute for the polarisability α in terms of refractive index and concentration of the solution, we obtain

$$R_\theta = KMC$$

where K is a constant for a given solution incorporating the refractive index of the pure solvent, Avagadro's number, the incident wavelength and the rate of change of refractive index with concentration. Thus if R_θ is measured, and the concentration C is known, the molecular weight can be determined. When the sample is a heterogeneous mixture of several molecular species of concentrations C_1, C_2...C_i, then, instead of a single molecular weight, the light scattering experiment will yield

$$\Sigma_i M_i C_i$$

where M_i is the molecular weight of the component i. The weighted average molecular weight is then obtained as

$$<M_w> = (\Sigma_i M_i C_i)/\Sigma_i C_i$$

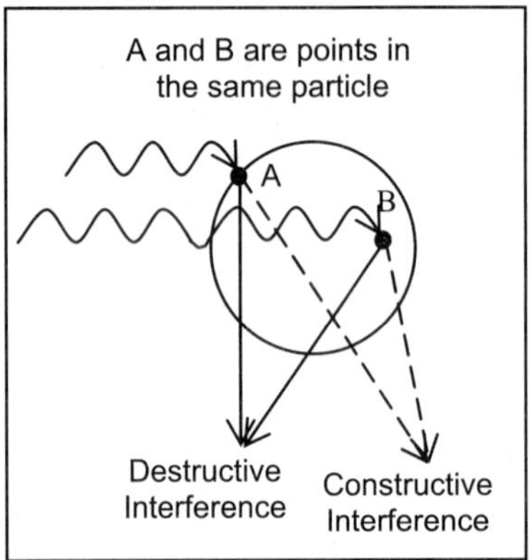

When the particle size was small, as in the above case, we could assume that the dipoles were oscillating in phase for all the scattering elements within the particle. The other situation is when the molecules in the solution have a size that is comparable to the wavelength of the incident radiation. In this case, different parts of the molecule are considered to scatter independently and thus the radiation emanating from the different portions undergo interference (Figure 3.10). This effect is treated by applying a correction factor P, which is a function of θ, to the Rayleigh scattering ratio. $P(\theta)$ can be expressed in terms of radius of gyration R_G of the molecule. The radius of gyration gives information about the shape of the molecule. Thus

$$R_\theta = KMC \times P(\theta) = KMC(1-16\pi^2 R_G^2 \sin^2\theta/3\lambda^2)$$

At the limit of infinite dilution (and since $1/(1-x)$ is approximately equal to $(1+x)$ for small value of x),

$$(KC/R_\theta) = (1/M)(1+16\pi^2 R_G^2 \sin^2\theta/3\lambda^2)$$

KC/R_θ is measured for various values of the scattering angle θ, and plotted against $\sin^2(\theta/2) + k'C$, where k' is an arbitrary constant used for scaling (Figure 3.11). This is known as the Zimm Plot. The concentration data at each angle is extrapolated to C = 0. Similarly the angle-dependent scattering data is extrapolated to $\theta = 0$. Both the extrapolations intersect on the y axis giving 1/M. The radius of gyration can be obtained from the slope of the C = 0 curve. The value of R_G depends on the shape of the particle. For a sphere of radius r, the radius of gyration is given by $R_G^2 = (3/5)r^2$, whereas for an ellipsoid of semi axes a and b, $R_G^2 = (a^2 + 2b^2)/5$. For a long rod of length l, $R_G^2 = l^2/12$. In this manner one can obtain information about the molecular weight and the radius of gyration, and therefore the shape of the molecule, from the light scattering experiment.

Figure 3.10: Interference in light scattered from particles with size comparable to λ, the wavelength of the incident light.

3.10: Small angle X-ray scattering

This technique is analogous to that of light scattering. For an average small protein having a maximum dimension of 100 Å, light scattering can give information about the molecular weight, but not about the shape of the protein, since the correction term $P(\theta)$ is close to unity. In other words, proteins appear spherical when probed with visible light. In order to obtain more precise information on the shape of

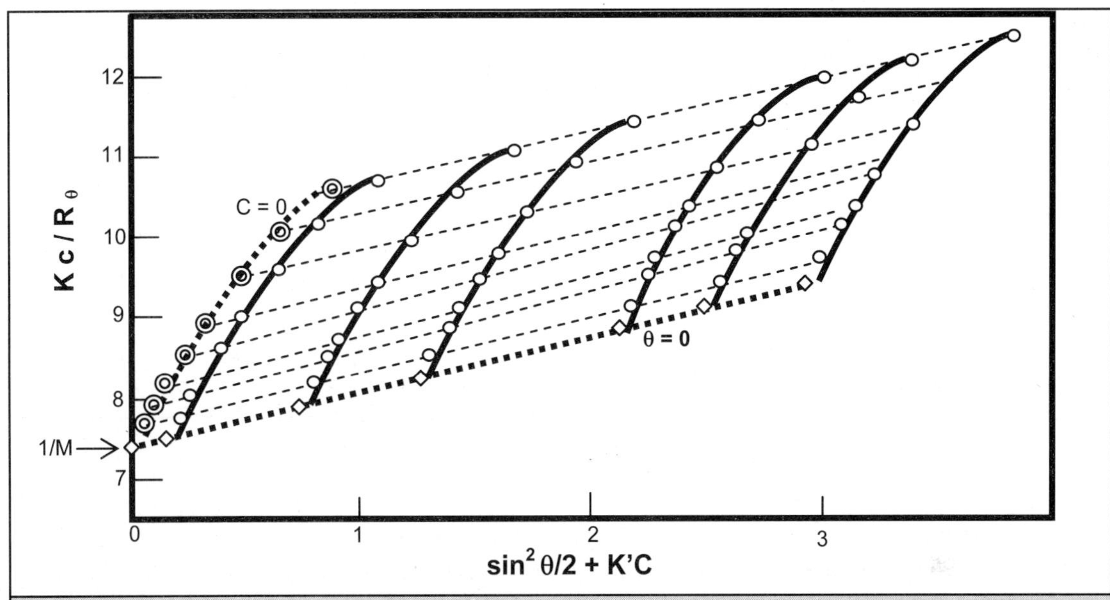

Figure 3.11: Zimm Plot for light scattering from a biomolecular solution. Experimental data (O), extrapolations to zero concentration (◎) and extrapolations to θ = zero (◇), K' is a constant.

these molecules, radiation with a smaller wavelength, such as X-rays, needs to be used. Small angle X-ray scattering or SAXS is a diffraction technique. In order to understand the need for measurements at small angles, the angle of diffraction for an inter-planar spacing of 100 Å can be calculated from the diffraction condition or Bragg's law to be

$$\sin\theta = \lambda/2 \times 100 = 0.0077$$

for CuK_α radiation of wavelength 1.5418 Å. This gives $\theta \sim 0.45°$. For a spacing of 1000 Å, the same expression gives θ as $0.045°$. Thus, the angle (with respect to the direct beam) at which the measurements will have to be made to obtain information at such spacing or such resolution is very small.

The SAXS experiment consists of allowing a narrow X-ray beam of known wavelength to be incident on a dilute solution of the sample, and then measuring the scattered intensity as a function of the angle. Since the scattering angle of interest is only a few degrees, at the most, from the direct beam, great care must to taken to collimate the beam to make it as thin as possible, so that the portion of the direct beam passing through the sample unaltered does not mask the scattered beams. Also the direct beam must be carefully and completely excluded in the measurement of the scattered intensity. Since the solution is sufficiently dilute, we assume, as in the light scattering experiment, that the molecules scatter independently. For each molecule, the molecular structure factor is defined as the sum of individual scattered X-rays from all the atoms in the molecule. Taking into consideration the difference in scattering between the different relative orientations of neighbouring molecules and integrating the structure factor over all possible relative orientations, we obtain the following expression for the scattering intensity.

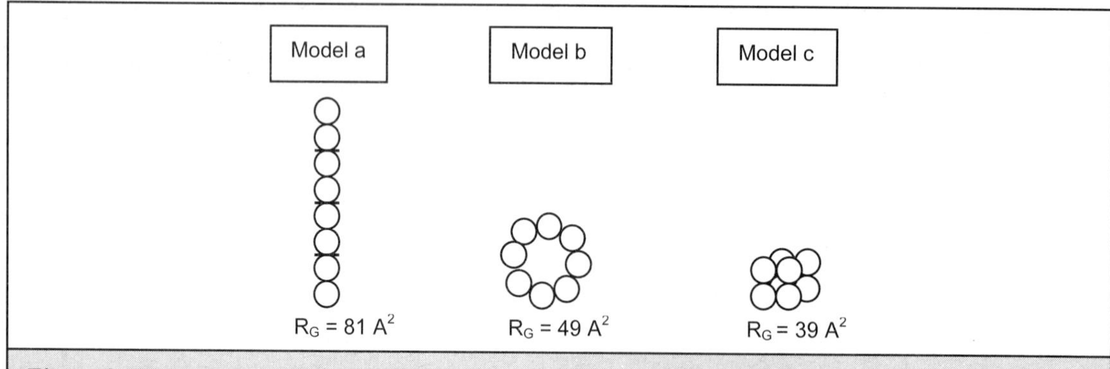

Model a   Model b   Model c

$R_G = 81\ A^2$ $R_G = 49\ A^2$ $R_G = 39\ A^2$

Figure 3.12: Models for the quaternary structure of α-lactalbumin at pH 4.5.

$$\langle I(\theta)\rangle = n_\varepsilon^2(1-16\pi^2 R_G^2 \sin^2\theta/3\lambda^2)$$

for small values of θ. R_G is the radius of gyration and n_ε is the number of electrons in the molecule (as obtained from knowledge of the molecular weight). This equation indicates that the zero-angle X-ray scattering from the solution is approximately proportional to the square of the molecular weight of the

particle. A plot of the intensity versus $\sin^2\theta$ yields the radius of gyration. A more common way of using this expression is to rewrite it as

$$\ln[I(\theta)/I(0)] = -(4\pi R_G\sin\theta/\lambda)^2/3$$

Now the small angle scattering can be plotted as $\ln[I(\theta)/I(0)]$ versus $\sin^2\theta/\lambda^2$. This plot is called the Guinier plot and it yields the radius of gyration even in the absence of information about the molecular weight. R_G in turn gives information on the shape of the molecule.

As an example of the application of SAXS consider the experiments on α-lactalbumin. At pH 4.5 these gave the radius of gyration as 34 Å. From previous experiments it was known that this protein exists in solution as a tetramer of four dimeric units. Various models for the arrangement of these units (Figure 3.12) were considered and the radius of gyration calculated for each was compared with the experimental value. From these comparisons it was clear that model c in Figure 3.12 is likely to be the correct one.

The scattering of neutrons from solutions can also be used to analyse the shape and size of molecules in solution. The experiment is called SANS, and the techniques and the theory are similar to SAXS. The major difference is that neutrons are scattered by the atomic nucleus, while X-rays are scattered by the electrons around the nucleus. Therefore, the two techniques may be used in a complementary fashion. In SAXS and other diffraction experiments using X-rays, the intensity of the radiation

Table 3.5: Values of the neutron scattering lengths for some biologically relevant elements.

Element	Neutron scattering length ($b \times 10^{-13}$ cm)
H	-3.74
D	6.67
C	6.65
N	9.40
Fe	9.51
Pt	9.50

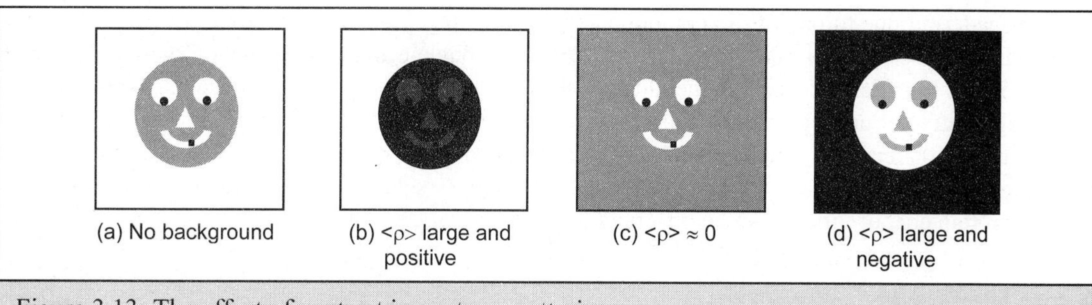

Figure 3.13: The effect of contrast in neutron scattering.

scattered is a measure of the electron density in the sample and is proportional to the atomic scattering factor. On the other hand, the intensity of neutron scattering is proportional to a factor called the scattering length of the atomic nucleus. Each element has a different value for the scattering length (Table 3.5). This value depends on the precise nature of the interactions of the incident neutrons with the nucleus. It therefore does not show any relationship with the atomic number or atomic weight. This is unlike the case for X-ray scattering, which is related to the atomic number. The scattering length is a measure of the scattering power. Neutron scattering lengths can be calculated from absolute scattering intensity measurements. Note that the scattering length for hydrogen is negative, and has a magnitude comparable to heavier atoms such as carbon. Thus unlike as in the case for X-rays, the contribution of hydrogen atoms to neutron scattering is clearly distinguishable. Neutron scattering can give supplementary information on hydrogen atoms which is not available from X-ray scattering.

Another interesting feature is that the neutron scattering from deuterium is positive and of higher magnitude compared to scattering from hydrogen. By replacing different proportions of water (H_2O) with deuterated water (D_2O) in the solvent, its scattering density can be varied at will. In this way the contrast between solute and solvent can be manipulated to observe details of interest. The contrast $\langle\rho\rangle$ is the mean difference in scattering density between the solute and the solvent (Figure 3.13). Thus

$$\langle\rho\rangle = \langle\sigma_{macromolecule}\rangle - \langle\sigma_{solvent}\rangle$$

where $\langle\sigma\rangle$ represents the average neutron scattering intensity of the appropriate component of the solution. When there is no background, the object will appear as in Figure 3.13(a). When $\langle\rho\rangle$ is positive and large, the object will appear as in Figure 3.13(b). When $\langle\rho\rangle$ is zero then the shape of the object will not be seen (Figure 3.13(c)) but internal details will be visible. When $\langle\rho\rangle$ is large and negative we see the object as in Figure 3.13(d). Thus contrast plays an important role in determining the amount of detail that can be obtained.

Deuterium can also be covalently substituted for hydrogen in cases where the sample is composed of two components. One of the components is labeled with H and the other with D. Then by adjusting the $H_2O - D_2O$ composition of the solvent, the following situation can be arrived at: Either

$$\langle\sigma_1\rangle \cong \langle\sigma_{macromolecule}\rangle$$

or

$$\langle\sigma_1\rangle \cong \langle\sigma_{solvent}\rangle$$

and the scattering due to one of the components will be blacked out. For example, when dealing with icosahedral viruses, the protein coat encapsulating the nucleic acid molecule of the viral genome usually makes it difficult to observe the latter in scattering experiments. If we consider that the neutron scattering length per unit volume for fully protonated protein is 3.1×10^{14} cm/Å^3, while that of nucleic acid is 4.4×10^{14} cm/ Å^3, and that for fully deuterated protein it is 8.5×10^{14} cm/Å^3, and for fully deuterated nucleic acid it is 7.4×10^{14} cm/ Å^3, we see that by selective labelling, the nucleic acid core and the protein coat of viruses can be observed independent of each other.

3.11: Mass spectrometry

The mass spectrometer is an important tool that uses mass/charge ratios to identify the atoms and molecules in a sample. According to electrodynamic theory, an electrically charged particle (i.e. an ion) moving in a magnetic field will be acted upon by a force given by

$$F = q(\mathbf{v} \times \mathbf{B})$$

where q is the charge on the particle, $\mathbf{v}$ is its velocity and $\mathbf{B}$ is the magnetic field. ($\mathbf{v}$ and $\mathbf{B}$ are both vectors and the term within brackets is their cross product. The resultant force is perpendicular to both $\mathbf{v}$ and $\mathbf{B}$.) If, in addition, an electric field $\mathbf{E}$ is used to accelerate the ion, the total force is given by

$$F = q(\mathbf{E} + \mathbf{v} \times \mathbf{B})$$

Substituting this expression into Newton's second law, we obtain

$$(m/q)\mathbf{a} = \mathbf{E} + \mathbf{v} \times \mathbf{B}$$

The magnitude and direction of the acceleration $\mathbf{a}$ thus depends not only on the electric field $\mathbf{E}$ and the magnetic field $\mathbf{B}$ but also on the m/q ratio of the ion. One may thus separate, detect and measure ions with different m/q ratios, by varying $\mathbf{B}$. This is the principle of all mass spectrometers. In other words, the amount of deflection of the ion from its straight-line path depends on the charge and mass of the particle. The higher the charge (and/or the lighter the particles), the greater will be the deflection. A mass spectrometer is an extremely sensitive instrument and even trace amounts of ionized molecules may be detected.

A mass spectrometer consists of a sample injector, an ionizer, a mass analyzer and an ion detector (Figure 3.14). These chambers are kept in vacuum to avoid collisions of the sample with gas molecules. The sample is vaporized and is then injected into the ionization chamber. This chamber consists of an electrically heated metal coil, which emits electrons. The electrons are accelerated towards the positively charged electron trap. These electrons bombard the atoms/molecules in the sample and knock off one or more electrons out of the sample particles transforming them into positively charged ions. These positive ions are repelled by the ion repeller, which also carries a slight positive charge. The ions are accelerated by the electric field and formed into a finely focussed beam by electrostatic lenses and ion guides. The ions then pass through the magnetic field created by the electromagnets, and are deflected to different extents depending on their mass/charge ratios. The detector is placed so as to detect the ions only in the central stream. In order to detect ions in the other streams, the magnetic field has to be decreased or increased. The current produced in the detector is proportional to the number of ions reaching the detector. The instrument can be calibrated to record current versus m/q. The resolution of a mass spectrometer is given by the following expression.

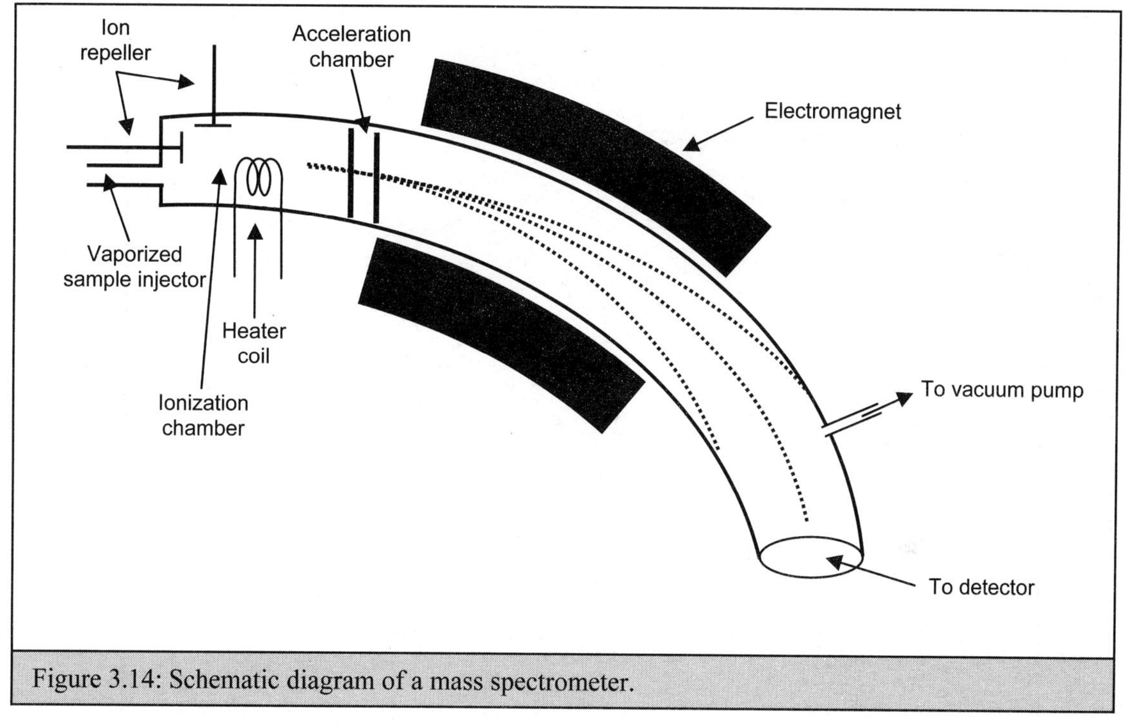

Figure 3.14: Schematic diagram of a mass spectrometer.

$$R = m/\Delta m$$

where m is the ion mass, and Δm is the difference between two resolvable peaks in a mass spectrum.

There are several types of mass spectrometers (MS), and the following are some of the commonly used ones: ion trap MS; magnetic and/or electric sector MS; time of flight MS (TOF); quadrupole MS; and Fourier transform MS (FTMS). Of these, the TOF mass spectrometer, in conjunction with the Matrix Assisted Laser Desorption and Ionisation technique (MALDI-TOF) has proved to be of immense utility in molecular biology.

3.11.1: MALDI-TOF:

In molecular biology, MALDI-TOF is used for oligonucleotide and amino acid sequencing. The sample is either injected directly into the ionization chamber, or routed through a chromatography column. In the latter case, the MS is attached to a HPLC (high-pressure liquid chromatography), gas chromatography or capillary electrophoresis separation column so that the sample is separated into a series of compounds, which are individually analyzed by the MS. Methods to ionize the sample include electrospray ionization (ESI), matrix assisted laser desorption ionization (MALDI), thermospray ionization, field ionization, etc. ESI and MALDI are the most frequently used methods. Most of the ionization methods produce both positive and negative ions, and the user must decide whether to detect the positively charged or negatively charged ions. Photomultipliers, electron multipliers and micro channel plates are the most commonly used detectors. Multiwire proportional counters are extremely precise, but also fragile and expensive.

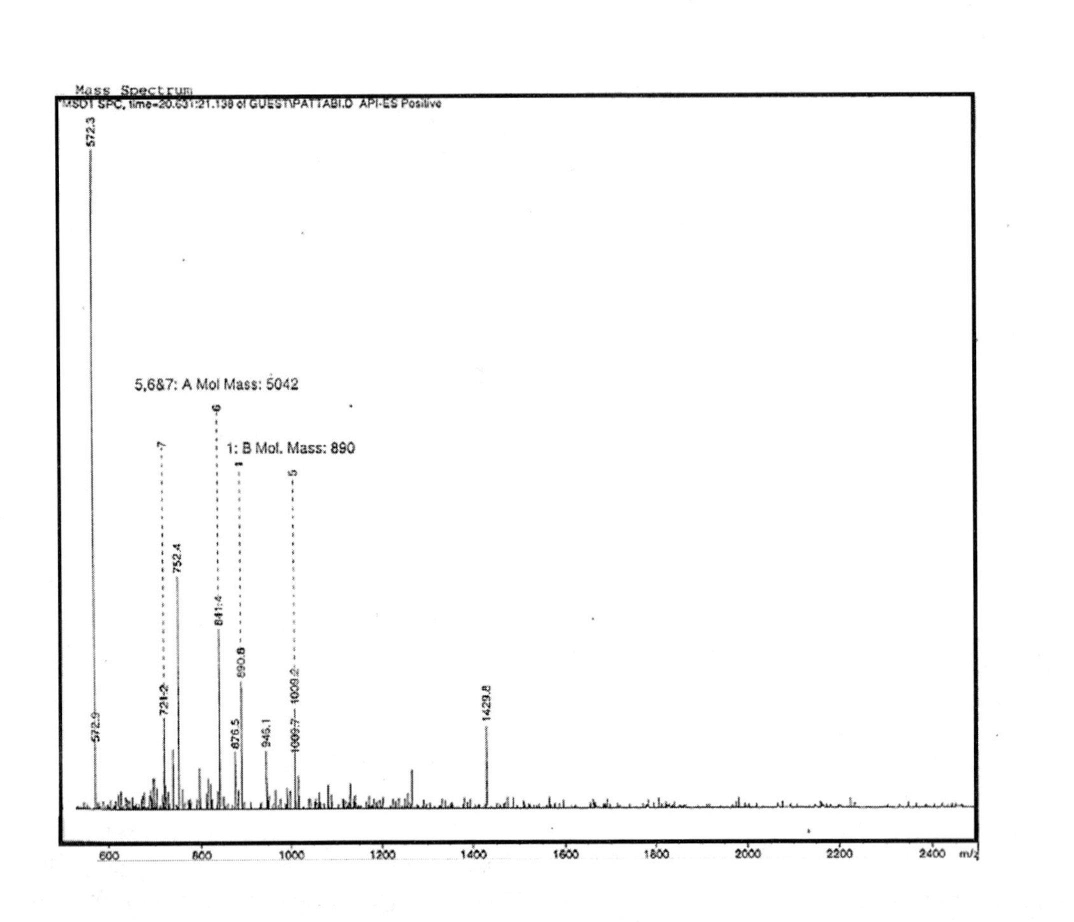

Figure 3.15: Mass spectrum of α-trypsin - nonapeptide complex. The peak B (molecular mass 890 daltons) corresponds to the nonapeptide.

In ESI the sample is dissolved in a polar, volatile solvent and pumped through a narrow stainless steel capillary. A high voltage of 3 to 4 kV is applied to the tip of the capillary. Due to the strong electric field at the tip of the capillary, and the sample is dispersed into the chamber as an aerosol of highly charged droplets.

MALDI deals with thermostable, non-volatile, high molecular mass organic compounds, and is used for the analysis of proteins, peptides, glycoproteins, oligosaccharides and oligonucleotides. The sample is mixed with a matrix compound that absorbs light energy very efficiently and converts it to excitation energy to ionize the sample molecules present. A laser beam that is made to incident on the matrix supplies the light energy. In this process, ions of the analyte (together with the other compounds in the matrix) are sputtered from the surface of the matrix. Sinapinic acid and alpha-cyano-4-hydroxy cinnamic acid are commonly used as the matrix. The sample is dissolved in a volatile solvent and 1-2 µl of this is mixed with equal volume of solution containing excess of matrix compound. A time-of-flight analyzer is used, in which the time taken for the ions to travel through field-free region is measured. Heavier ions (i.e. those with larger m/q values) are slower than the lighter ones. The m/q scale of the spectrometer is calibrated with a known sample before the experimental measurements are made. MALDI-TOF is specifically useful for biological samples as the ionization method used is relatively soft. Singly charged molecular and related ions are produced, and the spectra are easy to interpret (Figure 3.15).

Summary

- Physical properties such as sedimentation, viscosity, osmotic pressure and light scattering give information about the size and shape of macromolecules.

- Macromolecules act in solution and hence hydration, friction and diffusion play a very important role in many of the biological events.

- Macromolecules are generally hydrated. They have both bound and unbound water, which affects their thermodynamic properties. Hydrodynamic properties like diffusion, viscosity and sedimentation in turn are affected by friction.

- Coefficient of friction for spherical particles is given by Stokes law $f_{sph} = 6\pi\eta r$

- The frictional force opposes the motion of the particle through a medium. When the movement of the molecule is due to concentration gradient, it is called diffusion. If it is due to force of gravity or the centrifugal force, then sedimentation takes place.

- According to Fick's first law, the rate of diffusion across a boundary is proportional to the concentration gradient, i.e. $dn/dt = -DA(\delta c/\delta r)$

- An ultracentrifuge is an instrument used to measure sedimentation coefficients, which helps in molecular weight determination.

- Svedberg's equation is used to determine the molecular weight of a macromolecule using the sedimentation method.

- The density gradient technique can also be used for molecular weight determination.

- Viscosity like friction offers resistance to liquid flow.

- Poiseuille's equation governs the viscosity of a liquid flowing through a capillary.

- Molecules in solution undergo both translational and rotational motion due to Brownian motion of the solvent. The rotational movement is characterised by rotational diffusion coefficient.

- Flow birefringence and electric birefringence methods are based on the difference between the refractive index in different directions produced by oriented molecules in solution.

- These measurements can be used to determine the rotational diffusion coefficient.

- Molecular weight of macromolecules in solution can also be determined using light scattering.

- For an average small protein having a maximum dimension of 100 Å, light scattering cannot give information about the shape of the molecule. Instead, small angle x-ray scattering (SAXS) is used.
- The radius of gyration, and hence the shape of the molecules can be obtained from this experiment.
- Neutron scattering from solutions can also be used to study the shape and size of the molecules in solution.
- Neutron scattering length for hydrogen is negative and can be effectively used to locate the hydrogen positions.
- Neutron scattering is complementary to that of X-rays.
- The mass spectrometer is an important tool in the identification of atoms and molecules in a sample using their mass/charge ratio.

Chapter 4

Spectroscopy

4.1: Introduction

Electromagnetic energy depends on the wavelength (or frequency) of the radiation. Different wavelengths are useful in probing different aspects of materials. Table 4.1 indicates how the various regions of the electromagnetic spectrum are used.

In a spectroscopic experiment electromagnetic radiation of a particular frequency or a range of frequencies is allowed to fall on the sample under study. The radiation coming out of the sample is then analysed in terms of the intensity at different frequencies. This indicates which particular frequencies are absorbed (or emitted) by the molecule, thus giving a picture of the molecular energy levels. These, in turn, can be interpreted in terms of the chemistry or stereochemistry of the molecule. Under ordinary circumstances, all the atoms and molecules would exist in the ground state energy

Table 4.1: The electromagnetic spectrum and the corresponding physical experiments.

Name	Wavelength (metres)	Frequency (Hz)	Use
X-rays	$10^{-12} - 10^{-8}$	$10^{20} - 10^{16}$	Diffraction, small angle scattering
Ultraviolet (UV)	$10^{-8} - 4\text{x}10^{-7}$	$10^{16} - 7.5\text{x}10^{14}$	Electronic structure of molecules
Visible	$4\text{x}10^{-7} - 7.5\text{x}10^{-7}$	$7.5\text{x}10^{14} - 4\text{x}10^{14}$	Electronic structure of molecules
Infrared (IR)	$7.5\text{x}10^{-7} - 10^{-3}$	$4\text{x}10^{14} - 10^{11}$	Vibrational and rotational spectra of molecules
Microwave	$1\text{x}10^{-3} - 1$	$10^{11} - 10^{8}$	Rotational spectra
Radio waves	$1 - 10^{3}$	$10^{8} - 10^{5}$	NMR

level. When, due to the incident radiation, electrons in the outer shell are raised by one energy level, the phenomenon is known as resonance absorption and the technique of atomic absorption spectroscopy is based on this. The energy required lies in the ultraviolet region of the electromagnetic spectrum. Ultraviolet absorption spectra in molecules appear as wide bands. This is because the molecule usually contains sets of closely spaced electronic energy levels, and the transitions between these levels, as indicated by the spectra, are difficult to resolve. Molecular spectra have additional complexities arising from vibrational and rotational motion. While transitions between electronic levels are found in the ultraviolet and visible regions, those between different vibrational levels but within the same electronic level are in the infrared region, and those between rotational levels are in the infrared and microwave regions. Fluorescence spectroscopy of molecules also makes use of the absorption phenomenon but in a different way. A part of the energy gained by a molecule by absorption of electromagnetic radiation may be dissipated away, for example, as heat. The remaining portion of the energy is radiated as the molecule returns to its ground state. Since the energy radiated is now less than that originally absorbed, the frequency of emitted radiation is less than that of the incident radiation. This process is called fluorescence. For example, many organic compounds fluoresce in the visible region after being irradiated with ultraviolet radiation. Many molecules have the ability to rotate the plane of polarisation of plane-polarised light. This property is called optical activity. Its variation with wavelength of the light is called optical rotatory dispersion. A related phenomenon is circular dichroism, where the polarised light is split into left and right circular polarisation and the difference in the absorbance of the two components is measured. All these spectroscopic techniques give a greater or lesser amount of information about the stereochemistry and conformation of the molecules in the sample. They are of great utility in studies of biological molecules. We will now treat each of them in greater detail.

4.2: Ultraviolet/visible spectroscopy

If a beam of light of intensity I_0 falls on a cell containing the sample, the emergent radiation intensity I is less intense than the incident radiation, since a portion of it is absorbed by the sample. The absorption is different at different wavelengths and is characteristic of the sample. This characteristic quantity is called the absorbance and may be calculated from the Beer-Lambert principle. The Beer-Lambert law states that in a sample, each successive portion along the path of the incident radiation, containing an equal number of absorbing molecules absorbs an equal fraction of the radiation that traverses it. If we consider an infinitely thin slice of the sample (dl), the light traversing it may be considered a constant. Then the fraction of the light absorbed is simply proportional to the number of absorbing molecules,

$$-dI/I = C\varepsilon' dl$$

where C is the concentration of the molecule in moles per litre, ε' is a proportionality constant called the molar extinction coefficient and the negative sign indicates a diminution of the intensity of the radiation as it traverses the sample. ε' is a constant for a given wavelength for a given molecule and is independent of the concentration. Integrating this equation over the entire sample thickness l, we obtain

$$A = \log(I_0/I) = C\varepsilon(\lambda)l \tag{4.1}$$

where $\varepsilon = \varepsilon'/2.303$, i.e. molar extinction coefficient converted to log base 10. It is expressed as a function of wavelength λ. The term $\log(I_0/I)$ is called the absorbance A. In an experiment, if A is measured at each wavelength λ, $\varepsilon(\lambda)$ can be calculated and tabulated. Usually, the wavelength λ_{max} at the maximal extinction and the extinction coefficient ε_{max} at this wavelength are used to characterise the sample. Sometimes, especially for biopolymers, the molar extinction coefficient is difficult to measure. Instead the average extinction per residue is calculated. In the actual experimental set up, in order to eliminate the effect of the solvent, the effect of the path length, etc., what is measured usually is the difference in absorbance between a cell containing the sample, and one containing the solvent alone. A double beam spectrophotometer has a beam splitter that splits the beam into two paths of exactly equal length. The sample is placed in the path of one half of the beam and the reference in the other.

UV/Vis absorption spectroscopy is used frequently in biophysical research for simple tasks such as determining the concentration of a sample, as also for answering complex structural questions. Electronic transitions between molecular orbitals are responsible for the spectrum and each spectrum characterises the quantum mechanical state of the molecule. The most commonly observed transitions in biomolecules are: (a) Between the ground state π orbital and the excited π^* orbital, i.e. the $\pi \rightarrow \pi^*$ transition. (b) The nonbonding n orbital and the π^* orbital, i.e. $n \rightarrow \pi^*$. Though the σ orbitals do not usually participate, $n \rightarrow \sigma^*$ transitions have also been observed. At any given wavelength, one particular chemical group in the molecule will dominate the absorption spectrum. Such groups are called chromophores. In biological molecules such as nucleic acids and proteins the chromophores absorb light at wavelengths less than about 300 nm[6]. Biologically interesting macromolecules cannot be studied in the gas phase and the presence of the solvent itself becomes a constraint since the chromophores of the solvent will dominate at certain wavelengths. Thus, the most common solvent, water, absorbs very strongly below 170 nm, restricting useful study to wavelengths above this value. Protein molecules have three types of chromophores, namely the peptide group, the amino acid side chain groups and any prosthetic groups that may be present. The peptide group has an $n \rightarrow \pi^*$ absorption band at 210 – 220 nm. The intensity of this band is much smaller than the intense $\pi \rightarrow \pi^*$ transition at 190 nm, which may be considered a 'signature' absorption of the peptide group. A third transition corresponding, tentatively, to the $n \rightarrow \sigma^*$ transition in the peptide, is also sometimes observed. As far as the amino acid side chains are concerned, many of them absorb in the same region as the peptide group. In the case of the aromatic amino acids tryptophan, tyrosine and phenylalanine, the dominant band is between 230 and 300 nm, i.e. above the peptide bands. The most intense is the 280 nm absorption of tryptophan. Tyrosine also has an absorption band at almost the same wavelength, though somewhat weaker. Phenylalanine is much weaker and at approximately 260 nm. It is usually observed only in the absence of significant amounts of tryptophan or tyrosine in the protein. The absorption bands of the side chains change with the pH of the solvent. This is mainly due to the change in the protonation of the chromophores. For example, the removal of the OH proton in tyrosine at pH > 10.9 shifts the absorbance band from 274 nm to about 295 nm. Disulphide formation changes the spectrum of cysteine, but the absorption is weak and difficult to observe. The other possible chromophores in protein arise from the prosthetic groups such as the heme group, flavin, etc. Routine measurements of the concentration of protein solution assume that 1 mg per ml of protein will have an

---

[6] nanometre = 10^{-9} metre

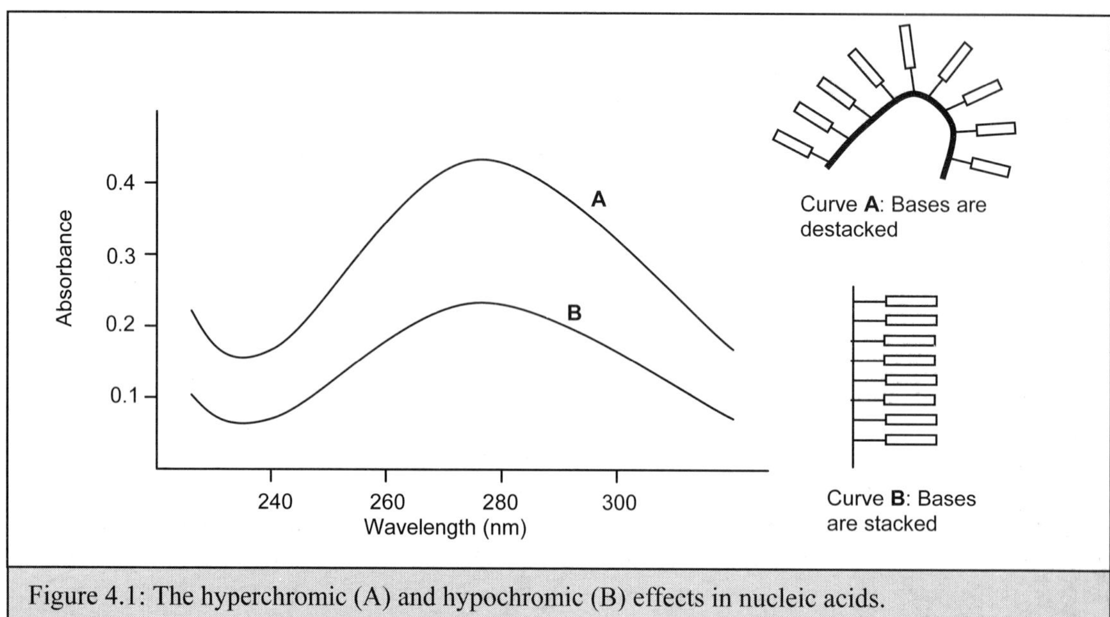

Figure 4.1: The hyperchromic (A) and hypochromic (B) effects in nucleic acids.

absorbance of 1.0 at 280 nm, i.e. $A_{280} = 1$. This number was arrived at after a series of experiments on proteins with a known number of tyrosines and tryptophans. If the measurement is made at 230 nm, $A_{230} = 3$ for a solution containing 1 mg of protein per ml of solution.

The UV absorption of nucleic acids is almost entirely due to the nucleotide bases. The sugar-phosphate backbone has little or no contribution at wavelengths greater than 200 nm. The spectra of the individual bases are quite complex. But they are all marked by a strong, broad band near 260 nm. Each of these strong bands corresponds to several $\pi \rightarrow \pi^*$ and $n \rightarrow \pi^*$ transitions. Some unusual bases, such as the Y base ($\lambda_{max} = 325$ nm) and 4-thiouridine ($\lambda_{max} = 340$ nm) that are found in tRNA, have the absorption maxima quite far away from the peak positions for the normal bases. This leads to their use in monitoring the formation and unravelling of tRNA structure and the interactions between various groups as the structure in formed. Nucleic acid polymers show dramatic effects called hypochromism and hyperchromism. These are effects that arise due to aggregation of chromophores. Quantum mechanical theory says that groups of chromophores that aggregate together have different effects on the absorption spectrum. This difference is due to interactions between the electric dipoles present in the chromophores. In a polynucleotide if the bases are stacked one on the other, as for example when the polymers takes up the double helical structure, the quantum mechanical effects leads to a diminution of the intensities of absorption. The effect is most noticeable at the 260 nm band, and is called hypochromism. If on the other hand the bases are arranged in almost the same plane, i.e. in an end-to-end arrangement, the absorption intensity is increased, leading to hyperchromism (Figure 4.1). Hypochromism is commonly used to follow the denaturation behaviour or the melting behaviour of the DNA double helix. At higher temperatures or at other disruptive conditions, when the base-base stacking is disturbed and the double helix is denatured, the intensity of absorption at 260 nm increases.

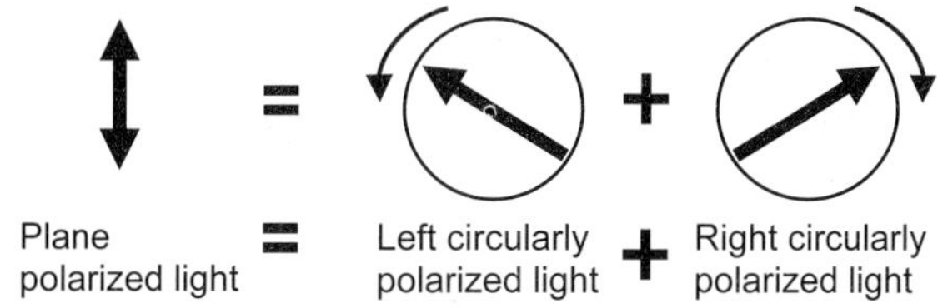

Figure 4.2: Plane polarized light can be resolved into two components.

4.3: Circular dichroism (CD) and optical rotatory dispersion (ORD)

Most biological molecules or systems are optically active; i.e. they rotate the plane of polarised light. If the rotation is clockwise when seen looking towards the light source, it is said to be dextro-rotatory. If the rotation is counter clockwise, it is levo-rotatory. For any compound, the extent of rotation depends on the number of molecules in the path of the polarised light, or in other words, for solutions, on the concentration and on the path length of the beam through it. It also depends on the wavelength of the radiation and the temperature. Optical activity is quantified by the specific rotation

$$[\alpha]_\lambda^t = \frac{\alpha^0}{dc}$$

i.e. the specific rotation $[\alpha]$ at a temperature t and for wavelength λ is the measured rotation α^0 divided by the concentration c in grams per cubic centimetre and the light path d in centimetres. Optical activity may also be quantified as the molar rotation

$$[M]_\lambda^t = \frac{\alpha M}{dc}$$

where M is the gram molecular weight of the substance. For polymers, it is common to use the mean residual rotation

$$[m]_\lambda^t = \frac{\alpha M_0}{dc}$$

where M_0 is the mean residue molecular weight, i.e. the average molecular weight of the monomers. Rather than the rotation at a single wavelength, measurement of the change in optical activity with change in wavelength yields more useful structural information. This is called Optical Rotatory Dispersion (ORD). A closely related phenomenon is Circular Dichroism (CD). Plane polarised light can be resolved into two beams, each circularly polarised, but in opposite senses (Figure 4.2). The refractive index for the right circularly polarised light, n_R may be different from that for the left circularly polarised light, n_L. Similarly the absorbance A_R for the right circularly polarised light may be different from A_L. When n_R and n_L are unequal, each component will be retarded to a different extent as it passes through the sample. This leads to a change in the phases of the two components when they make up the plane-polarised light, leading to a rotation of the plane of polarisation. This is in fact the theoretical basis for the phenomenon of optical activity. Since the refractive index is always dependent on the wavelength, the angle of rotation is also wavelength-dependent. At a given wavelength, λ, the angle of rotation is given by the following equation.

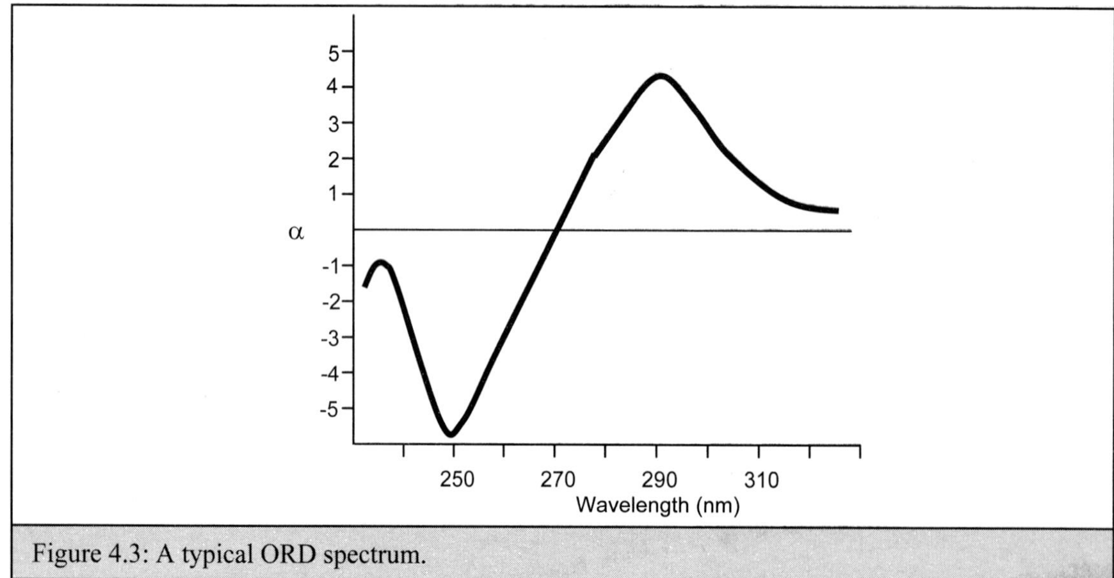

Figure 4.3: A typical ORD spectrum.

$$\alpha_\lambda = (180d/\lambda)(n_L - n_R)$$

where d is the path length. α, or more commonly the specific rotation $[\alpha]_\lambda$ can be plotted as a function of wavelength (Figure 4.3). This is called the Optical Rotatory Dispersion (ORD) spectrum. The difference in the absorption between the right and the left components does not introduce a phase difference between the two, but instead produces a difference in intensity, i.e. a difference in magnitude of the wave. The resultant beam is no longer plane polarised but is elliptically polarised (Figure 4.4). The difference in absorbance $A_L - A_R$ is called the circular dichroism or CD. The ellipticity of the ellipse is also a measure of the differential absorbance, known as the ellipticity θ.

$$\theta(\lambda) = 2.303(A_L - A_R)180/4\pi = 33(A_L - A_R)$$

where the absorbance A is defined as in equation (4.1) for a particular wavelength λ.

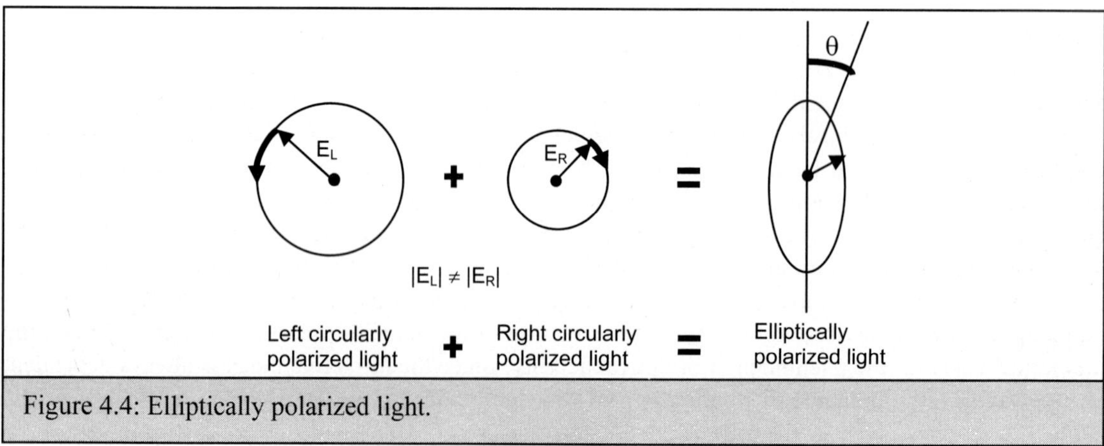

Figure 4.4: Elliptically polarized light.

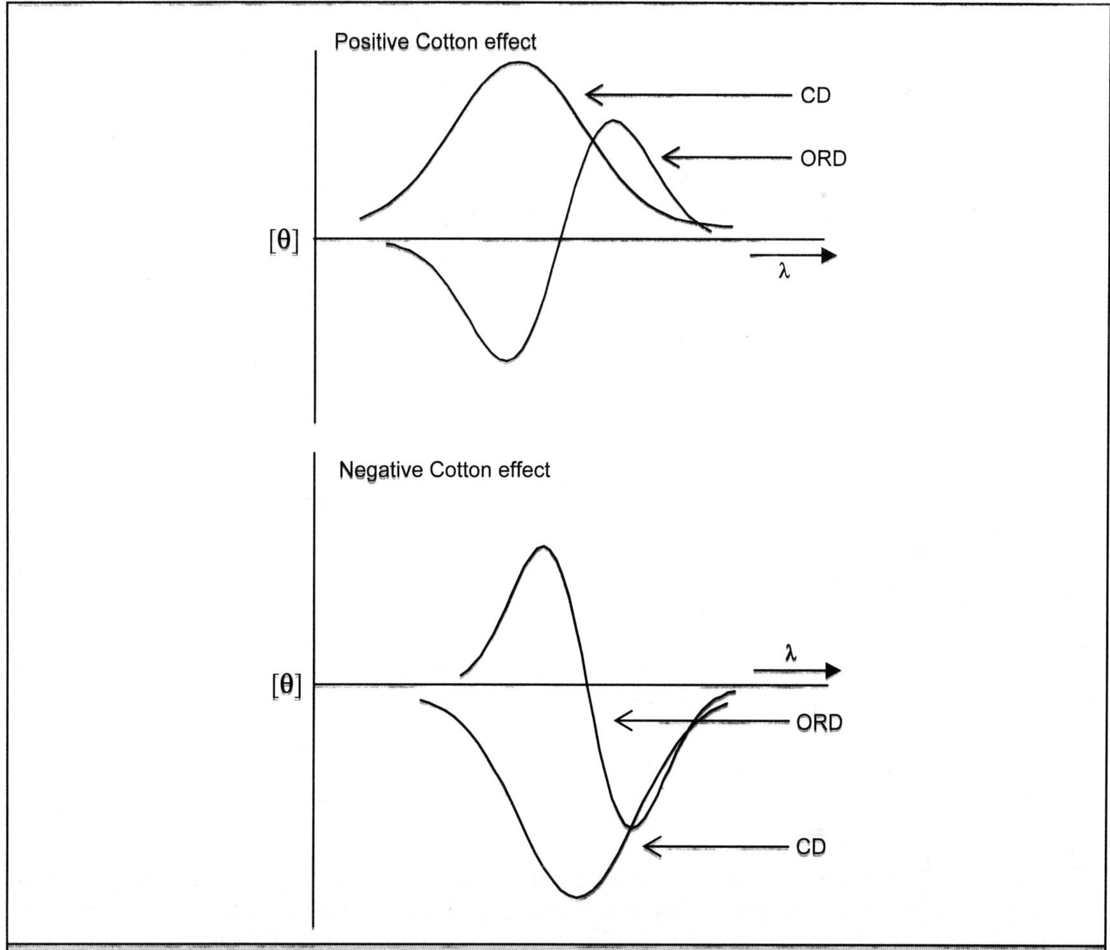

Figure 4.5: The positive and negative Cotton effects illustrated by means of both CD and ORD curves.

The molar ellipticity is given by the following expression.

$$[\theta(\lambda)] = M\theta(\lambda)/10dc$$

where M is the molecular weight, d is the path length in centimetres and c is the concentration in grams/ml. $[\theta(\lambda)]$ is measured in degrees. A plot of the difference in the refractive index, i.e. $(n_L - n_R)$ versus the wavelength is called the ORD curve and a plot of the difference in absorbance, i.e. $(A_L - A_R)$ versus the wavelength is called the CD curve. As seen in Figure 4.5, the two phenomena are related, and together are called the Cotton effect. Both CD and ORD curves can be negative or positive and go by the names 'negative Cotton effect' and 'positive Cotton effect' respectively. Rigorously, the measurement of either CD or ORD implies the other. However, in practice, especially with the availability of appropriate instrumentation, CD is the measurement of choice.

The CD spectrum of a polymer is different from that of a monomer, the difference reflecting the three dimensional arrangement. CD, therefore, is immensely useful to study three-dimensional structures of biopolymers such as proteins and nucleic acids. The $\pi \rightarrow \pi^*$ transitions discussed previously dominate the spectra, and in principle, quantum mechanical calculations of the CD of the biopolymers can be performed. This would allow the interpretation of the spectrum in terms of the structure. This however is an extremely complex task and, in practice, the existing optical activity data for polypeptides, proteins and nucleic acids with known structure are used in conjunction with semi-empirical equations to analyse unknown systems.

For proteins the aim is to analyse the CD spectrum to obtain secondary structural features such as the percentage of α-helical regions or β-sheet regions or random coil regions. The protein CD spectrum is dominated by the polypeptide backbone chromophores and the effects of the side chains are usually negligible. It can be considered as a linear combination of the spectra of each of the regular secondary structural regions. One may thus write

$$[\theta(\lambda)] = k_\alpha[\theta_\alpha(\lambda)] + k_\beta[\theta_\beta(\lambda)] + k_\gamma[\theta_\gamma(\lambda)]$$

where $[\theta(\lambda)]$ is the measured ellipticity, k_α, k_β and k_γ are the fractional compositions of α-helix, β-sheet and random coil regions respectively, and $[\theta_\alpha(\lambda)]$, $[\theta_\beta(\lambda)]$ and $[\theta_\gamma(\lambda)]$ are the measured CD of polypeptides in the conformation indicated. This equation can be used to estimate k_α, k_β and k_γ for an unknown protein, provided $[\theta_\alpha(\lambda)]$, $[\theta_\beta(\lambda)]$ and $[\theta_\gamma(\lambda)]$ are known. There are two methods of obtaining this information; both of them giving equivalent results when applied to unknown proteins. In the first method homopolypeptides that are known to possess one of the above secondary structures are chosen and the CD spectrum is measured for each. These spectra are used as the basis for calculating the fractional compositions. The other semi-empirical method obtains the basis values of the $[\theta_\alpha(\lambda)]$, $[\theta_\beta(\lambda)]$ and $[\theta_\gamma(\lambda)]$ from the CD spectra of proteins of known three-dimensional structure. The second method is theoretically more attractive as the spectra will automatically include effects of length of the polypeptide on tertiary interactions.

In the case of nucleic acids, the most intense chromophores are the bases. One cannot therefore neglect the sequence and consider the backbone alone. A further contribution to the intensity arises from the stacking of the bases. Semi-empirical theoretical calculations of the CD spectrum of di- or tri-nucleotides involve treating the ellipticity due to the molecule as the sum of those due to the individual bases, plus a strong contribution from the base-base stacking interactions. Thus for a trinucleotide with the bases B_1, B_2 and B_3

$$[\theta_{B_1 B_2 B_3}(\lambda)] = 1/3[\theta_{B_1}(\lambda)] + 1/3[\theta_{B_2}(\lambda)] + 1/3[\theta_{B_3}(\lambda)] + I_{B_1 B_2} + I_{B_2 B_3} + I_{B_3 B_1}$$

where $I_{B_1 B_2}$...etc., are due to the base-base interactions. Note that the term $I_{B_3 B_1}$ represents a 'next-nearest neighbour' interaction, rather than the 'nearest neighbour' interaction of the other two terms. Normally one needs to consider only the 'nearest neighbour' interactions. This approach of building up the CD spectrum of a polynucleotide from the spectrum of the monomer can be extended quite successfully to longer chains, though the number of contributions then increases. For DNA the base-paired double-helical structure must also be taken into account. Accurate determinations of the average structures of several different synthetic polynucleotides with several different structures have been made. A dramatic example of the use of CD was when an inversion of the high salt spectrum of poly d(CG), lead to the prediction of a novel left-handed double helical structure.

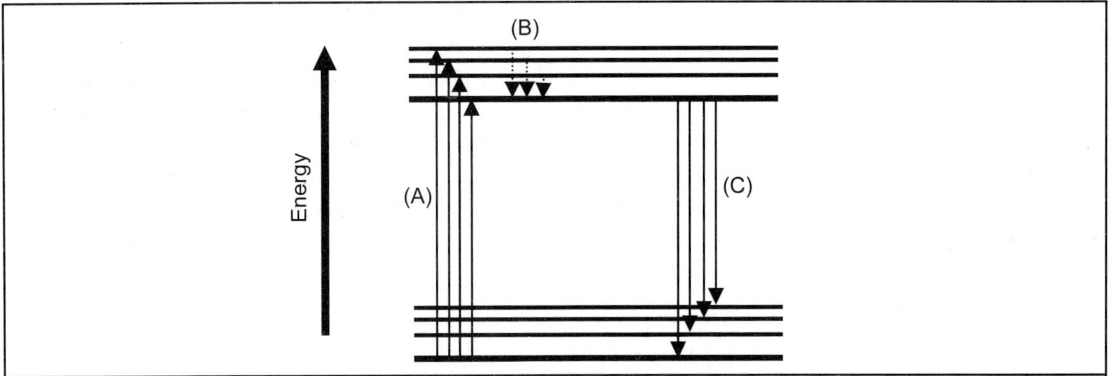

Figure 4.6: The electronic energy levels in a molecule, indicating the fine structure, and possible transitions. (A) Transitions from ground to excited state. (B) Non-radiative loss of energy. (C) Radiative transitions to lower state.

4.4: Fluorescence spectroscopy

Fluorescence is a phenomenon by which light is absorbed by a system at one wavelength and emitted by it at a different, and longer, wavelength. Measurements of fluorescence give information about molecular conformation and dynamics and form an important part of the techniques used in biophysics. The basic principle of fluorescence can be understood in terms of the electronic energy levels. According to quantum mechanics, the molecule is usually found in one of the energy levels (Figure 4.6). Each of these electronic levels themselves has a so-called 'fine structure' due to the vibrational motion of the molecules. In other words, each electronic level is not a single energy value, but consists of a band of closely spaced energies, which arise due to the different vibrational modes. The molecule usually exists in the ground state. When a beam of light (visible or UV) falls on it, some of the energy is absorbed and electrons jump up to the excited state where they can go to any of the vibrational levels. So far the experiment is the same as UV/visible absorption spectroscopy. However, now the electron in the upper electronic level can lose some of its energy in many ways. One of these is by going to a lower vibrational level within the same electronic level. When now the electron jumps back to the ground state, it does so by emitting radiation, but the radiation is of less energy and therefore longer wavelength than the incident light. The molecule is said to fluoresce. A study of the intensity of the fluorescent radiation as a function of the wavelength is known as fluorescence spectroscopy. Fluorescent radiation is spontaneously emitted radiation, as opposed to stimulated emission. It is governed by a rate constant k_F

$$k_F = A_{ba} = 1/\tau_F$$

where τ_F is called the radiative lifetime of the excited electronic state b of the molecule. A_{ba} is the Einstein coefficient of spontaneous transition from state b to state a, which is the lower, ground state. A_{ba} can be evaluated in terms of the frequency of the incident light and the strength of electric moment dipole corresponding to the states a and b. The molecule can also go from the excited state to a lower state by several other mechanisms that do not involve emission of radiation. For example, the excitation energy may be lost through collisions with the solvent molecules or by increasing the vibration of the molecules. The rate at which energy is lost through this process is called k_{ic} for 'internal conversion'. Another way in which the energy can be dissipated is by a phenomenon called quenching, resulting from the formation of complexes with solute molecules. The rate constant governing this process is called k_Q. A third process is when the excited state is converted to another, excited state of a lower energy though the transition that may be normally disallowed by quantum mechanics. The new excited state then dissipates its energy by internal conversion or by emission at a much longer wavelength than the fluorescence. The rate constant for this process is designated as k_{is} for 'intersystem crossing'. Thus the actual proportion of the absorbed incident light which is reemitted as fluorescent radiation is small and is called the quantum yield. It is given by the expression

$$\phi_F = k_F/(k_F + k_{ic} + k_Q + k_{is})$$

Chromophores or UV/visible absorption groups in biological molecules like proteins and nucleic acids have rather low quantum yields and very short radioactive lifetimes. For example, the most intensely absorbing group among the constituents of proteins is the tryptophan side chain. Its absorption maximum is at 280 nm while its fluorescence emission maximum is at 348 nm. It has $\phi_F = 0.20$ and $\tau_F = 2.6$ nanoseconds. The tyrosine side chain absorbs at 274 nm, emits at 303 nm, $\phi_F = 0.14$, $\tau_F = 3.6$ nanoseconds. Phenylalanine absorbs at 257 nm, emits at 282 nm, has a very poor quantum yield of 0.04 and a lifetime of 6.4 nanoseconds. Other side chains have yields much lower than this. In nucleic acids, the bases have yields of the order of 10^{-4} and lifetimes less than 0.02 nanoseconds. It may be noticed that the strongest fluorescent chromophores are rigid planar moieties, since in such groups, the energy dissipation by vibrational and rotational motion is much less. The flexibility of the nucleic acid backbones and of the other side chains and backbones of proteins is responsible for the fact that virtually no fluorescence is detected from them.

Nevertheless, since fluorescence is extremely sensitive to the environment of the chromophore, it is an effective technique to study conformational changes and ligand binding. The studies are enhanced

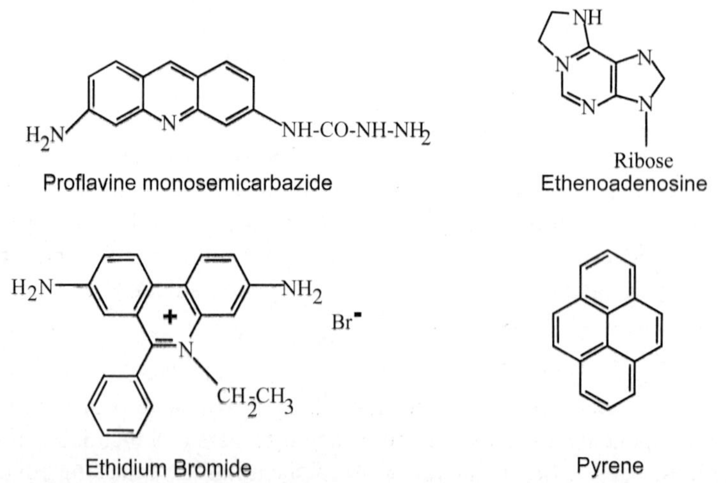

Figure 4.7: Some common fluorescent probes.

where a is a constant such that am_I is the value of the local field and H^O_{res} is the field strength at which resonance would occur if a = 0. Figure 4.15 shows the situation for a hydrogen atom with its single electron and the single proton in the nucleus, I=1/2. The dashed lines show the case when a = 0. The solid lines are the two transitions that actually occur, giving rise to the indicated hyperfine splitting. In general, if the spin quantum number of the nucleus is I, the ESR spectrum is split into 2I+1 lines.

ESR spectroscopy yields useful information in biological systems. The binding of paramagnetic metal ions such as Cu^{2+}, Mn^{2+}, etc., (i.e. those with an unpaired electron) to proteins or other molecules can be studied by the splitting that the binding induces in the absorption lines. Both the common isotopes of nitrogen, i.e. N^{14} as well as N^{15}, will lead to a splitting of the absorption line of Cu^{2+} when the metal ion binds. The resonance frequency of a paramagnetic species also depends on the orientation of the spin with respect to the external magnetic field. In a liquid, the tumbling motion will average out all the orientations. But in a crystalline solid, the ESR spectrum contains information regarding the orientation. This property was used, for example, in determining the angle of the heme group, containing paramagnetic iron, with respect to the crystal axes. Also, the orientation effect leads to an inverse correlation between the tumbling motion of the molecule and the width of the absorption line – the wider the line, the slower the tumbling. This has helped in the study of fluidity of membranes and the dynamics of proteins.

In the case of molecules that do not have intrinsic paramagnetic atoms (as in most biological macromolecules), ESR studies are made by attaching one or more organic or other radicals. Such paramagnetic groups are known as spin-labels and act as a reporter or probe. For most studies, spin-labels containing the nitroxide group are used. The ESR spectrum of the spin label can then be analysed in terms of the width of absorption line and the hyperfine splitting, to yield information about the conformation and dynamics of the biological molecules. This technique is especially useful in the study of membrane bilayers.

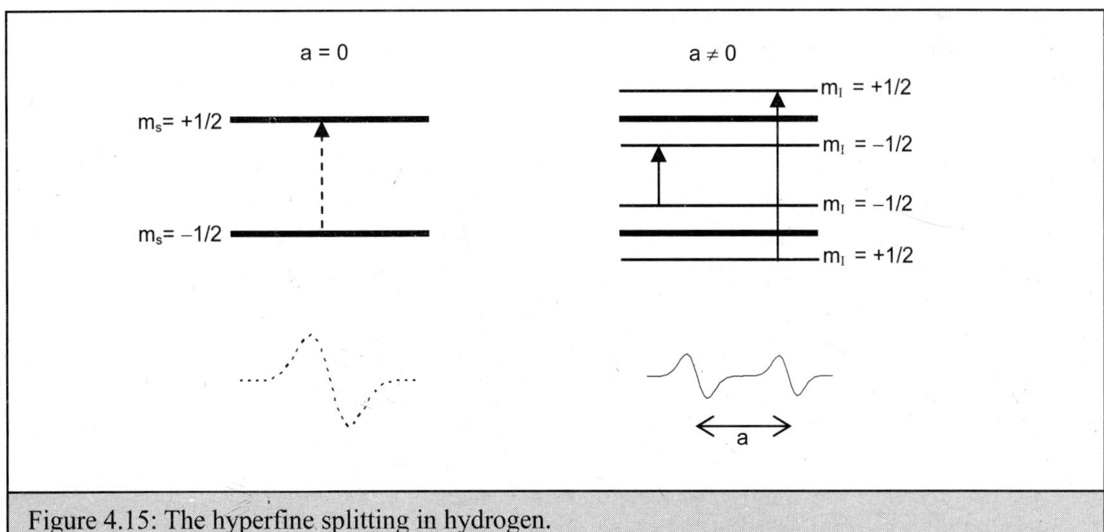

Figure 4.15: The hyperfine splitting in hydrogen.

Summary

- Different wavelengths of electromagnetic radiation are useful in probing different aspects of materials.

- In a spectroscopic experiment electromagnetic radiation of a particular frequency or a range of frequencies is allowed to fall on the sample under study. The radiation coming out of the sample is then analysed in terms of the intensity at different frequencies.

- Ultraviolet absorption spectra studies transitions between electronic states of atoms in the molecule.

- The Beer-Lambert law states that in a sample, each successive portion along the path of the incident radiation, containing an equal number of absorbing molecules absorbs an equal fraction of the radiation that traverses it.

- UV/Vis absorption spectroscopy is used frequently in biophysical research for simple tasks such as determining the concentration of a sample, as also for answering complex structural questions.

- Protein molecules have three types of chromophores, namely the peptide group, the amino acid side chain groups and any prosthetic groups that may be present.

- The UV absorption of nucleic acids is almost entirely due to the bases.

- Nucleic acid polymers show dramatic hypochromism and hyperchromism. These are effects that arise due to aggregation of chromophores.

- Circular dichroism and optical rotatory dispersion spectra are based on the optical activity of the moelcules, i.e. on the direction of the rotation of plane polarised light.

- Optical activity is quantified by the specific rotation.

- The measurement of the change in optical activity with change in wavelength is called Optical Rotatory Dispersion (ORD).

- The difference between the absorbance of the right circularly polarised light and the left circularly polarised light is called Circular Dichroism (CD).

- The CD spectrum of a polymer is different from that of a monomer, the difference reflecting the three dimensional arrangement. CD, therefore, is immensely useful to study three-dimensional structures of biopolymers such as proteins and nucleic acids.

- Fluorescence is a phenomenon by which light is absorbed by a system at one wavelength and emitted by it at a different, and longer, wavelength.

- Measurements of fluorescence give information about molecular conformation and dynamics and form an important part of the techniques used in biophysics.

- Chromophores are UV/visible absorption groups in biological molecules like proteins and nucleic acids.

- Since fluorescence is extremely sensitive to the environment of the chromophore, it is an effective technique to study conformational changes and ligand binding.

- Infrared spectroscopy is a probe of the vibrational motion of the molecules.

- Vibrational absorption spectra consist of bands rather than lines. The spread in the energies is usually due to conformational differences in the molecules and free rotation about the bonds

- The amide I and the amide II bands are the principal infrared absorption bands of the peptide group in proteins.

- The Raman effect arises from a coupling of the electronic and vibrational states of the molecule.

- Raman scattering is a probe of the vibrational states of the molecule, just like IR spectroscopy, except that the incident radiation is not infrared, but visible light and what is measured is not the absorbance but the frequencies of the scattered light.

- In the application of Raman spectroscopy to proteins the amide I and the amide III bands are used to probe the structure.

- The Raman spectra of nucleic acids has two distinct classes of lines; those arising from the bases, and those due to the sugar-phosphate backbone.

- Raman spectra of liquids and membranes are used to follow order-disorder transitions, through intermediate liquid crystalline phases.

- The utility of Electron Spin Resonance (ESR) in chemistry and biology comes from the variation that the 'splitting factor' undergoes in different chemical environments.

- The binding of paramagnetic metal ions such as Cu^{2+}, Mn^{2+}, etc., (i.e. those with an unpaired electron) to proteins or other molecules can be studied by the splitting that the binding induces in the ESR absorption lines.

- In the case of molecules that do not have intrinsic paramagnetic atoms (as in most biological macromolecules), ESR studies are made by attaching one or more organic or other radicals, known as 'spin-labels'.

Chapter 5

Light Microscopy

5.1: Introduction

Many of the interesting features of biological systems are too small to be seen with the naked eye. When molecular detail is not required, the light microscope is an ideal, and hence essential, instrument for a biologist. Electron microscopy can also be used to elicit highly detailed information, but suffers from several limitations, including complex operating procedures. Light microscopy is inexpensive and easy to use. An additional, crucial advantage it has over the electron microscope is its ability to focus at different levels within a three-dimensional object. In addition, the new methods and applications, which are even now being constantly developed, have helped to retain the importance of the light microscope in biological studies, in spite of the availability of advanced instruments of all kinds based on a variety of other physical principles.

Light microscopes are built up from a stack of lenses, usually made of glass, and in this chapter the optical principles that govern their arrangement are discussed first. A discussion of the resolving power of microscopes follows. Finally we discuss the newly developed methods in microscopy.

5.2: Elementary geometrical optics

Optical lenses can be either concave or convex. Convex lenses (Figure 5.1(a)) are also called converging lenses. Concave lenses (Figure 5.1(b)) are known as diverging lenses. Each lens has two focal points as shown. In the figure the focal lengths, defined by OF_0 and OF_1 are equal. This is not always so and depends on the relative refractive indices of the lens and the medium on either side of it. A thin concave lens always gives a minified (i.e. made smaller) virtual image of any object viewed through it (Figure 5.2). Virtual images are images that cannot be captured on a screen or on photographic film, except with the help of additional lenses. Concave lenses are thus not normally used in microscopes expect in combination with other lenses. A real image, in contrast, can be captured on a screen. Convex or converging lenses may form real or virtual images, depending on the positions of the object with respect to the focal points of the lens. Also the images may be minified, magnified, or

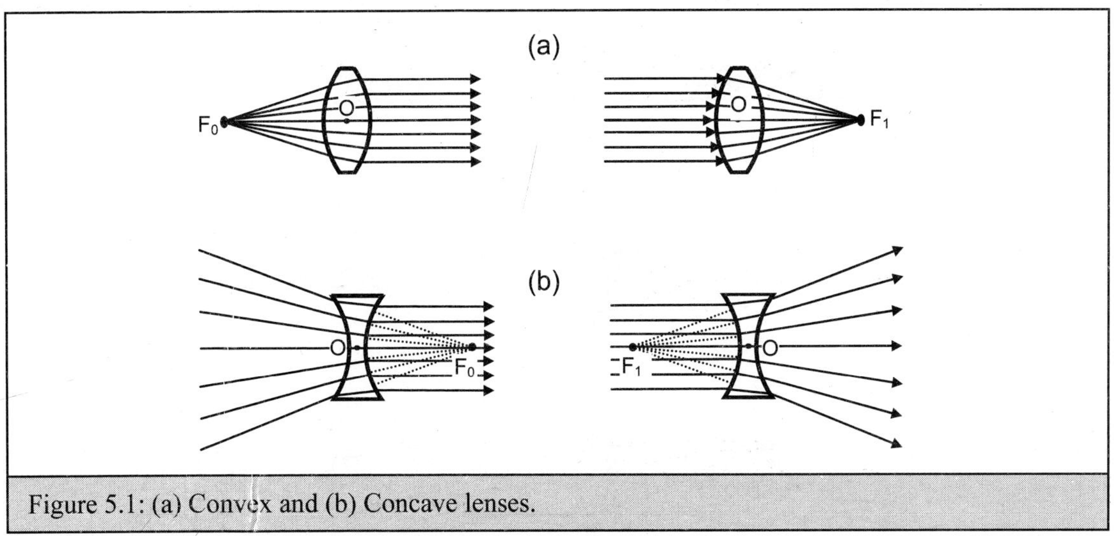

Figure 5.1: (a) Convex and (b) Concave lenses.

of the same size. Figure 5.3 illustrates the various possibilities. When the object is at a distance greater than twice the focal length from the lens, a real, inverted, minified image occurs at a point between one and two times the focal length on the other side of the lens (Figure 5.3(a)). When the object is at a point exactly twice the focal length, the image is real, inverted and the same size (Figure 5.3(b)). When the object is at a distance between one and two times the focal length, the image is real, inverted and magnified, and occurs at a distance greater than twice the focal length (Figure 5.3(c)). When the object is at the focal point, the image occurs at infinity (Figure 5.3(d)). The basic compound microscope uses two convex lenses in the arrangement shown in Figure 5.4. The lens closest to the object is called the 'objective'. The lens closest to the eye is called the 'eyepiece'. The microscope is adjusted so that the object (a) is at a distance between one and two times the focal length. The image (b) formed is therefore real, magnified and inverted. This real image acts as the object for the second lens, i.e. the eyepiece. The eyepiece magnifies this intermediate image and produces a large virtual image (c), which then becomes the object for the lens in the eye or for the camera lens. The final real image (d) is formed on the retina of the eye or the photo film. The relative positions of the eyepiece and the objective are kept constant, and focusing the microscope requires movement of both as a single element. The overall magnification of the instrument is the product of the magnifications of the

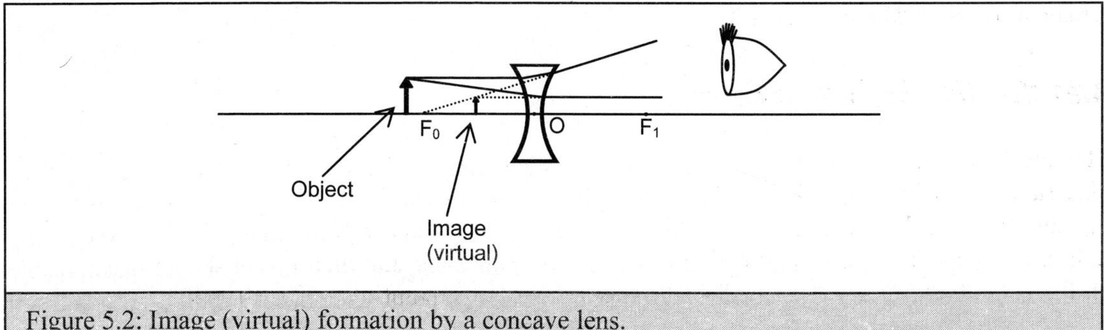

Figure 5.2: Image (virtual) formation by a concave lens.

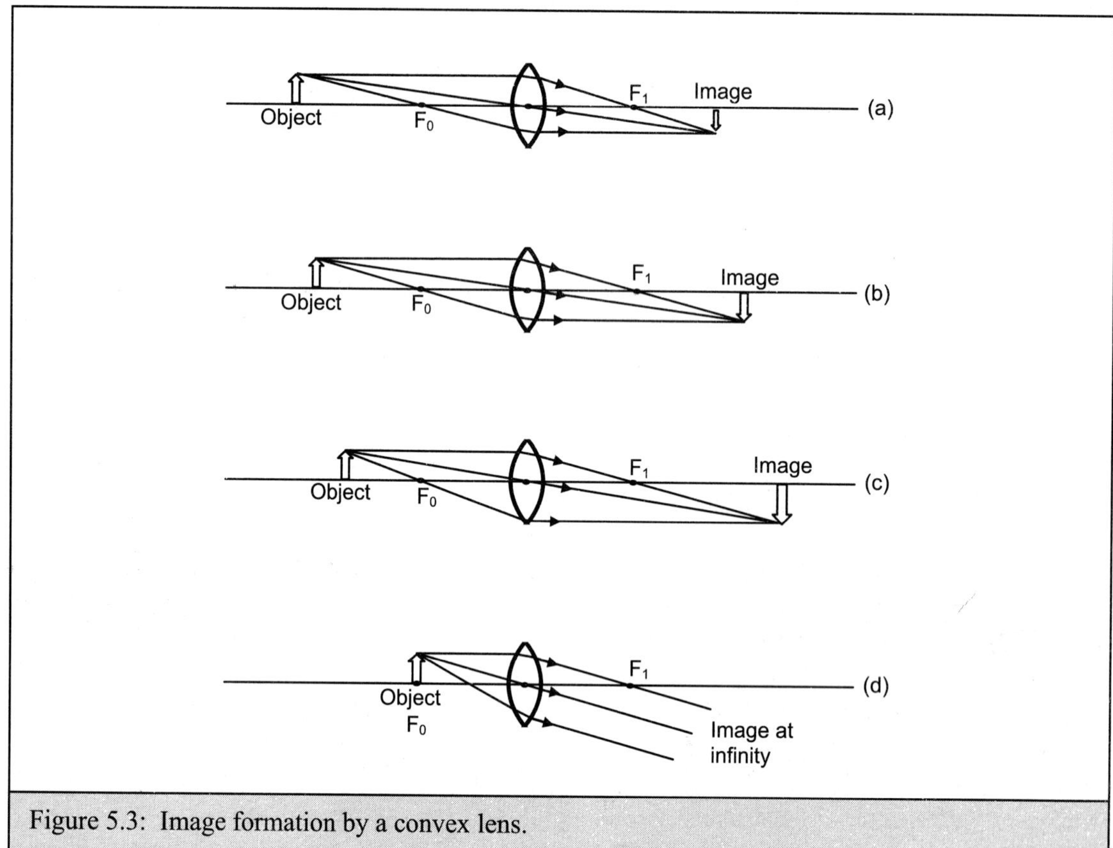

Figure 5.3: Image formation by a convex lens.

objective and the eyepiece. In most microscopes these are changeable and various combinations of the two lenses can be used to obtain the required magnification. In some microscopes, the magnification can be 'zoomed' up and down in a continuously variable fashion. Lenses with aspherical surfaces are also available, to reduce optical aberrations. In fact, even the simplest of compound microscopes available today will not be as simple as the one shown in the figure. The objective is a combination of convex and concave lenses to produce the desired magnification and eliminate distortions. Similarly the eyepiece is built up of many lenses. Nevertheless, the principles underlying the magnifying systems remain as illustrated here.

5.3: The limits of resolution

The resolution of a microscope is its ability to distinguish closely spaced objects. Heisenberg's Uncertainty principle (see Chapter 1) specifies the theoretical limits to the amount of magnification that an object can be subjected to and this figure is far, far greater than the magnifying power of any light microscope. For the magnification to make sense, however, the final image should make visible all the individual features. The smallest distance at which two point-like objects can be placed and still

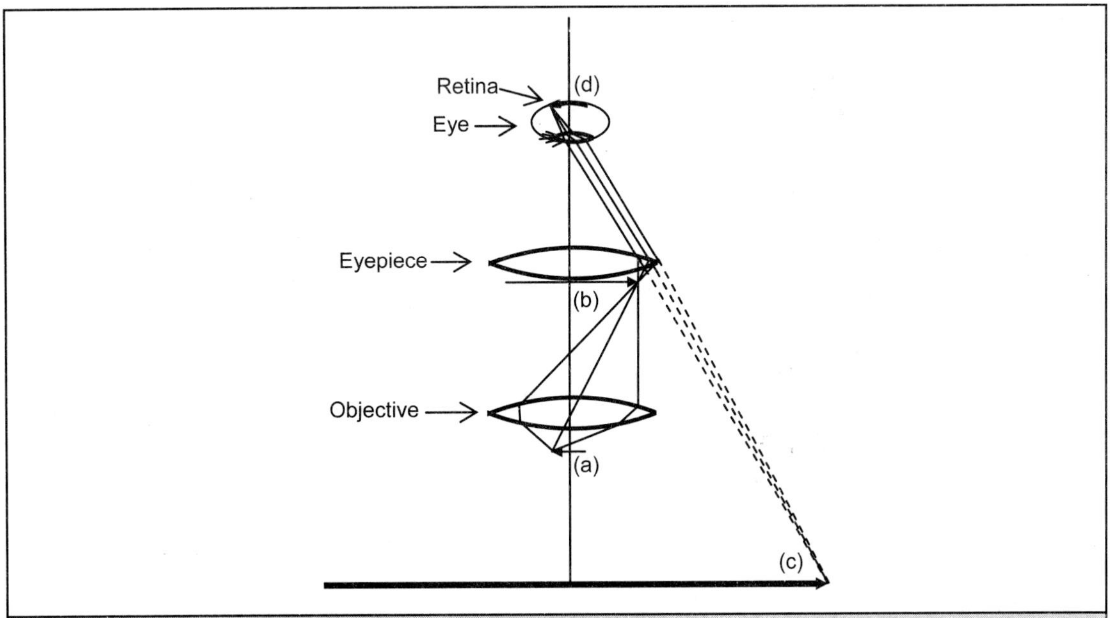

Figure 5.4: Schematic diagram of a compound microscope.

appear as two distinct objects governs the actual, effective magnifying power of the microscope. The smaller this distance, the higher is the resolution of the microscope and, therefore, the higher is the magnification possible. The resolution limit of an instrument can be defined on the basis of Rayleigh's criterion, which, in turn, comes from an analysis of the diffraction of light. Light waves coming from different points of the object form the final image after travelling through slightly different distances. As explained in the chapter on X-ray diffraction (Chapter 7), the waves can interfere constructively or destructively to form a diffraction pattern. Figure 5.5 shows the image, i.e. the diffraction pattern that is formed by a small pinhole. A central bright image is surrounded by concentric rings, which rapidly become fainter. The central image is called the first maximum, and the dark ring immediately surrounding it is the first minimum. The diffraction pattern due to two pinholes close to each other will overlap to different extents depending on the distance between the holes. Rayleigh suggested that two such objects could be considered as resolved, if in the diffraction pattern, the first maximum of one object overlapped with the first minimum of the other (Figure 5.6). Rayleigh's criterion is especially useful in deciding the resolution limit when dealing with self-luminous objects, such as for example, stars viewed through a telescope.

Another obvious limit to the magnifying power of a microscope is the wavelength of the light used for illumination. If the spacing between the objects is less than about 1/2 the wavelength of the light, then the images of the two objects will almost exactly superimpose on each other and they cannot be resolved. For the usual type of light used in microscopes this limit is close to 2 50 nm. Taking into consideration the

Figure 5.5: Diffraction pattern from a pinhole.

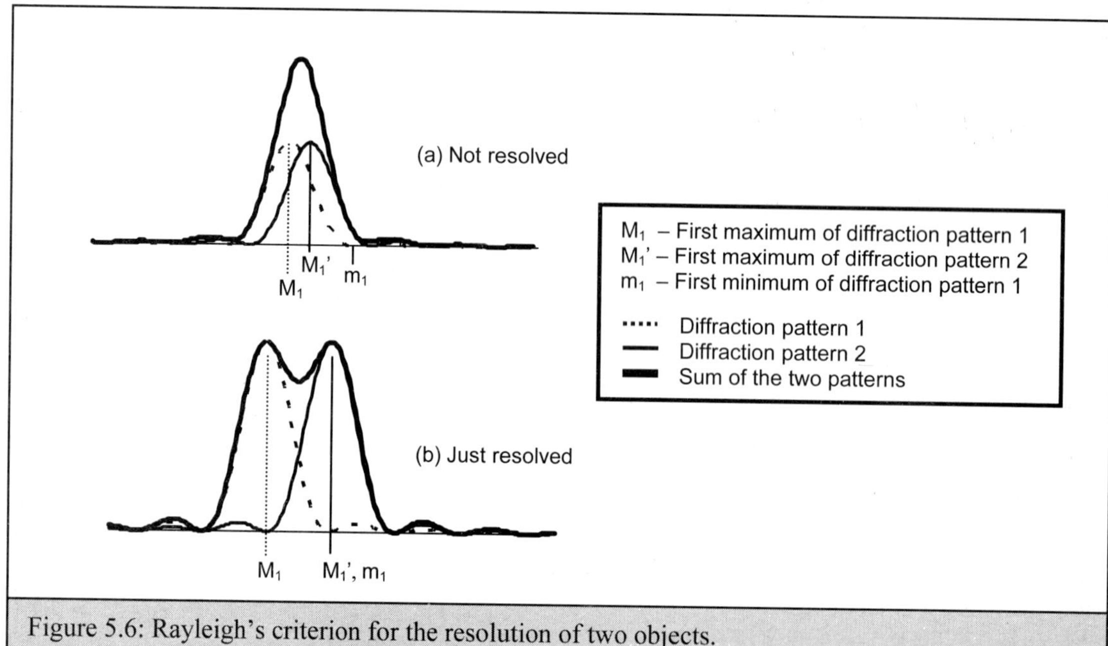

Figure 5.6: Rayleigh's criterion for the resolution of two objects.

fact that the eye (or the photographic film) imposes its own limits, the total magnification that is usually possible with light microscopes using ordinary light is about X800 (X800 or 800X means 800 times). While the Rayleigh criterion is simple, there are certain disadvantages when applied to images formed by microscopes. Apart from being strictly applicable only to self-luminous objects, there is also the problem of analysing diffraction from a complex object. The approach of Abbe is probably more suited to microscopes. Abbe realized that the objective lens in a microscope acted essentially as a Fourier summation device. Fourier showed that any regular periodic function could be written as a sum of simultaneously varying waves of different wavelength and amplitude. When light falls on an object, the effect of diffraction is to split the wave into Fourier components. A summation of all these components will give back the original image, and this task is performed by the objective lens. However, owing to the large angles at which most of the components are scattered, the objective usually is able to capture only the low order components. This leads to a fuzzy representation of the objects. The angles at which the various components are scattered also depends on the spacing of the features in the object – the smaller the spacing, the larger the angle. Thus features which are very close together are not well resolved since fewer of the Fourier components of the diffraction pattern from these features are used by the object lens in the Fourier summation. Abbe suggested that a pair of features on the object be considered as well resolved if both the zeroth and the first orders of the Fourier components reach the objective. This criterion leads to essentially the same result as the Rayleigh criterion, for the resolving power of a microscope and the highest meaningful magnification. It has the additional advantage of providing a rigorous understanding of the similarities between the microscope lens system and the mathematical operations of Fourier transformation.

5.4: Different types of microscopy

5.4.1: Bright field microscopy:

Bright field microscopy is obtained in the normal method of illuminating the object, when thin, more or less transparent samples (such as cells or bacteria) are being viewed. The illumination is usually as illustrated in Figure 5.4. The light is transmitted through the sample and the image is formed by the absorption of this light. Thus the image will appear darker than the background, which is the bright field. In other words, the image is formed due to the contrast in the amplitude of the light waves coming from the object, and those that are transmitted unchanged through the object. This imaging technique may be called amplitude contrast.

5.4.2: Dark field microscopy:

In the case of dark field microscopy the background illumination is cut off completely, leaving the field of the image dark. An extra condenser lens is added so that the rays of light fall obliquely on the object. The direct beam then passes straight through and is cut out by means of a diaphragm. Only the light scattered by the object is collected by the objective and sent on to the eyepiece. The light coming from the source itself has its central portion cut off, and only an annular ring of light falls on the condenser lens. This technique requires a very powerful source of illumination, since only the scattered light is used in the image formation. Nevertheless it is one of the oldest kinds of microscopy, and still finds application when viewing very small objects. Bacterial flagella, for example, with diameters of only about 20nm can be followed in motion using a very strong source of light and dark field microscopy. This is also a form of amplitude contrast microscopy.

5.4.3: Phase contrast microscopy:

Every wave has a characteristic phase. This is illustrated in Figure 5.7 where wave 'b' has the same phase as wave 'a' while wave 'c' has a different phase. In a microscope if the phase of the rays scattered from the sample can be made to have a different phase from the unscattered rays, then the image of the specimen will be different in light intensity as compared to the background and will be more clearly visible. This is because when two waves (e.g. the scattered rays and unscattered rays) are

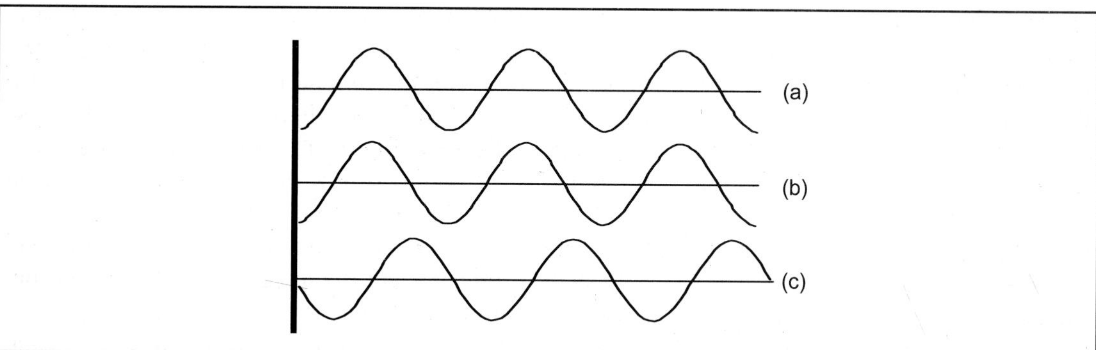

Figure 5.7: Waves of different phases. (a) and (b) are in phase, (c) is 90° out of phase with the other two.

Summary

- When molecular detail is not required, the light microscope is an ideal instrument for a biologist to view biological objects.

- Light microscopes are built up from a stack of lenses, usually made of glass.

- Virtual images are images that cannot be captured on a screen or on photographic film, except with the help of additional lenses.

- A thin concave lens always gives a smaller virtual image of any object viewed through it.

- A real image can be captured on a screen.

- Convex or converging lenses may form real or virtual images, depending on the positions of the object with respect to the focal points of the lens.

- The basic compound microscope uses two convex lenses. The lens closest to the object is called the 'objective'. The lens closest to the eye is called the 'eyepiece'.

- The image formed by the objective is real, magnified and inverted.

- A large virtual image is formed finally by the eyepiece at the lens of the observer's eye.

- The resolution of a microscope is its ability to distinguish closely spaced objects. The resolution limit of an instrument can be defined on the basis of Rayleigh's criterion.

- Rayleigh suggested that two objects could be considered as resolved, if in the combined diffraction pattern of the two objects, the first maximum of one object overlapped with the first minimum of the other.

- Another limit to the magnifying power of a microscope is the wavelength of the light used.

- The total magnification that is usually possible with light microscopes using ordinary light is about 800 times.

- Bright field microscopy is obtained in the normal method of illuminating the object, when thin, more or less transparent samples (such as cells or bacteria) are being viewed.

- In the case of dark field microscopy, only the light scattered by the object is collected by the objective and sent on to the eyepiece.

- The basic principle of phase contrast microscopy is that the light not scattered by the object has its phase shifted inside the microscope before being allowed to recombine with the light scattered by the object.

- Fluorescence is the phenomenon by which certain substances absorb light of a particular wavelength. After a very short interval of time the light is re-emitted with its wavelength altered.

- Fluorescence microscopy uses this property to specifically image materials of interest.

- In a polarising microscope, light polarised by a polariser is allowed to fall on the object. Light scattered from the object is then viewed through an analyser.

- A polarising microscope helps to determine the structures of ordered domains within the sample.

Chapter 6

Electron Microscopy

6.1: Introduction

Electrons are negatively charged subatomic particles. Their use in imaging can be explained in terms of their quantum mechanical wave-like property, which allows a beam of electrons to be considered in exactly the same way as a beam of electromagnetic radiation such as light. Electrons possess two additional advantages for use in microscopy. The first of these is that according to quantum physics, the wavelength of the waves associated with an electron varies inversely as its energy. Thus by accelerating the electron between charged plates with a very high applied voltage, the wavelength of the 'radiation' used in the microscope could be made very small. As explained earlier (Chapter 5), this implies a concomitant increase in the resolution and hence in the possible magnification. The second advantage, which actually allows the use of electrons for microscopy, is that, unlike electromagnetic radiation of short wavelengths such as X-rays, electrons can be focussed by means of magnetic lenses. This is dependent on the well-known principle in physics that a charged particle in a magnetic field is deflected in a direction perpendicular to its direction of motion. This principle has been used in the construction of electron microscopes.

In this chapter we will first study some elementary electron optics including the principle components of the electron microscope. Next we describe the different types of microscopic techniques, the preparation of the specimen, and some applications. We go on to an introduction to advanced image reconstruction techniques and electron diffraction. We end with a discussion of tunnelling electron microscopy and the atomic force microscope.

6.2: Electron optics

The wavelength of the electron beam depends on its energy, which is in turn dependent on the voltage used to accelerate the electrons. Modern electron microscopes use accelerating voltages in the range 1000 volts to 1000 kilovolts. The most popular microscopes use about 100 kilovolts. According to the

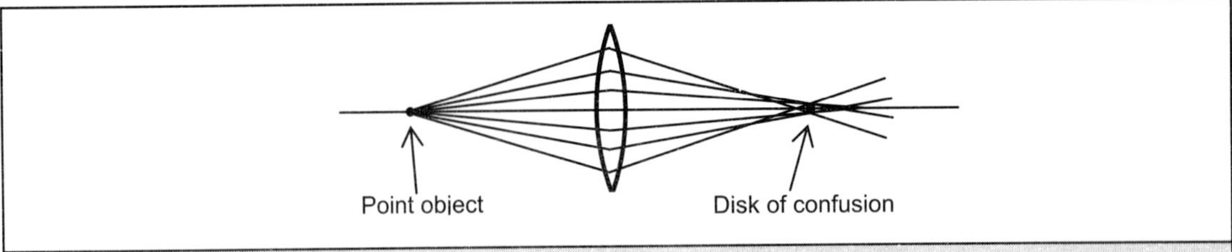

Figure 6.1: Spherical aberration in electron optics, caused by imperfections in the magnetic lens.

de Broglie equation (see Chapter 1), this voltage will correspond to a wavelength of about 0.03 Å. This should, in principle, allow the imaging of details on an atomic scale. However several factors come in the way of reaching this high resolution. Vibration, damage to the specimen and contamination are some of the problems encountered in microscope design. The most important factor limiting the resolution is the inability of the magnetic lenses to focus the beam accurately. This is called the aberration of the lens. Aberration is chiefly of two types. The first, called spherical aberration arises from the fact that all the electron beams scattered from the specimen which fall on the lens should be focussed on to the same spot but in practice are not. Figure 6.1 gives an illustration of how the images from a point will be spread out into a series of images. Thus the image of the point will be seen as a disk, called the disk of confusion. The spherical aberration can be reduced by cutting out some of the higher angle beams scattered from the point by means of a stop as illustrated in the figure. But obviously there is a limit to the narrowness of the aperture beyond which diffraction effects will come into play. The second important type of aberration, called chromatic aberration, arises from the fact that the focal length of the lens is different for different wavelengths. Since the source cannot produce a uniformly energetic electron beam, therefore the image of the point again spreads out into a disk of confusion, this time caused by the chromatic aberration.

The final quality of the image depends not only on the resolution but also on the contrast. Contrast is the difference in intensities between the object and the background as a fraction of the background intensity. For visibility, this ratio should be larger than 0.2. The contrast can be enhanced if the object is dense as compared to the background. Usually, the support for the specimen is a carbon film and biological materials, especially, have about the same density. The use of electron microscopy in biology is thus especially problem prone. As will be described later, one can use staining methods to increase the contrast. It is also possible to increase the contrast by increasing the intensity of the incident electron beam. This is because the background intensity varies as the square root of the incident intensity while the image intensity will vary directly as the incident intensity. However, increasing the intensity of the incident beam leads to another severe problem in the study of biological materials, viz. radiation damage. This will destroy the sample leading to very poor quality images.

Thus because of the above problems and also because of others not fully discussed here, the actual practical resolution and magnification in biological electron microscopy is below what in principle is possible. Nevertheless, electron microscopy is far superior to the light microscope in its resolution and magnification. Objects such as cells, viruses, DNA strands, ribosomes, etc. can be imaged clearly at resolutions of a few tens of Angstroms, at magnification of 100000X or more.

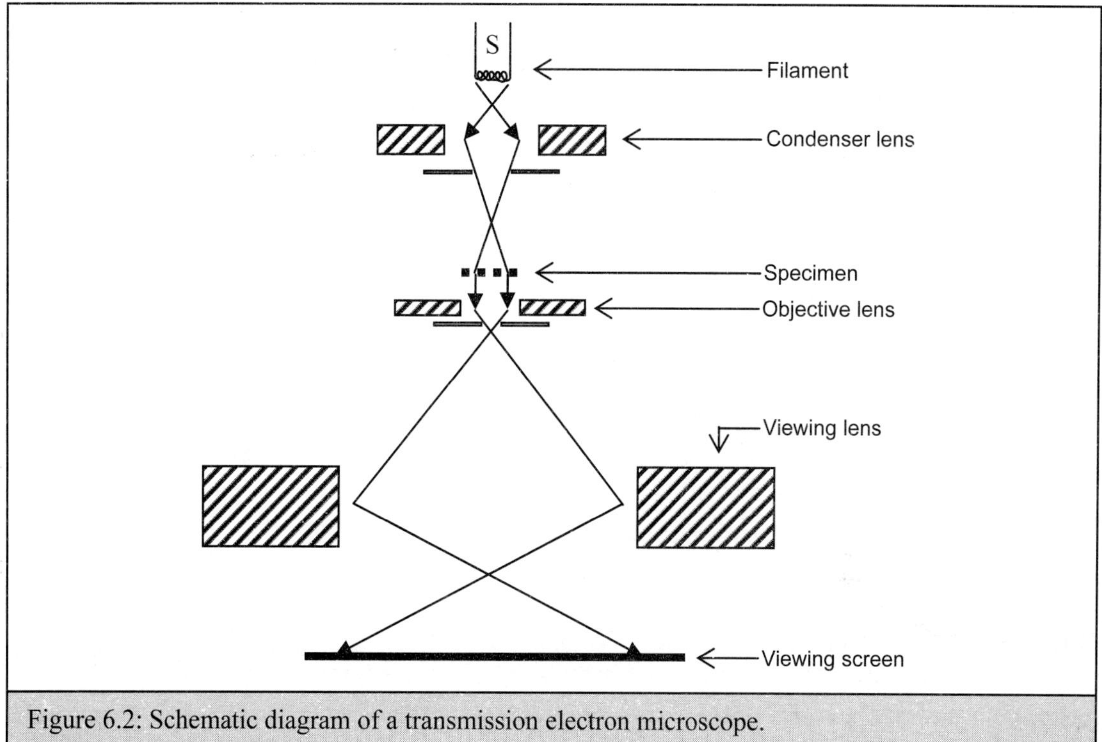

Figure 6.2: Schematic diagram of a transmission electron microscope.

6.3: The transmission electron microscope (TEM)

This is one of the most commonly used instruments. Its geometry is shown in Figure 6.2. The source 'S' of the electron beam may be a tungsten filament, but other materials like lanthanum hexaboride are also used. The beam path, the placement of the lenses, the specimen, the aperture etc, follow the plan of the light microscope as seen in the figure. The entire arrangement, including the specimen, has to be placed in high vacuum to avoid extraneous scattering and absorption of the electrons by air. The lenses are magnetic lenses in the electron microscope, and unlike as in the light microscope, the image is not viewed directly through an eyepiece, but is projected on to a fluorescent screen on which the electrons form the image. An extra aperture called the objective aperture is placed in the electron microscope and not normally in the light microscope. This is because the image forming process in the TEM is different from that in light microscopes. In a light microscope, light is transmitted or absorbed by the specimen. This creates a contrast between the different parts of it, and hence creates the image. In the TEMs the situation is quite different and most of the incident beam passes through the specimen unimpeded, and very little is absorbed. Figure 6.2 shows that only a small portion is actually scattered in the forward direction. It is this forward-scattered fraction that is used to form the image. Contrast between different features in the image, e.g., raised features against flat ones, is created by differentiating between electrons scattered into wide angles from those that are scattered into small angles. The objective aperture cuts off the large angle electrons, and the amplitude of the electron

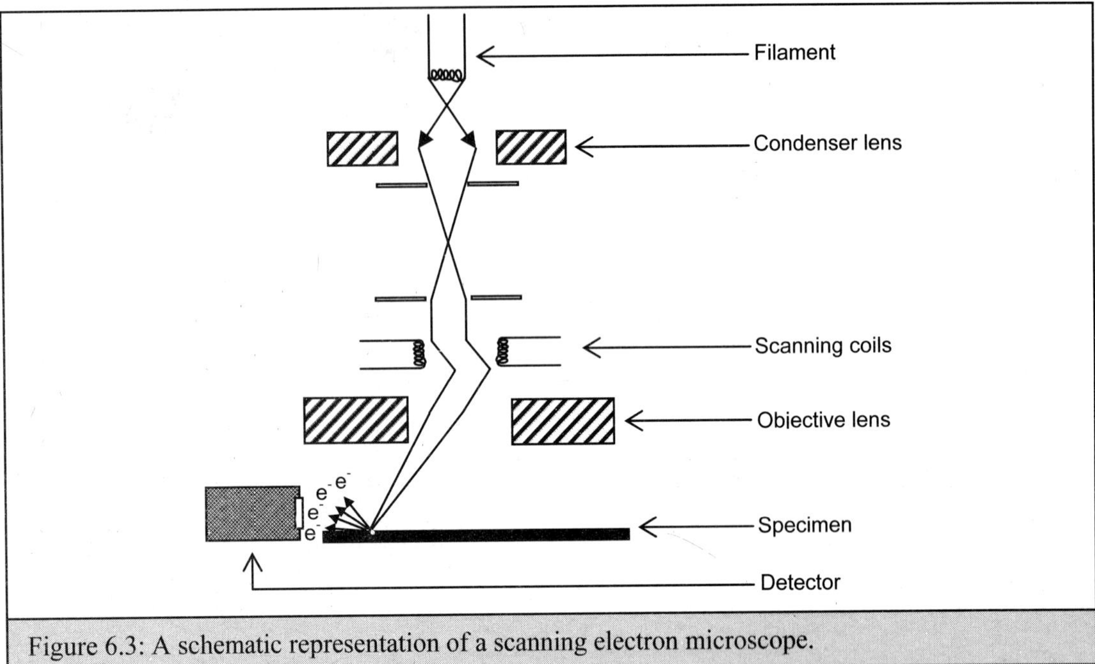

Figure 6.3: A schematic representation of a scanning electron microscope.

beam from different portions of the specimen is different. This leads to amplitude contrast, which creates the image. Most biological materials however do not have enough contrast to be viewed unless they are stained. Dark field imaging is also possible and is achieved by cutting out the un-scattered beam.

6.4: The scanning electron microscope (SEM)

A SEM is very useful in obtaining images of the surface of thick specimens. In the scanning-transmission mode (STEM), the SEM can also be used to study thin specimens and in this respect has some advantages over the conventional TEM. Figure 6.3 is a schematic illustration of the construction of the SEM. The most obvious difference as compared to a TEM is that the electron beam, after passing through the condenser lens is deflected in a raster pattern over the specimen stage, similar to the pattern in a television picture tube. The objective lens is split into two parts. One part is placed between the condenser lens and the specimen. This can in fact be regarded as an additional condenser lens, which focuses the electron beam onto a small spot on the specimen. Image signals can be collected by the detector, which is placed on the same side of the specimen as the source and the raster coils. This detector is used in the surface-scanning mode and collects the secondary electrons knocked out of the specimen by the primary beam, as well as the back-scattered electrons reflected from the surface of the sample. Image signals can also be collected in the scanning transmission mode by the detector after passing through the second half of the objective. These are the forward-scattered electrons.

Images obtained in the STEM mode are very similar to those obtained in the fixed beam TEM. The surface SEM image is quite different. In particular, the contrast in the surface imaging arises partly from the different scattering powers of the different atoms in the specimen, and partly from variations in the topography. In both the STEM mode as well as the surface SEM modes the images are generated in the following way: As the electron beam scans the specimen, the scattered electrons from each position is measured by the detector, one after the other in time. The measured intensities are displayed on the imaging screen one after the other in space, in the same type of raster scan as the electron beam. Thus, if the scattering is high at particular point during the scan, then the corresponding point on the viewing screen is bright. If the scattering is low, the corresponding point is dark. This creates an image of the object. The resolution of the image depends on the size of the electron spot used to scan the specimen and this may be as small as 50 Å. Both surface SEM and STEM have changed the use of electron microscopy quite dramatically and have made quantitative studies possible. The scanning principle has also been used more recently in the construction of the atomic force microscope and the scanning tunnelling microscope. These will be discussed later.

6.5: Preparation of the specimen for electron microscopy

Preparation of the specimen is one of the most important steps in the process of imaging by electron microscopy, and often takes longer, and requires greater skill, than the actual observation. This is especially true in the case of biological samples. This is because biomolecules and assemblies are prone to disintegration in the high vacuum condition under which the imaging is done. Secondly, without staining with heavy atoms, the lighter atoms such as carbon, hydrogen, nitrogen and oxygen, which are the principle components of biological samples, do not offer sufficient contrast against the support, and cannot therefore be observed. Finally, unless protected, the intense beam of electrons can easily destroy the sample.

The specimen preparation involves the following steps: - a) Preservation: The specimen has to be preserved in the relevant configuration. Care should be taken to avoid artefacts being introduced into the specimen during the preparation. b) Staining: In order to increase the contrast between the sample and the background, the specimen has to be stained with a heavy, electron-rich metal. c) The specimen has to be dried, i.e. water has to be removed in order that this does not interfere with the imaging. d) The specimen has to be made into a form suitable for observation by replication and/or microtomy.

(a) Preservation: The two main methods of preserving the sample are by freezing and by chemical fixation. Freezing the sample has to be done with care to avoid formation of ice crystals that may radically alter the structures of interest. Rapid freezing is one way of reducing ice crystal formation, and a cooling rate of about 5000 degrees per second is sufficient to freeze whatever water there maybe in the sample into a glass-like structure, which does not affect the sample. Though the sample is usually cooled down to only liquid nitrogen temperatures, the required high cooling rates may be obtained if the whole process is accomplished in a few milliseconds. Once the specimen is frozen and is retained at the liquid nitrogen temperature or below, the structure, morphology and constitution do not alter for even months. The sample can also be preserved by chemical fixation. Chemicals, most commonly osmium tetraoxide and gluteraldehyde, are used to preserve the structure and shape of the specimen. Both chemicals work by creating covalent bonds between different molecules, especially proteins or lipids. Osmium tetraoxide has the additional advantage of providing a heavy metal stain, doing away with the necessity of applying the stain separately. However, unlike gluteraldehyde, it does not act on a wide range of tissues. The great advantage of chemical fixation, as compared to freezing is that the procedure is simpler. Moreover, chemicals can be used to preserve much thicker samples since the penetration is greater. Handling and storage of chemically fixed samples is also easier.

b) Staining: As mentioned earlier, both the biological samples studied and the carbon films on which they sit consist of light atoms, and without contrast enhancement the samples will be invisible with respect to the background. Thus staining the sample must be resorted to, in spite of the fact that this will result in a substantial loss of resolution. The technique of positive stain adds a specific heavy atom (uranium to DNA samples, or osmium to lipids) by chemically complexing them. Another technique of positive staining is called shadowing. Here heavy atoms, such as tungsten are evaporated from a filament onto the sample. If the sample is bombarded at an angle, then walls of the metal atoms pile up on one side of vertical features in the sample, while no metal atoms are deposited in the shadow of the wall. This allows the easy visualisation of the contours of the raised features in the sample. In the technique of negative stain, the heavy metal stain fills the area where the molecules are not present. Thus the outline of the molecules or molecular complexes will be visible.

c) Dehydration: For wet specimens the dehydration has to be carried out gradually in order that the sudden loss of water does not distort the structure. In the case of frozen specimens such sudden change in the water content is less likely, and dehydration is carried out by maintaining the sample in an atmosphere of very low pressure. A common method of dehydrating wet samples is to immerse it in alcohol solutions, slowly increasing the concentration of alcohol, such that the water is ultimately replaced completely by alcohol. In some cases the heavy metal stain can act also as the dehydrating agent. When a thin layer of the staining solution such as potassium phosphotungstate or ammonium molybdate or uranyl acetate is added to a thin specimen exposed to air, the solution takes on an amorphous, glassy shape and protects the morphology of the specimen during dehydration.

d) Replication, microtomy, etc.: This step in the specimen preparation process renders it suitable for viewing in the microscope. Biological macromolecules such as DNA, etc., can be deposited directly onto a very small (~1 mm diameter) fine grid made of copper, which is then treated with the staining solution and inserted into the microscope for viewing. Replication is a method used to prepare samples when only the surface features are of interest. A thin layer of metals such as gold, tungsten, palladium or platinum, or a thin plastic film, is cast over the sample and allowed to set. The specimen can then be discarded, as a replica of it is now preserved in the metal or plastic film. If the interior of the specimen is of interest, sectioning becomes necessary. In this case the specimen is embedded in a resin such as epoxy resins ('araldite') or polyester resins. It is then sectioned using a microtome. A microtome is a device that uses glass or diamond knives to produce very thin sections of the embedded samples. The thickness of the sections can be as small as 100 nanometers. These sections are then placed on the copper grid for viewing. The heavy metals that are used in the staining process also act to conduct away the excess negative charge that accumulates on a non-conducting sample. For very thin samples viewed on the copper grid, the grid will act to ground the excess charge. In cases where a thick non-conducting sample is used without staining, e.g. for surface SEM studies, it should be coated with a thin metal film solely to act as the conductor.

6.6: Image reconstruction

Very often the electron micrographs of biological samples, such as viruses or ribosomes or other large molecular assemblies, consist of several repeated indistinct images of the object. The image can be made more distinct by reconstruction procedures. There are chiefly two such procedures. The first involves simply superposing all the images, such that the 'true' parts of the images, or the signal, gets reinforced, while the background 'noise' will be cancelled. The selection of the identical parts of the image for superposition can be accomplished visually. A more rigorous method is to calculate a superposition function in two dimensions. The superposition function or the cross correlation function will have a maximum value when one image is rotated and translated to superpose on another image optimally. Such optimisation can be accomplished on the computer if the images are digitised and fed to it. A more accurate way to accomplishing a clear image from several indistinct images of the same object is possible if the images are arranged in a regular repeating two-dimensional pattern. If the image is used as a kind of imperfect diffraction grating and is placed in the path of a source of light, a diffraction pattern will be formed. The diffraction pattern can then be filtered to allow through only those parts which correspond to the periodic lattice. The filtered diffraction pattern is in turn used as the grating and placed in the path of a light beam. The image formed will correspond to a much clearer version of the original object. Mathematically, the procedure is equivalent to performing a Fourier transform, cleaning up the coefficients by filtering, and then performing a reverse Fourier transform. The method of image reconstruction has been widely used to study crystalline materials. Virus structures (such as the stem of the T phage), membranes and two-dimensional arrays of ribosomes have all been studied by this method. The method can be extended to three dimensions by taking several two-dimensional images obtained by tilting the specimen at different angles. In fact a single micrograph of a suspension of particles is bound to contain images of the particle in various three-dimensional orientations. Image reconstruction techniques can be fruitfully used to put together these images to form a 3-D picture of the particle.

6.7: Electron diffraction

The electron microscope can be also be used with a few modifications, to obtain electron diffraction patterns from two-dimensional crystals. Instead of analysing these patterns through geometrical optics, it is possible to use Fourier theory to mathematically transform the diffraction pattern into a three-dimensional image of the particle. Since the electron wavelength is much smaller (~0.03 Å) than that of X-rays, it may be thought that a much higher resolution will be obtained. However several obstacles come in the way of attaining such high resolution. Since electrons are easily absorbed, only very thin samples can be used. Electron diffraction is ideal for use in those cases where three-dimensional crystals do not grow but two-dimensional arrays form easily, such as to study membrane proteins. From such a thin `crystal', almost the entire diffraction pattern will be in a single plane and most of the three-dimensional pattern required to perform a complete Fourier transform is not available. Another problem, also present in the case of X-ray diffraction, is the loss of phase information and the consequent inability to perform the Fourier analysis. One way in which this problem is got over is by using the electron micrograph to obtain some information about the approximate phases and using this information in combination with the measured electron diffraction intensities. In spite of these

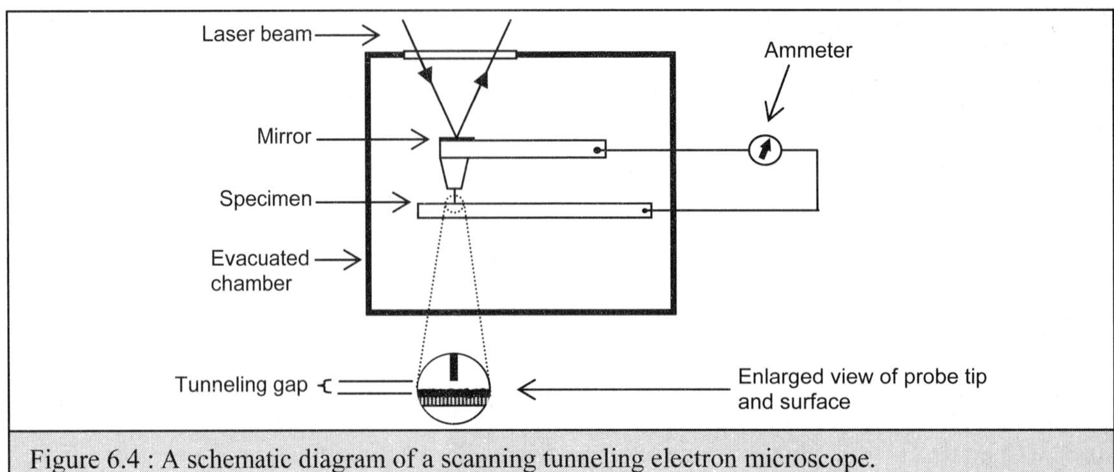

Figure 6.4 : A schematic diagram of a scanning tunneling electron microscope.

difficulties, electron diffraction is a popular technique to obtain low-resolution images of large particles that do not easily form crystals suitable for X-ray diffraction.

The experimental techniques of preparing the sample and obtaining the diffraction pattern are not very different from those for obtaining electron micrographs. Since the electron beam can be focussed on very small areas, the ordering in the sample needs to extend only up to a few microns (or less) to obtain the pattern. The intensities are then extracted from the electron diffraction photograph by densitometry. The intensities and the calculated or estimated phases are combined to perform a Fourier transform to obtain the electron density images of the particles. It is possible to obtain the diffraction patterns from tilted crystalline samples and calculate three-dimensional images. These will improve the quality of the final results.

6.8: The tunnelling electron microscope

In the last decade, the tunnelling electron microscope has become an integral part of any laboratory interested in the study of surfaces. Its use in biological imaging is somewhat less widespread, though in recent years it has been used to produce images of DNA molecules that have enabled precise measurement of the pitch of the double helix, as well as of proteins and macromolecular assemblies.

The operation of the tunnelling electron microscope is based on the well-known quantum mechanical principle of electron tunnelling. The equations of quantum mechanics assign a finite non-zero probability for an electron to go from one region of low potential energy, to another such region, even if it does not possess sufficient energy to overcome the barrier between the two regions. This phenomenon may be equivalently visualised as a football going from one well into which it has been dropped, into another one nearby, without coming to the surface or making a hole between the wells. In the tunnelling electron microscope, this phenomenon is used in the following way. A very, very sharp needle is brought close to the surface to be imaged. The distance is of the order of a few Angstroms. At such a distance electrons from the surface will tunnel across the gap and set up a measurable tunnelling current in the needle. This current is a precise indicator of the distance between the tip and the surface. The needle can now be moved (in a raster pattern) across the surface, at each point its height being adjusted to maintain the tunnelling current at a constant value. Thus a record of the vertical portion of the needle at each position becomes equivalent to a record of the surface topography and can be converted into an easily observable picture.

Figure 6.4 is a schematic representation of a scanning tunnelling electron microscope. The tip of the needle can be made so fine that at the very end there is only one single atom. Also, modern electrical measurement technology allows even nano-amperes of tunnelling current to be measured. These technological developments have been instrumental in making possible the visualisation of individual atoms on the surface of materials.

The use of this microscope in the study of biological samples is still not widespread for several reasons. The chief among these is that non-conducting or semi-conducting materials have to be stained with heavy metals before they can be imaged, leading to a loss of resolution. The preparation of the tip and its positioning above the surface also has to be more precise and controlled in the case of biological materials.

6.9: Atomic force microscope

The tunnelling microscope has been modified in the following way. Instead of keeping the needle tip at a small distance from the sample the needle is pushed right against the surface. The force present in the tip is kept constant and the tip is scanned across the surface like a phonograph needle running in the groove of a record. A record of the tip's vertical motion is made by reflecting a laser beam from a mirror fixed to the top of the needle and using an interferometer to accurately measure the distance travelled by the laser beam, and hence the position of the tip. This record can be converted and displayed as an image of the surface. Since the atomic force microscope does not depend on a current, it can be used to visualise the surfaces of conductors as well as non-conducting materials. Atomic force microscopes or AFMs are now routinely used tools in the fast-developing field of nano-biotechnology. The AFM is sensitive to forces as small a piconewton. This makes it possible to measure the forces acting on, or exerted by, single biomolecules. For examples AFM has been used to study the movement of a RNA polymerse on the DNA, the forces at antibody-antigen binding, the forces required to produce supercoils in a single strand of DNA, etc. The use of AFM, along with the technique of trapping molecules using finely focussed laser beams has contributed to the rapid development of the field of single-molecule biophysics, or the study of physics at the level of single biological molecules.

Other microscopes have been built which are based roughly on the same principle of a tip scanning the surface. In each different type of microscope a different force is measured. Thus we have friction force microscopes, magnetic force microscopes, electrostatic force microscopes, and scanning thermal microscopes. The last may be described as the world's smallest thermometer. A thermocouple attached to the needle tip measures the temperature at each position. Since heat is essentially motion at infrared wavelengths, it may be possible to measure the infrared spectrum of a single molecule!

Summary

- The use of electrons for imaging can be explained in terms of their quantum mechanical wave-like property, which allows a beam of electrons to be considered in exactly the same way as a beam of electromagnetic radiation such as light.
- A major advantage that actually allows the use of electrons for microscopy, is that, unlike electromagnetic radiation of short wavelengths such as X-rays, electrons can be focussed by means of magnetic lenses.
- The wavelength of the electron beam depends on its energy, which is in turn dependent on the voltage used to accelerate the electrons.
- The actual practical resolution and magnification in biological electron microscopy is below what is possible in principle.
- Electron microscopy can produce of images of objects such as cells, viruses, DNA strands, ribosomes, etc. at resolutions of a few tens of Angstroms, i.e. at magnification of 100000X.
- The transmission electron microscope (TEM) is one of the most commonly used instruments.
- The lenses are magnetic lenses in the electron microscope, and unlike as in the light microscope, the image is not viewed directly through an eyepiece, but is projected on to a fluorescent screen on which the electrons form the image.
- In the TEM most of the incident beam passes through the specimen unimpeded, and very little is absorbed. The small forward-scattered fraction is used to form the image.
- Contrast is created by differentiating between electrons scattered into wide angles from those scattered into small angles.
- The most obvious difference of a scanning electron microscope (SEM) as compared to a TEM is that the electron beam is deflected in a raster pattern over the specimen stage.
- As the electron beam scans the specimen, the scattered electrons from each position is measured by the detector, one after the other in time.
- The measured intensities are displayed on the imaging screen one after the other in space, in the same type of raster scan as the electron beam. This creates an image of the object.
- The resolution of the image depends on the size of the electron spot used to scan the specimen, and this may be as small as 50 Å.
- Preparation of the specimen is one of the most important steps in the process of imaging by electron microscopy.
- The specimen preparation involves the following steps: - a) Preservation. b) Staining. c) Dehydration. d) Replication and/or microtomy.
- The two main methods of preserving the sample are by freezing and by chemical fixation.
- In order to increase the contrast between the sample and the background, the specimen has to be stained with a heavy, electron-rich metal.
- A common method of dehydrating wet samples is to immerse it in alcohol solutions, slowly increasing the concentration of alcohol, such that the water is ultimately replaced completely by alcohol.
- Replication is a method used to prepare samples when only the surface features are of interest.
- A microtome is a device that uses glass or diamond knives to produce very thin sections of the embedded samples.

- Image reconstruction uses many indistinct images of the sample to produce a single clear image.
- The electron microscope can be also be used with a few modifications, to obtain electron diffraction patterns from two-dimensional crystals.
- Electron diffraction is a popular technique to obtain low-resolution images of large particles that do not easily form crystals suitable for X-ray diffraction.
- Atomic force microscopes or AFMs are routinely used tools in the fast-developing field of nano-biotechnology.

Chapter 7

X-ray Crystallography

7.1: Introduction

In June 1913 Laurence Bragg determined the structure of NaCl using X-ray analysis. Since then, in the intervening 95 years, continuous development of theory, experimental practice and technology has made possible the determination of the three dimensional structures of about 350,000 organic and biological molecules. This chapter discusses the powerful technique of X-ray crystallography in detail. Crystallography is the study of crystals. Since biological materials are only rarely crystalline in the natural state, one may wonder what biophysics has to do with crystallography. The answer lies in the fact that, except for NMR, X-ray crystallography is the only available technique of obtaining a detailed, accurate and unbiased picture of a molecule.

The method of X-ray crystallography (Figure 7.1) can be described very briefly as follows: Crystallise the molecule, subject it to irradiation by a beam of X-rays and make a record of the 3-D diffraction pattern. Analysis of this data through computer calculations results in a model of the molecule, which can then be refined against the observed X-ray data. At the end of the procedure one obtains a set of three-dimensional coordinates, which describe the position in space of each atom in the molecule. Such a refined model can be displayed and manipulated either as plastic or wooden models, or on the computer graphics terminal. In the following portions of this chapter we will look into each of these aspects in detail.

7.2: Crystals and symmetries

A crystal is a regular, repeating three-dimensional array of atoms or

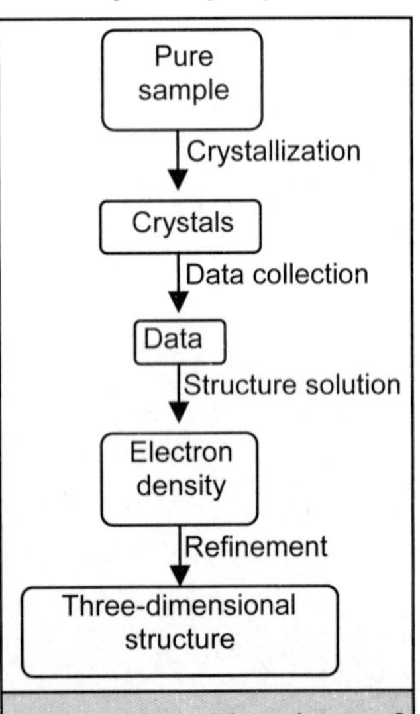

Figure 7.1: Flowchart of the X-ray crystallographic method.

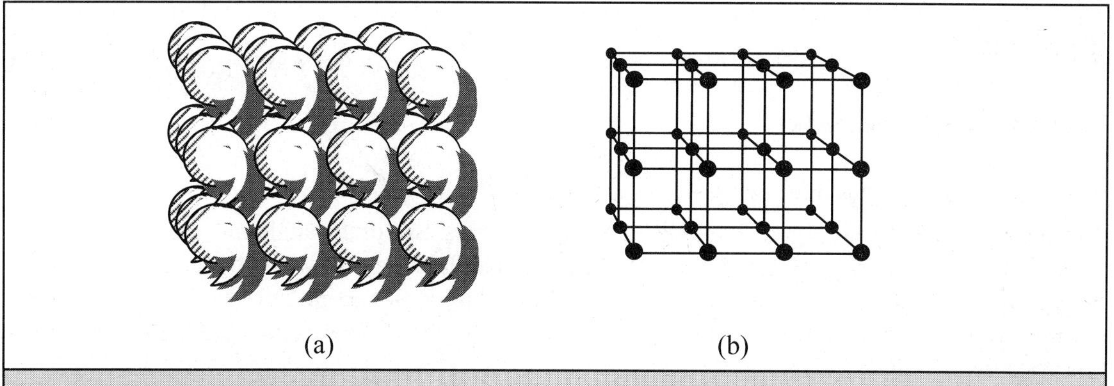

Figure 7.2:   (a) A regular periodic arrangement of three-dimensional objects. (b) The underlying lattice.

molecules or groups of molecules. If one imagines a point as representing each repeating unit, the array of points that represents the crystal is called a lattice (Figure 7.2). There are 14 different symmetrical patterns in which the points can be arranged in space and, in addition, be consistent with the requirement that the patterns be repeated in all three dimensions regularly to form the crystal. In other words, whatever other symmetry the lattice may possess it has to be consistent with the translational symmetry that defines a crystal. These 14 patterns are called Bravais lattices.

The Bravais lattices exhibit mirror symmetry and rotational symmetries. A lattice possesses mirror symmetry, if there exists in the crystal, a plane across which the lattice can be reflected without in any way changing the pattern of arrangement of the lattice points. Similarly, an n-fold axis of rotational symmetry exists if the lattice can be rotated about the axis by $360/n^\circ$ without any change in the pattern. Consistent with translational symmetry, only one-fold (360° rotation), two-fold (180°), three-fold (120°), four-fold (90°) and six-fold (60°) axes of rotational symmetry are possible.

Each Bravais lattice may be characterised by three axes, which are defined by joining a point to three neighbouring points that are not in the same plane. A unit cell is formed by completing the box

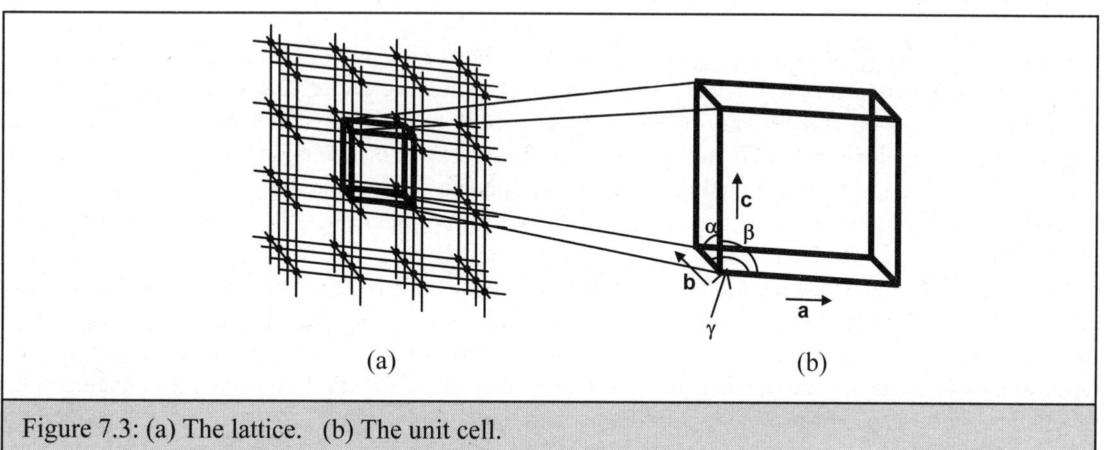

Figure 7.3: (a) The lattice. (b) The unit cell.

and is described by the length of the three sides a, b and c and the angles between them, α, β and γ (Figure 7.3).

7.3: Crystal systems

A triclinic unit cell has three unequal axes and angles (Figure 7.4a).

A monoclinic unit cell (Figure 7.4b) has three unequal axes, two of the angles as right angles, and no limitation on the third one. This condition can be obeyed both in the case where the lattice points are at the corners of the unit cell, as well as when there is an extra lattice point on one of the faces. This is conventionally taken to be the ab plane or the 'C' face. There are thus two possible Bravais lattices in this system.

An orthorhombic unit cell (Figure 7.4c) has three unequal axes, and all three angles are right angles. There are four possible Bravais lattices; with the lattice points at the corners; with an extra lattice points on the ab plane; with an extra lattice point at the body centre of the unit cell; and finally with extra lattice points on all six faces.

A rhombohedral system (Figure 7.4d) has three equal sides and three equal angles.

A tetragonal system (Figure 7.4e) has two equal sides and all angles as right angles. There are two possible Bravais lattices.

A hexagonal unit cell (Figure 7.4f) has two equal axes, and one angle as 120°, the other two as right angles.

A cubic system (Figure 7.4g) has equal sides and right angles. There are three Bravais lattices.

The Bravais lattices represent a complete set. No others are possible in a crystal. For example, a tetragonal lattice could have points at the centres of all the faces, like the face-centred cubic lattice in Figure 7.4g. However, this is not another tetragonal lattice, for a rotation of 45° around the c axis will give back the body-centred lattice of Figure 7.4e.

7.4: Point groups and space groups

Crystals can be classified in terms of the symmetry between their faces. In this case we consider the crystal as a whole and ignore for the present the lattice structure underlying it. Each crystal can be identified by the group of symmetry operations that pertain to it. The symmetry operations can be n-fold rotation, and mirror planes. In addition we may have centres of inversion – each point in space is matched by a similar one at an equal distance on the opposite side of the centre. Rotatory inversion is another symmetry operation that the crystal as a whole may possess – the operation consisting of rotation about an axis combined with inversion about a centre.

Each group of symmetry operations that identify a crystal is called the symmetry point group of that crystal. Since the point group of the crystal is also the point group of the underlying lattice, only symmetry operations consistent with lattice translations are allowed. With this restriction, a total of 32 point groups is possible for crystals and is divided among the seven crystal systems. For example, the monoclinic system has always one two-fold axis, and therefore the point groups 2, (i.e. it has one two-fold axis alone), m (i.e. one mirror plane alone) and 2/m (i.e. a two-fold axis and a mirror plane

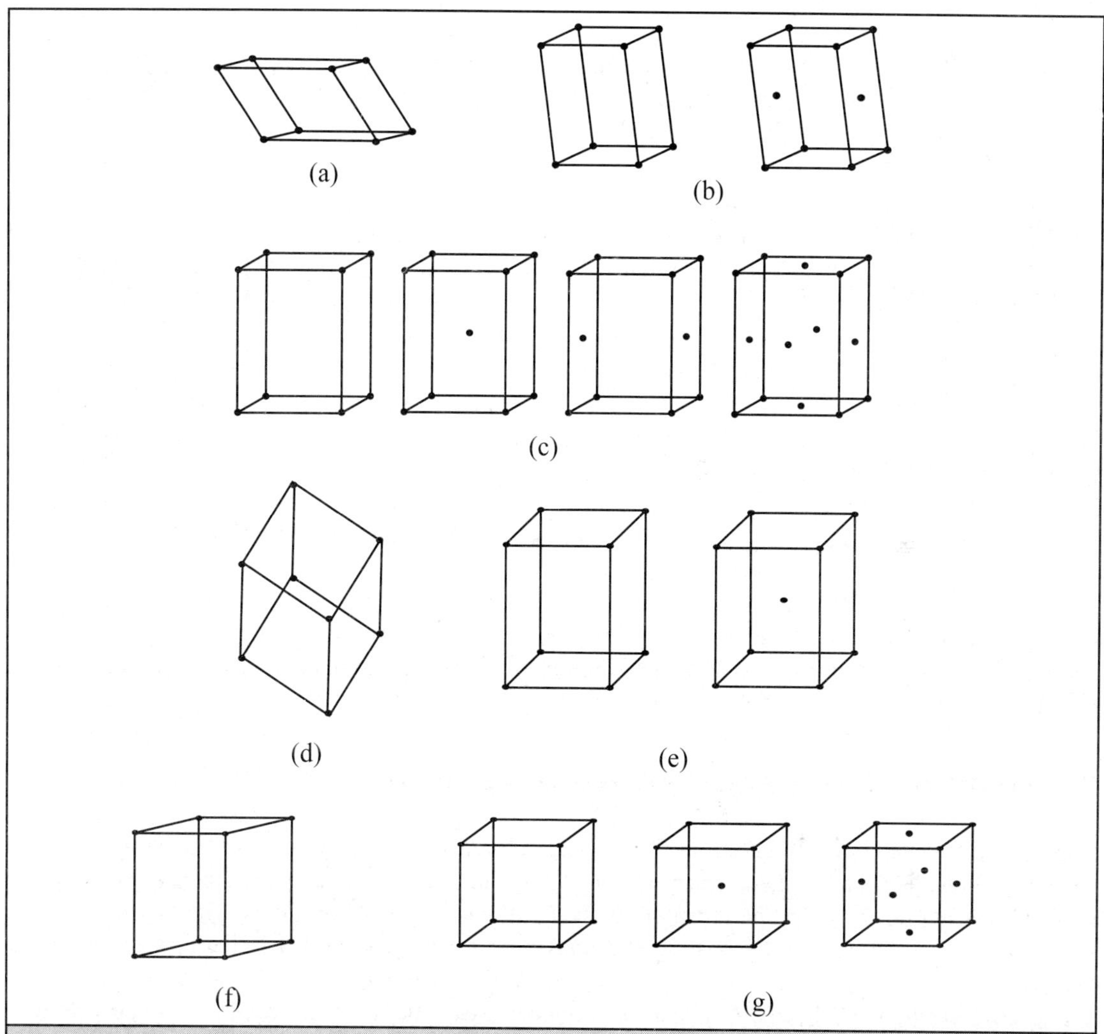

Figure 7.4: The seven crystal systems and the fourteen Bravais lattices. (a) Triclinic. (b) Monoclinic. (c) Orthorhombic. (d) Rhombohedral. (e) Tetragonal. (f) Hexagonal. (g) Cubic.

perpendicular to it) all fall in the monoclinic system. The point group 222 (three two-fold axes perpendicular to each other) requires three perpendicular axes and hence will fall in the orthorhombic system (Figure 7.5). Other point groups are similarly assigned to other systems.

In addition to the lattice, when we take into consideration the structural components of crystals, i.e. the atoms or molecules, additional symmetry elements, combining translation with rotation or reflection come into play. One of these is the screw axis, which corresponds to a rotation and a translation, and is designated n_m, which implies an n-fold rotation, followed by a translation parallel to the rotation axis, by a distance m/n of the unit cell axis. Thus, for example, a 3_1 screw about c implies a 3-fold rotation about the c axis and a translation of 1/3 times the length of c along the c axis (Figure

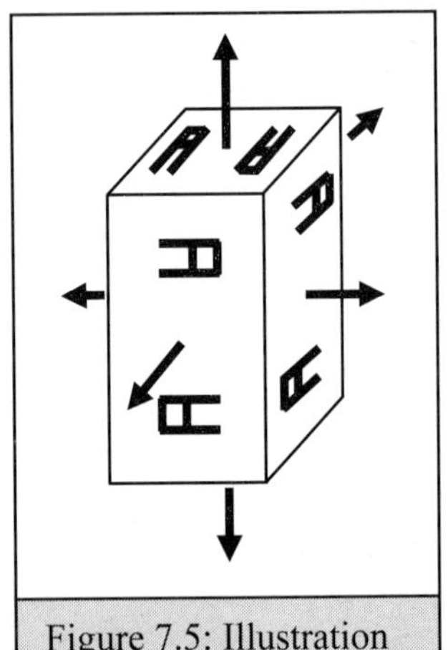

Figure 7.5: Illustration of the 222 point group symmetry.

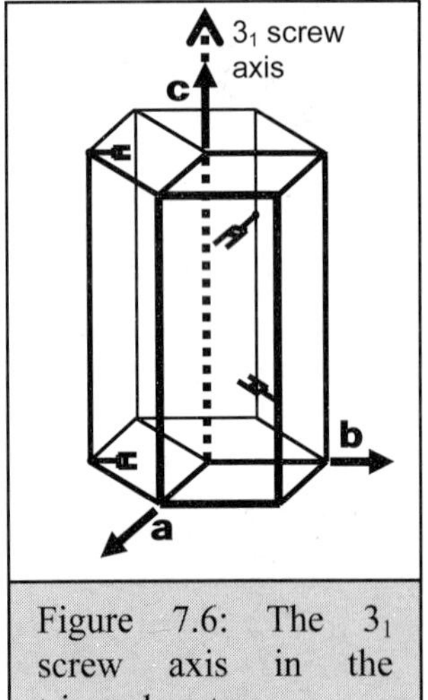

Figure 7.6: The 3_1 screw axis in the trigonal system .

7.6). Similarly a glide plane arises when a mirror reflection is followed by a translation of 1/2 along a unit cell axis (a, b, or c glides) or a unit cell diagonal (n glide).

The three dimensional assembly of planes, axes and centres of symmetry is called a space group. The symmetry operation of a space group, together with the unit cell translations, will transform a single object into a crystal. Just as there are only seven distinct crystal systems, fourteen Bravais lattices and 32 point groups, there are only 230 space groups, or 230 distinct ways in which an object, such as a molecule, can be arranged in three dimensional space to form a crystal.

A description of the 230 space groups is beyond the scope of this book. The International Tables for Crystallography[8] contain all the details. The symmetry of the crystal, i.e. its space group, can be deduced from its X-ray diffraction pattern and is an essential first step in the X-ray analysis. Before we can subject a crystal to X-ray analysis, however, we need to grow it.

7.5: Growth of crystals of biological molecules

The chief difference between crystals of biological substances and those of other important materials such as silicon is that, in the former, the repeating units are invariably molecules or groups of molecules, while in the latter the repeating units are atoms or groups of atoms. This implies that the forces of bonding between one repeating unit and another are much weaker in the former, being van der Waals forces or hydrogen bonds. In the latter, on the other hand, the crystals are built up from strong covalent or ionic bonds. Crystals of biological materials are thus usually soft and brittle as compared to inorganic substances.

Another difference is the size. Biological and organic molecules usually form very small crystals (approximately 1 mm^3) as compared to inorganic materials (about 1 cm^3). Bio-crystals are usually grown from solution and may have a large solvent content. This is especially true of crystals of macromolecules such as proteins and nucleic acids where the solvent can take up as much as 80% of the crystal. The growth of bio-crystals is a multi-parameter process. Table 7.1 gives a list of factors that could affect crystallisation.

To grow crystals of any compound from solution, the molecules have to be brought to a supersaturated state. When the solution returns to equilibrium, the substance will precipitate out,

---

[8] *The International Tables for Crystallography, Vol A,* T. Hahn, (ed.), 1987. D. Reidel Publishing Company, Dordrecht, Holland.

hopefully as crystals, and not as an amorphous precipitate. Supersaturation can be achieved by slow evaporation. Also, by varying one of the parameters listed in Table 7.1 one may change the solubility and hence the saturation level of the molecule.

In order to obtain good, defect free crystals of a size suitable for X-ray analysis, the equilibration needs to be very carefully controlled, and techniques have been developed for this. Slow evaporation (Figure 7.7) is conceptually the simplest of techniques. It is chiefly used to crystallise small organic molecules. The solution is placed in a small glass beaker and sealed with an airtight cover, except for one or two small apertures through which slow evaporation of the solvent takes place, bringing the solute to supersaturation and hence to crystallisation.

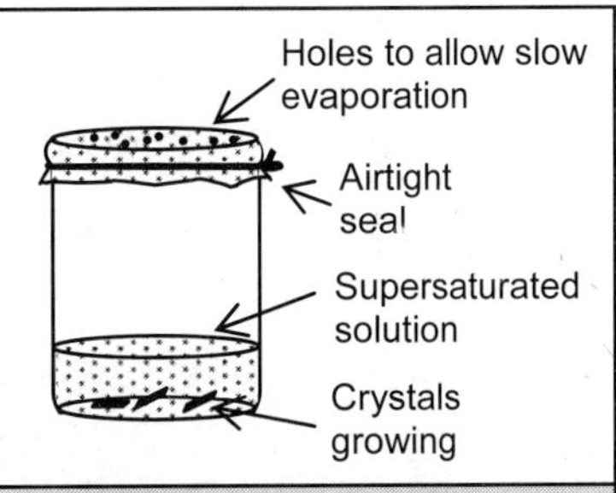

Figure 7.7: Slow evaporation technique to grow crystals from supersaturated solutions.

For biological macromolecules, especially, dialysis can be used to obtain crystals. The solution of the molecule is separated from a large volume of the solvent by a semi–permeable membrane, which gives small molecules (ions, additives, buffer, etc.) the freedom to circulate, but prevents the movement of the macromolecule. The macromolecule solution is gradually built up to supersaturation by changing the concentrations of the small molecules in the large volume of the solvent. This has the effect of gradually changing the pH or the ionic strength or any similar parameter.

Vapour diffusion is another commonly used technique (Figure 7.8). A droplet of solution containing the molecule is equilibrated against a reservoir containing the precipitant, i.e. the crystallising agent, at a higher concentration than in the drop.

In batch methods, the molecule to be crystallised is mixed with the crystallising agent at a high concentration so that supersaturation is immediately reached. Though this may lead to a large number of tiny crystals, it frequently gives rise to good crystals.

Interface diffusion (Figure 7.9) is used for small molecules as well as large biological macromolecules. The precipitant solution is gently layered over the solution of the molecule. Crystals usually grow at the interface.

Table 7.1. Factors affecting crystallisation of organic molecules and biological macromolecule.
-- purity of the sample.
-- concentration of the sample
-- temperature, pH of the solution
-- time, rate of equilibration
-- ionic strength
-- convection currents
-- volume of the crystallisation solution
-- pressure, vibrations
-- additives, their character and concentration
-- solvents (especially for small organic molecules)

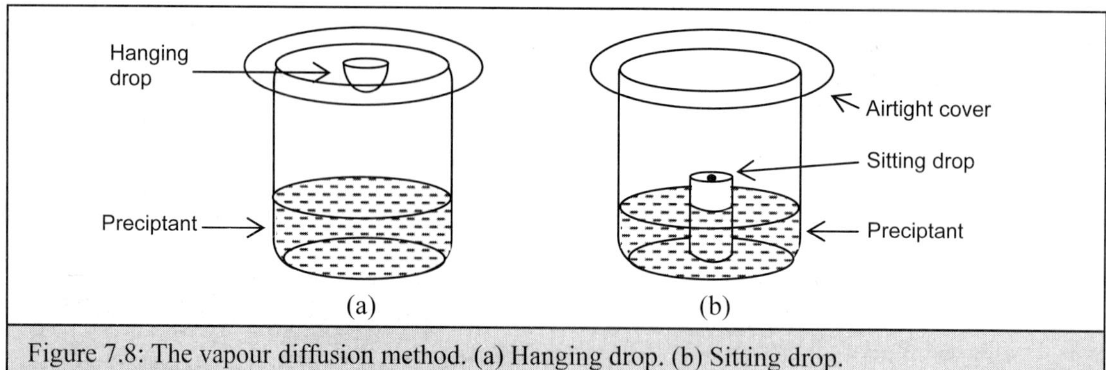

Figure 7.8: The vapour diffusion method. (a) Hanging drop. (b) Sitting drop.

Sometimes it is necessary to introduce seeds into the solution to induce the crystals to grow. These seeds are usually obtained from previous less successful crystallisation attempts, and may be extremely tiny crystallites. Table 7.2 gives some of the solvents, additives and precipitants used in growing crystals of organic and biological molecules.

7.6: X–ray diffraction

Diffraction is a well-known phenomenon that occurs when two or more waves are combined. If the waves combine in such a way that the peaks of one wave fall on the peaks of the other, the total effect is to enhance the intensity. On the other hand if the combination is such that the peak of one wave falls on the trough of the other, the effect is one of cancelling each other out and the intensity of the combination is zero. If an incoming wave is split into two by a pair of slits, and the two parts of

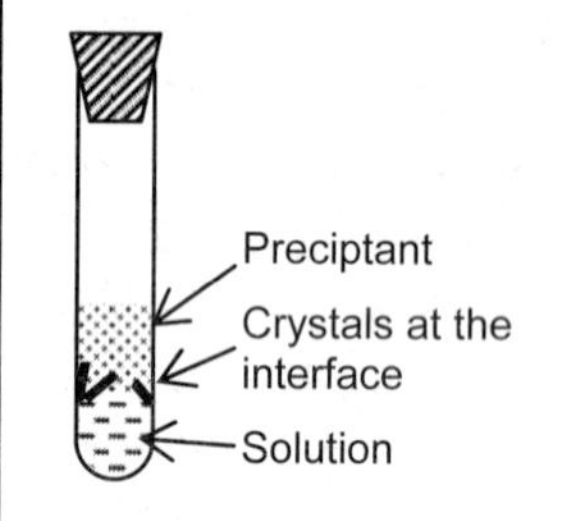

Figure 7.9: The interface diffusion method.

Table 7.2: Compounds required to crystallise biomolecules.

Molecule	Solvents	Additives	Precipitants
Small organic Molecules	Organic solvents (e.g. CCl_4, ethyl acetate, etc.), alcohols, water	Metal ions, salts	Alcohols, organic Solvents
Proteins	Aqueous solutions	Co-factors, sugars, metal ions, ion complexes, polyamines, salts	Ammonium sulphate, polyethylene glycol (PEG), Methylpentanediol (MPD)
Nucleic acids	Aqueous solutions	Polyamines, metal ions, salts	MPD, PEG, Isopropanol, other Alcohols

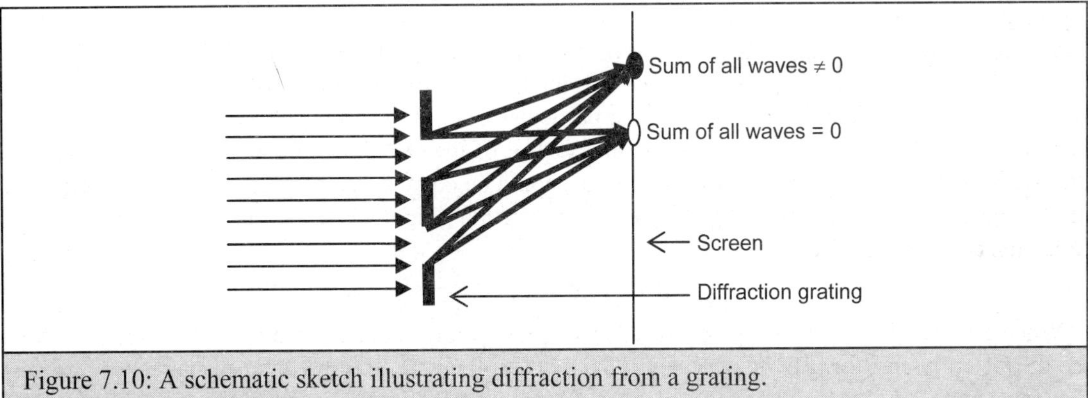

Figure 7.10: A schematic sketch illustrating diffraction from a grating.

the wave are allowed to interfere with each other, then a pattern is produced on a screen placed on the other side of the slits. Whether the two beams interfere constructively or destructively at a particular point on the screen, is decided by the difference in the length of the paths that each of the waves take from the slit to the point on the screen. If the difference corresponds to a whole number of wavelengths, the combination of the two waves is perfectly constructive and the intensity is greatly enhanced. A regularly spaced series of slits has the same type of effect and is called a diffraction grating (Figure 7.10). It is possible to have two and three-dimensional diffraction gratings. A crystal lattice is nothing but a three-dimensional grating (Figure 7.2(b)). However, it cannot diffract visible light. This is because for the diffraction effect to be noticeable, the spacing between the slits in the grating should be of the same order of magnitude as the wavelength of the incident radiation. In a crystal the distance between the lattice points corresponds to the dimensions of atoms and molecules i.e., a few Angstroms. X-rays are radiation with wavelengths of this size, and hence diffraction from crystals is observed with X-rays. The beginning of a theory of X-ray diffraction from crystals was developed by Max von Laue. A few years later, W.L. Bragg came up with the idea of treating diffraction from a crystal as reflection from the lattice planes and put the theory on a firm footing. A

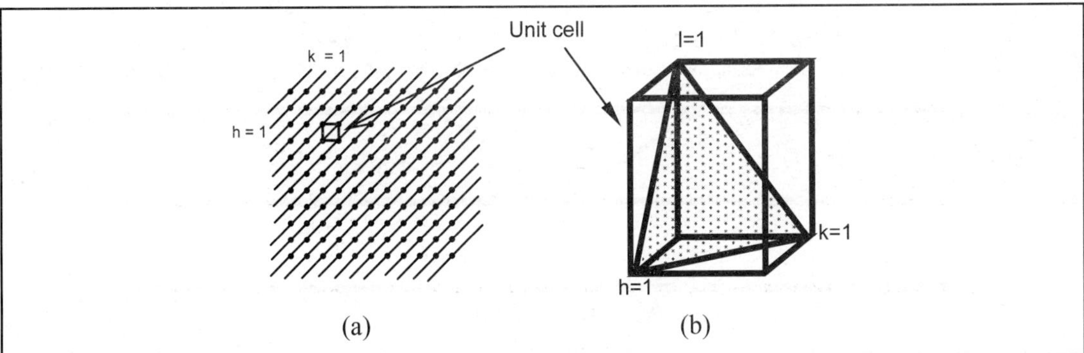

Figure 7.11: Lattice planes and Miller indices. (a) A two-dimensional lattice showing the set of 'planes' (lines) with the Miller indices h=1,k=1 or (1,1). (b) A three-dimensional unit cell showing the plane (1,1,1).

crystal can be characterised as having lattice planes, each of which is identified by a set of integers h, k and l (Figure 7.11). These numbers are known as Miller indices. Miller indices help in identifying the plane from which reflection takes place. The interference of the beams of X-rays reflected from successive parallel planes corresponding to a particular triad of hkl values, leads to diffraction effects. For the interference to be constructive, as in the case of the one-dimensional diffraction grating, the difference in path lengths of the reflected beams should be equal to an integral number of wavelengths of the incident beam. A simple geometrical construction (Figure 7.12) leads to the famous Bragg reflection condition or Bragg's law

$$2d\sin\theta = n\lambda.$$

When a crystal is irradiated by an X-ray beam, Bragg's law is used to predict the angles θ at which the diffracted intensities may be expected. This requires knowledge of the inter-planar spacing 'd' and the wavelength λ of the X-rays. Conversely, if the angles of occurrence (and relative intensities) of the diffracted spots are measured in an experiment, the corresponding inter-planar spacing in the crystal may be calculated and the lattice structure can be determined. Therefore, the next step, after growing the crystal is to place it in a beam of X-rays and measure the angles and intensities of the diffracted spots.

7.7: X-ray data collection

The process of collecting a complete set of X-ray diffraction data from a crystal consists of measuring the intensity of the diffracted beam from each set of planes. The experimental set up, which is shown in Figure 7.13, consists of an X-ray source, a goniometer on which to mount the crystal, and a detector to measure the intensity of the reflected X-ray beam.

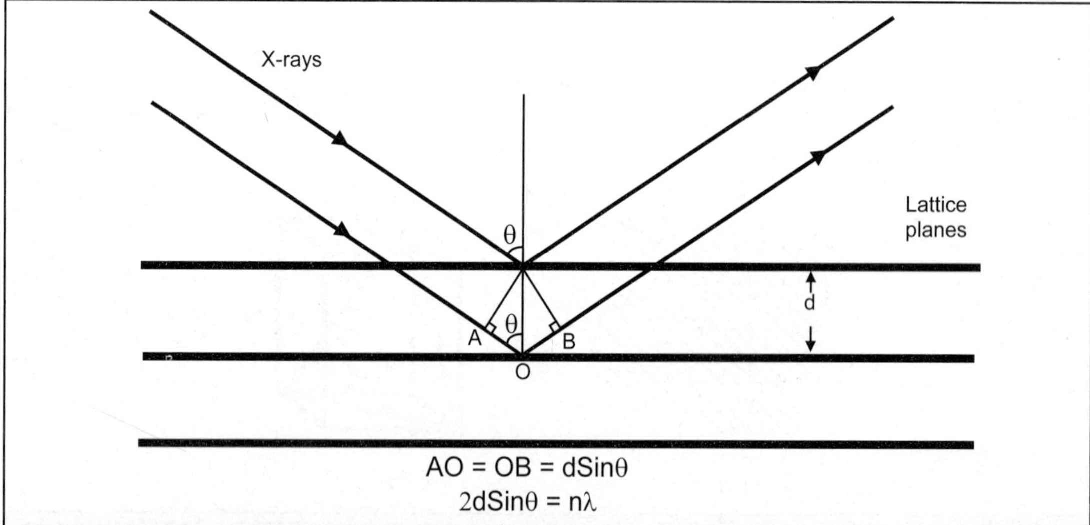

Figure 7.12: Bragg's Law. For constructive interference to occur, the difference in path lengths, given by AO+OB should be equal to an integral number of wavelengths or $n\lambda$.

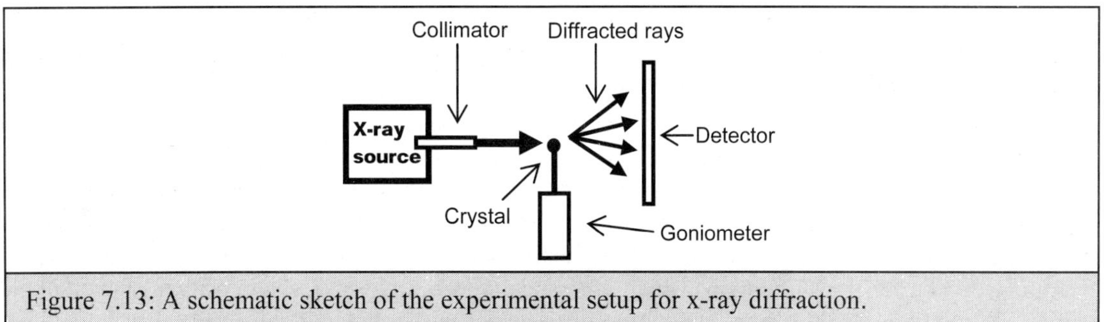

Figure 7.13: A schematic sketch of the experimental setup for x-ray diffraction.

The X-ray source is usually a sealed tube in which electrons are accelerated from one end and allowed to impinge at the other end on a metal target, usually copper or molybdenum for biologically relevant samples. This produces X-rays of wavelength 1.5418 Å (for Cu Kα) and 0.7107 Å (for Mo Kα). This process also produces a tremendous amount of heat and the tube has to be kept cool by circulating chilled water or by other means, such as rapidly rotating the metal target.

The collimated beam of X-rays then fall on the crystal mounted on a goniometer. This is a device that allows one to position the crystal in different orientations in order to bring different sets of planes into the reflecting position.

The detector of the diffracted X-rays may be a film, or a radiation counter. The first measurements of X-ray intensities used in the solution of the structure of NaCl, were made with an ionization chamber. This however allows the measurement of only one reflection at a time, since the crystal and the detector have to be set at different angles for each reflection. Films allow the recording of several reflections at a time, but suffer from the requirement of long exposure times, as well as imprecision in the conversion of the blackening of the film into a numerical value of intensity. Modern devices such as imaging plates or charge-coupled devices (CCDs) offer the advantages of both films and of counters and are now almost exclusively used in the collection of data from crystals of biological macromolecules.

The outcome of the data collection process is a list of angles, indicating the orientation of the crystal and the detector, and the intensity of the reflection from the oriented lattice plane at this position. Instead of the angles, it is usual to identify the set of planes that give rise to each diffracted spot by the Miller indices hkl and associate with each value of hkl the appropriate value of the intensity. From the list of intensities, the amplitude of each reflected wave is extracted by the simple process of taking the square root, since the measured intensity of a wave is the square of its amplitude. These amplitudes, represented as F_{hkl}, are now used in the analysis of the structure of the molecule that makes up the crystal.

7.8: Structure solution

The structure of the molecule is obtained by a Fourier transform of the observed amplitudes F_{hkl}. This statement, which forms the core of the X-ray crystallographic method, is amplified and explained below. In order to first understand the mathematical process of Fourier transformation, consider the periodic function shown in Figure 7.14. Fourier theory shows that such a function can be represented as a sum of a series of simple waves. Note however that it is not only necessary to know the amplitude (and wavelength) of the simple waves but also to know the relative positions, or phases of the waves in order to reconstruct the original periodic function.

A crystal which has the unit cell, with its complex contents, repeated periodically, can be considered to be made up of simple planes of electron density arranged in three dimensions. This periodic function can be represented as series of simple waves. The amplitudes of these simple waves are measured, i.e. F_{hkl}. In order to reconstruct the crystal as a three-dimensional superposition of these simple waves of electron density, it is necessary to know the phases. Once the phases are known the mathematical operation of Fourier transformation, which corresponds to the physical superposition of the simple waves, will yield a picture of the distribution of electrons within the unit cell of the crystal. In other words, we obtain a picture of the structure of the molecules in the unit cell. The problem of X-ray crystallography now translates into one of obtaining the phase information. This is called the phase problem in crystallography since, in the X-ray diffraction experiments, the phases cannot be measured directly, and the phases have to be determined by other indirect means. Some of these methods are described below.

7.8.1: The heavy atom or multiple isomorphous replacement (MIR) method:

This method requires that the crystal of the sample be grown in at least two ways, once in the presence of a 'heavy', usually metal, atom and once in its absence. An additional, rather stringent, requirement is that in both cases the crystals are 'isomorphous', i.e. they have the same unit cell dimensions and crystal symmetry. In such a situation the only difference in the arrangement of the atoms in the crystal would be the presence or absence of the heavy atom. Whether a particular atomic species is called 'heavy' or not would depend on the contribution it makes to the total electron density and hence to the X-ray diffraction intensities. Thus for small organic compounds of molecular weight about 1000, cobalt, copper, iron and bromine would qualify as heavy atoms. For proteins, crystals may have to be grown in the presence and absence of platinum, samarium, mercury, etc. The heavy atom method

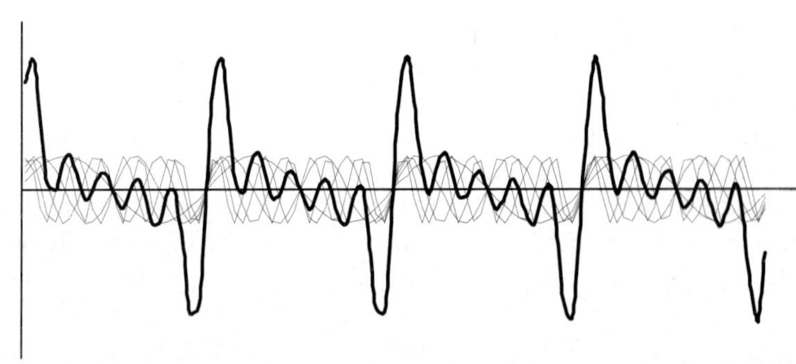

Figure 7.14: The Fourier decomposition of a complex periodic function (thick line) into its sine and cosine components (thin lines). As more and more component waves at different frequencies (wavelengths) are added, the resulting wave approaches a square wave more closely. The Fourier decomposition of a square wave itself would result in components of all possible frequencies.

depends on the assumption that, since a large portion of the scattering matter in the unit cell is concentrated into a single atom, a knowledge of the position of this atom and hence its contribution to the phases, would lead to a knowledge of the phases due to all the atoms in the unit cell, at least to a very good approximation. The position of the heavy atom itself is usually arrived at by a special type of Fourier analysis called the 'Patterson synthesis'. Once the position of the heavy atom is known, the phases can be calculated. On the basis of the above stated assumption, they are the phases due to all the atoms and a Fourier transform will give the approximate positions of the atoms. Usually, especially for macromolecules like proteins, there are ambiguities in assigning the phases, which can be overcome only by growing many crystals by substituting different heavy atoms and collecting and analysing the X-ray data for each crystal. This method is therefore usually called the 'Multiple Isomorphous Replacement' method.

7.8.2: Direct methods:

H. Hauptman and J. Karle pioneered a statistical analysis of the X-ray data to obtain the phases. It was realised that the values of the phases have to be such that the electron density obtained has to correspond to physical reality, i.e. there can be no 'negative' electron density in the crystal (the presence of an electron is considered as 'positive' electron density) and the density should correspond to separate or bonded atoms. In addition the phases for some of the reflections called semi-invariants can be restricted to a few values depending on the space group. These facts can be used iteratively in a computer program to generate the phases for all the reflections. A Fourier transform then gives the structure. The theory is very well developed for crystals with less than about 50 atoms in the unit cell and the technique is routinely used to determine small molecule structures. For larger molecules such as proteins, the method is still not fully successful.

7.8.3: Molecular replacement:

The molecular replacement technique is used when the unknown structure can be assumed to be similar to an already known structure. The problem then becomes one of first finding the orientation and position of the model molecule in the unknown unit cell, and next of finding the differences between the model and the actual structure. The first part of the problem is usually solved by means of a search procedure. The computer, in effect, generates the F_{hkl} for each position and orientation of the model in the unit cell and compares these calculated reflections with those actually observed in the experiment. The position and orientation corresponding to the best match is then taken up for the second part of the structure solution procedure, i.e. the refinement.

7.8.4: The anomalous scattering technique:

In this method the wavelength of the incident X-ray beam is adjusted so that the scattered intensities would be different when measured from opposite sides of the crystals. Normally the two sets of intensities would be the same, but in particular instances when the wavelength is close to the so-called absorption edge of one of the atomic species in the crystal, the intensities are different. This difference is then exploited to obtain the phases. Since the wavelength has to be tuned to match the element, this method became widely used after the advent of powerful, tunable (and very expensive!) sources of X-rays called synchrotron radiation sources.

7.9: Refinement of the structure

Once the approximate values of the phases are determined, the next step is to refine them. This usually more conveniently done by performing the Fourier transform and refining the approximate positions of the atoms obtained by including other known data such as stereochemistry.

7.9.1: Least squares methods:

This is a well-known mathematical technique used to fit a model to the observed data. Small changes are made in the model, and with each set of changes, the expected data are calculated. The differences between the observed and calculated data are once again used to determine the changes in model. This is an iterative process, and it terminates when further changes in the model do not lead to any further improvement in the calculated data. Since the differences can be either positive or negative, i.e. the calculated data points can either be more or less than the observed data points, the sum of the squares of the differences, which is always a positive quantity, is the term that is actually minimised in the least squares procedure. The whole procedure can be described in terms of a mathematical operation, with the changes to the model in each cycle being calculated automatically. The mathematical equations lend themselves to easy computerisation and the least squares method is widely used in many fields of science and engineering. As applied to crystallography, the function to be minimised is called the Residual or the R factor and is usually defined as

$$R = (\Sigma ||F_o|-|F_c||)/ (\Sigma |F_o|)$$

where F_c is the calculated structure amplitude for a particular reflection hkl, and F_o is the observed structure amplitude for the same reflection. The two summations in the equation are over all the reflections that have been measured. The changes are made to the coordinates to make the observed and calculated structure amplitudes equal each other as closely as possible. Note that only absolute differences between the calculated and observed data are taken into consideration. This is in keeping with the idea that the actual sign of the displacement is not relevant to the least squares technique. The calculation of the F_c involves a Fourier transform of the model and it is performed in each cycle after shifting the atoms by the (usually small) amounts indicated in the previous cycle. Though the amount of computation required is quite large, modern technology allows the complete refinement of structure of molecular weight of about 1000 in less than one day. In order to work, the least squares technique requires that the initial model be close (< 0.5 Å) to the actual final model. Furthermore a very large number of observations are required for reliable results. As a rule of thumb, between seven and ten observed data points are required for every parameter refined. For example, a molecule containing 20 atoms will have three positional parameters (x,y,z) and six parameters corresponding to the thermal vibration, a total of nine parameters for each atom. To ensure accurate refinement therefore, more than 1500 reflections will have to be measured. In the case of biological macromolecules, this condition is seldom satisfied. Owing to various factors, including high solvent content of the crystals, the molecules are not very well ordered and reflection planes with spacing of less than about 2.0 Å are usually not sufficiently regularly spaced to give measurable diffracted intensities. In other words the limit of resolution of the data in such cases is usually not beyond 2.0 Å. This means that the observation-to-parameter ratio is not good enough to allow refinement using the standard least squares techniques.

7.9.2: Constrained-restrained refinement:

There are two ways in which the observation-to-parameter ratio can be improved without having to collect additional X-ray data. The first is to increase the number of observations by including known chemical information. For example, in the extreme case, the whole molecule can be treated as a single rigid unit. The only variables are then the orientation and the position of the molecule, a total of six parameters. This is called constrained refinement. In the intermediate cases various parts of the molecule can be considered rigid units, which are then linked to each other by flexible bonds. Another method of improving the ratio based on available stereochemical information is called the restrained refinement technique, where the distances between the various atoms in the molecule are known. For example, the distance between two carbon atoms joined by a single bond cannot be very different from 1.54 Å. Similarly the angle at a sp^3 hybridised carbon atom has to be close to 109.5°. This chemical information is treated as additional observation points and for the same number of parameters, the ratio improves. The bond length and angle information, together with other stereochemical and physicochemical information such as Van der Waals contacts, hydrogen bonds, steric hindrances and intra-molecular electrostatic forces can also be programmed as an energy function which is to be minimised. This is then added to the residual function mentioned earlier and the whole is minimised by least squares or some other optimisation technique.

7.9.3: Maximum likelihood refinement:

Statistical theory showed that least squares refinement was based on the assumption that errors in the atomic coordinates were distributed normally about their mean values, and the refinement procedure facilitated in bringing them to their expected values. However, deeper analyses indicated that in fact the distribution of errors about the expected values was not the normal one, and in addition, each set of reflections or coordinates had a different distribution of the errors. These distributions could be estimated, and incorporated into the refinement in a procedure that utilised Bayesian statistics. It was possible therefore to calculate a precisely defined function called 'the likelihood' of a given set of coordinates, given a particular set of reflections, and the errors associated with them. The refinement procedures then consist of maximizing the likelihood by changing the coordinates in an iterative optimisation technique. One could also incorporate into these expressions constraints and restraints as described in the previous paragraph, such as optimal bond lengths, angles and torsion angles. These modifications and enhancements to the refinement techniques have proven to be very successful and are now incorporated in computer programs such as REFMAC, which are widely used to refine protein structures.

7.9.4: Computer graphics in refinement:

The advent of high-resolution computer graphics has made possible the development of several new techniques of refinement, and of utilising known chemical and biological information in the refinement process. Today much of the crystallographer's creativity is expressed at the computer graphics terminal. As mentioned earlier, once the approximate phases have been obtained, a Fourier transform using the amplitudes F_{hkl} as the Fourier coefficients produces a map of the electron density in the unit cell (Figure 7.15). The model is superposed on this map, and wherever the fit of the model to the electron density is not good enough, adjustments are made to improve the fit. After a cycle of fitting, or changing various parts of the model, the map can be recalculated using the changed positions of the atoms in the model to obtain the new set of phases. This process iteratively brings the model closer and closer into a perfect fit with electron density. Simultaneously a least squares or maximum likelihood refinement procedure is carried out and the conventional Residual or R factor is also used as a check on the correctness of the structure.

At the end of the refinement process we obtain a set of coordinates for each of the atoms of the molecule and another number (or set of numbers) representing the thermal vibrations of the atom. These can be subjected to further geometrical analyses to understand the chemistry or biology of the molecule.

7.10: Note on the resolution of an x-ray structure

The number of X-ray reflections that may be obtained from a crystal depends on the degree of order present in it. A greater degree of order implies that lattice planes with smaller inter-planar spacing are regularly arranged in the crystal and that X-ray reflections from these planes can also be measured. This implies that in the final structure, features at smaller distances can be resolved (i.e. seen as separate entities). The terminology of the resolution can be somewhat confusing. Since a higher resolution means smaller separation of the observable features, a resolution of 6 Å, is lower than a resolution of, say, 2 Å. When this number is larger, the resolution is lower, and vice versa. The resolution can also be expressed as the angular limit at which the intensity data can be observed. By Bragg's law

$$\sin\theta = n\lambda/2d$$

In other words, the lower the inter-planar spacing 'd' (i.e. the greater the resolution), the higher the

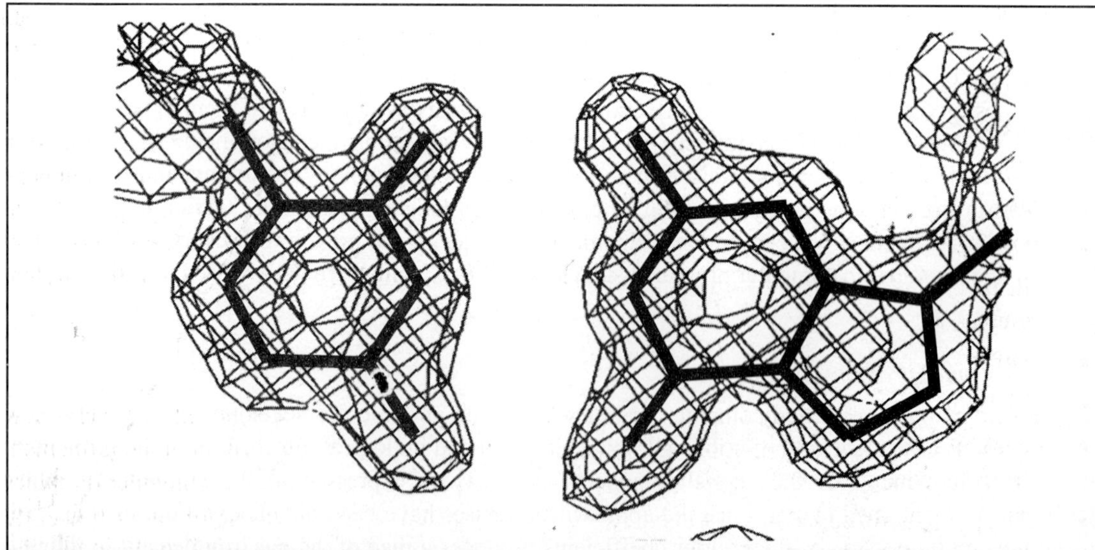

Figure 7.15: The electron density net corresponding to a G.C base pair in the crystal structure of a DNA fragment.

angle at which the diffracted beam corresponding to that set of planes is observed. The terminology in this case is more straightforward since $\sin\theta$ varies directly as the resolution. If we normalize with respect to the wavelength of the incident radiation, the resolution may be expressed as $\sin\theta/\lambda$ (or $1/2d$). Thus, the highest resolution achievable using radiation of a particular wavelength λ is $d = \lambda/2$ Å.

Summary

- Except for NMR, X-ray crystallography is the only available technique of obtaining a detailed, accurate and unbiased picture of a molecule.

- A crystal is a regular, repeating three-dimensional array of atoms or molecules or groups of molecules.

- The 14 different symmetrical patterns in which the points can be arranged in space in a crystal are called Bravais lattices. These are distributed among seven crystal systems, each defined by a particular set of symmetries and corresponding to a particular set of unit cell parameters.

- Each crystal can be identified by the group of symmetry operations that pertain to it, called the symmetry point group of that crystal.

- The three dimensional assembly of planes, axes and centres of symmetry is called a space group.

- There are only 230 space groups, or 230 distinct ways in which an object, such as a molecule, can be arranged in three-dimensional space to form a crystal.

- Crystals of biological materials are usually soft and brittle as compared to inorganic substances.

- Crystals of macromolecules such as proteins and nucleic acids are usually grown from solution and have a large solvent content.

- The growth of bio-crystals is a multi-parameter process. Some of the parameters are purity of the sample, temperature, pH of the solution, additives, etc.

- Vapour diffusion is commonly used technique to grow crystals of macromolecules.

- Sometimes it is necessary to introduce 'seed crystals' into the solution to induce the crystals to grow.

- Diffraction is a well-known phenomenon that occurs when two or more waves are combined.

- It is possible to have one-dimensional, two-dimensional and three-dimensional diffraction gratings. A crystal lattice is a three-dimensional grating.

- Diffraction from crystals is observed only with X-rays.

- The Bragg reflection condition or Bragg's law states $2d\sin\theta = n\lambda$.

- Bragg's law is used to predict the angles θ at which the diffracted intensities may be expected from knowledge of the inter-planar spacing 'd' and the wavelength λ of the X-rays.

- The process of collecting a complete set of X-ray diffraction data from a crystal consists of measuring the intensity of the diffracted beam from each of the potentially thousands of sets of Bragg planes.

- It is usual to identify the set of planes that give rise to each diffracted spot by the Miller indices hkl.

- The outcome of the data collection process is a list of Miller indices and the intensities of the reflections from the corresponding lattice planes.

- The amplitude F_{hkl} of each reflected wave is extracted by taking the square root of the measured intensity of a wave is the square of its amplitude.

- The structure of the molecule is obtained by a Fourier transform of the observed amplitudes F_{hkl}.

- However it is necessary to know the phases of the waves in order to perform this operation.

- The lack of such information experimentally is known as the 'phase problem' in crystallography

- Some of the methods of obtaining the phases are the heavy atom or multiple isomorphous replacement (MIR) method, Direct Methods, the molecular replacement technique and the anomalous scattering method.

- Once the approximate values of the phases are determined, the next step is to refine them.

- Some commonly used refinement techniques include the least squares methods, maximum likelihood refinement, and manual refinement using computer graphics.

- These techniques minimise the Residual or the R factor, usually defined as $R = (\Sigma||F_o|-|F_c||)/(\Sigma|F_o|)$, where F_c is the calculated structure amplitude for a particular reflection hkl, F_o is the observed structure amplitude for the same reflection, and the summation is over all the reflections.

- At the end of the refinement process we obtain a set of coordinates for each of the atoms of the molecule, which may be subjected to further geometrical analyses to understand the chemistry or biology of the molecule.

Chapter 8

NMR Spectroscopy

8.1: Introduction

Nuclear Magnetic Resonance spectroscopy is one of the most important experimental techniques currently in use in biophysics. Until the early 1970's NMR was of interest mainly to physicists and chemists. Physicists use NMR spectroscopy to study the nature and interactions of the atomic nucleus. Chemists use NMR spectroscopy to determine chemical structure of small molecules. With the development of techniques such as Fourier transform NMR, and two-dimensional and three-dimensional NMR, biochemists and biophysicists use the method to determine the structures of large molecules such as proteins and also to study the dynamic behaviour of such molecules. Magnetic Resonance Imaging is becoming increasingly popular in medicine as a means to obtain accurate three-dimensional images of the interior organs of the human body in an essentially safe and non-invasive manner.

NMR is a rapidly growing field and already it is so vast that only a few of the specialised techniques can be explained in the present book. In this chapter we first explain the basic principles of the NMR technique. NMR theory is then dealt with in somewhat greater detail including a description of the parameters measured in the different types of experiments. Finally we discuss the applications of NMR in chemistry, biology and medicine.

8.2: Basic principles of NMR

NMR is a spectroscopic technique. In common with other spectroscopic techniques, the NMR experiment consists of the measurement of the intensity of absorption (or emission) of electromagnetic radiation at each frequency over a specific range. In NMR this range is the radio-frequency range, i.e. from a few MHz to a few hundred MHz. At this frequency, the energy is absorbed or emitted when the nuclei of the atoms go from one

Figure 8.1: The spin of a nucleus.

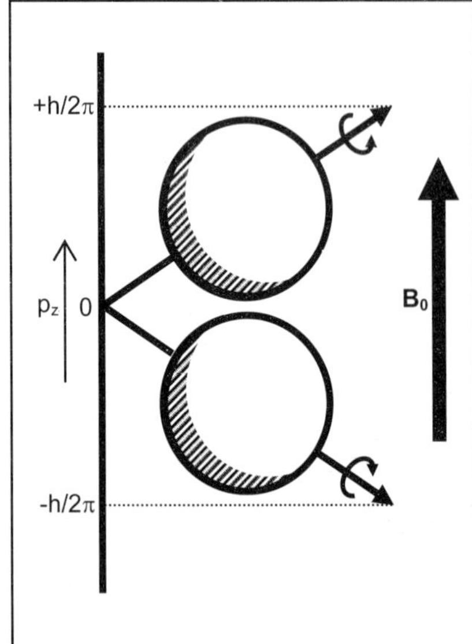

Figure 8.2: The two possible orientations of a nuclear spin in a magnetic field.

quantized spin state to another. Spin is a property possessed by only certain atomic nuclei such as the hydrogen nucleus H or the phosphorous isotope ^{31}P (Figure 8.1). A crude classical analogy to this quantum-mechanical property of spin is the rotation of a body about an axis that passes through its centre. Since the nucleus has a charge, the spinning motion will give rise to a magnetic field, and the spinning nucleus can be thought of as a small bar magnet. The axis of the bar magnet will be parallel to the axis of spin. When this nucleus is placed in a strong external magnetic field, then, by the laws of quantum mechanics and due to the interaction of the two magnetic fields, the axis of spin will orient itself in one of two possible orientations (Figure 8.2). Transitions between these two states can occur when another external oscillating magnetic field supplies sufficient energy for the spin system, i.e. the nucleus to go from the state or orientation of lower energy to the state of higher energy. The difference in energy of the two states corresponds to the energy of a radio-frequency field and hence a nuclear spin, placed in a static magnetic field, must be excited by radio-frequency electromagnetic radiation in order that the transition takes place. The energy absorbed from the radio-frequency field can be detected and this is the fundamental parameter measured in an NMR experiment. The frequency at which the transition occurs is called the resonance frequency. The resonance frequency of a particular type of nucleus, e.g. ^{1}H, will be different when it is placed in different chemical environments. For example the ^{1}H in a methylene group will have a distinctly different resonance frequency from one in a methyl group. By changing the frequency of the exciting radio-frequency field and measuring the intensity of the absorption at each frequency, it is possible to detect the nuclei and in each case identify their respective environments. It is this feature which is useful in identifying various chemical groups in molecules. If the magnetic field in which the spins are placed is very high, subtler effects become noticeable, i.e. the resolution of the spectrum increases. In this way it is possible to detect even conformational changes in the molecule.

Continuously changing the frequency of the radio field and measuring resonance is a tedious and time-consuming task. So much so, in the case of large molecules, where the changes in intensity are very small, formidably long data collection time would be required. Fourier transform NMR and multi-dimensional NMR have reduced the experimental time required immensely and have made it possible to use NMR to study large biological molecules and complexes.

Another aspect of the NMR experiment is used in medical imaging. After a nucleus makes a transition from a lower energy state to a higher energy state, it has a tendency to 'relax' back to the lower energy state by transferring the extra energy it has acquired either to another spin nearby, or to the other atoms and molecules, i.e. the 'lattice'. The time required for this relaxation will depend upon the nature of the lattice. In a living body, different tissues have different relaxation times. A measurement of the relaxation time of the spin nuclei in different parts of the body can be made. These measurements can then be displayed as a picture of the interior of the body, enabling medical practitioners to detect malfunctions or tumours etc. in the various organs.

8.3: NMR theory and experiment

Rigorous NMR theory requires knowledge of quantum mechanics, and such advanced explanations will not be attempted here. Some of the results that have been obtained will be stated without proof.

According to quantum mechanics the angular momentum of atomic nuclei or, in other words, the spin of an atomic nucleus, can only have certain discrete values characterised by the spin quantum number I. The value of I for any nucleus can either be an integer or a half-integer ($^1/_2$, $^3/_2$, $^5/_2$...), and depends on the number of protons (the atomic number) and the number of neutrons present. (The sum of the neutrons and protons is the mass number.) Table 8.1 gives the values of the spin for various possible combinations. For example the nuclei ^1H, ^{13}C, ^{31}P, and ^{19}F all have odd mass numbers and therefore a half-integral spin value, (In these particular cases the value is ½). The nuclei ^{12}C and ^{16}O have even atomic numbers (6 and 8 respectively) and even mass numbers and the spin is zero. Nuclei of deuterium, i.e. ^2H, and ^{14}N have odd atomic numbers (1 and 7 respectively) and even mass numbers. The value of the spin is 1. In chemistry and biology, the most useful nuclei are those with spin ½ and the following discussions will be restricted to such nuclei.

Table 8.1. The relation between the nucleon numbers and spin. Only nuclei with half-integral or integral spins are NMR-active.

Atomic number	Mass number	Spin quantum Number
Even or odd	Odd	Half integral
Even	Even	Zero
Odd	Even	Integral

When placed in a strong, uniform magnetic field, again according to quantum mechanics, the spin angular momentum vector can take up one of two positions (Figure 8.2) such that the component of the angular momentum along the direction of the applied magnetic field B_0 is

$$p_z = +½(h/2\pi) \text{ or } -½(h/2\pi)$$

where h is Planck's constant. The z component of the magnetic moment of such a nucleus is also quantized and is given by

$$\mu_z = \gamma p_z = +½(h/2\pi)\gamma \text{ or } -½(h/2\pi)\gamma$$

where γ is a parameter that depends on the particular nucleus and is called the magnetogyric (or gyromagnetic) ratio. The energy associated with these two positions due to the interaction of the magnetic moment with the applied magnetic field is

$$E = -\mu_z B_0 = \pm ½(h/2\pi)\gamma B_0$$

Thus the difference in energy between the two states is

$$\Delta E = \gamma(h/2\pi)B_0$$

Therefore, to induce the spin to make the transition from the lower energy state to the higher energy state, the applied oscillating magnetic field, i.e. the radio-frequency field should supply this energy. This implies that the frequency of oscillation should be ν where

$$\Delta E = h\nu_0$$

and

$$\nu_0 = \gamma B_0/2\pi$$

If we write $\omega_0 = 2\pi\nu_0$, where ω_0 is the angular frequency, then

$$\omega_0 = \gamma B_0$$

This is called the resonance condition and is the basic NMR equation.

Actual NMR experiments, of course, do not deal with single nuclei and single spins. The sample will contain a large number of nuclei, of the order of Avogadro's number ($\sim 10^{23}$). When the sample is placed in the static magnetic field, all the nuclei will not take up the lower energy state. Instead the ratio of the number of spins in the lower energy state to the number in the higher energy state is given by

$$\exp(-\Delta E/kT)$$

i.e. the Boltzmann formula, where k is the Boltzmann constant and T is the absolute temperature of the sample. Since

$$\Delta E = \gamma(h/2\pi)B_0$$

the ratio is

$$\exp(-\gamma(h/2\pi)B_0/kT).$$

This means that at room temperature and at medium magnetic fields, the energy difference between the two states is such that the population of spins in them differs by less than 1 part in ten thousand. This implies that the signal is very weak, since only a relatively few nuclei will absorb energy and make the transition. Note that if the applied static field is increased, the energy difference increases, leading to a larger population difference and hence a stronger signal.

8.4: Classical description of NMR

The NMR condition

$$\omega_0 = \gamma B_0$$

may also be derived from considerations of classical mechanics, as follows. If we consider the nucleus to be analogous to a spinning bar magnet, classical mechanics tells us that such a system placed in a magnetic field will precess about the direction of the external field with the so called Larmor frequency given by

$$\omega_0 = \gamma B_0$$

The large numbers of nuclei in a typical experimental sample will all precess about B_0 (Figure 8.3). This would mean that, while the net component of the magnetic moment in the x-y plane is zero, there would be a net magnetisation along the z-axis (presuming B_0 to be along z, as in Figure 8.3). In order to perform the NMR experiment, it becomes necessary to tilt the total magnetisation towards or onto the x-y plane as illustrated in Figure 8.4. (This is equivalent to inducing a transition of many nuclei, from the lower energy quantum state to the higher energy state). To understand this better, we may change our reference axial system from the static xyz system in Figure 8.3 to the rotating frame of

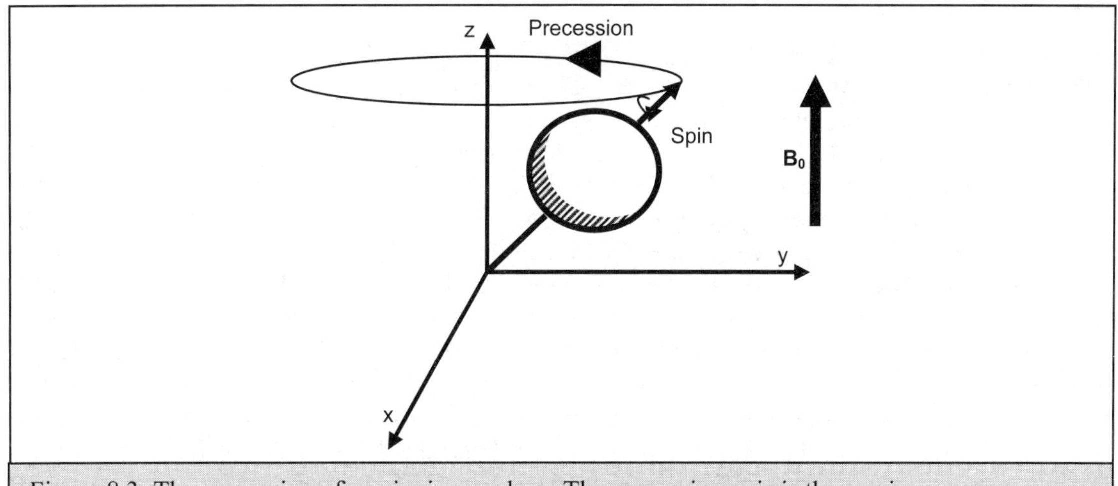

Figure 8.3: The precession of a spinning nucleus. The precession axis is the z axis.

reference given by x'y'z'. This reference frame is assumed to rotate about the z-axis with the same angular frequency as the Larmor frequency ω_0. Thus in this frame, the nuclear magnets do not precess and the apparent static magnetic field is zero. Now if a static magnetic field B_1 is applied along the x' direction in this field, then, just as the field B_0 caused a precession in the stationary reference frame, so will the field B_1 cause the nuclear magnets to precess about the x' axis, and the total bulk magnetisation will rotate about the x' axis. Of course the field B_1 is not actually a static field, but only appears so in the rotating frame. In fact B_1 is the radio-frequency field (of frequency ω_0 = Larmor frequency), which is applied in addition to B_0 in the NMR experiment. If the field B_1 is applied for a time t, then the magnetisation will rotate through an angle given by

$$\theta = \gamma B_1 t.$$

If the duration of the pulse of radio-frequency electromagnetic radiation is such that the magnetisation rotates by 90° onto the x-y plane, then such a pulse is called a 90° pulse. In the same way a 180° pulse will reverse the direction of the magnetisation. The pulse produces a net component of the magnetisation in x-y plane also. This component then produces an effect upon the detection coil in the NMR spectrometer, which is amplified and displayed as the NMR spectrum.

There are two main methods by which the NMR experiment is performed, viz. the continuous wave method and the Fourier transform method. In the continuous wave experiment the radio frequency field is continuously applied to the sample, and the magnetic field is changed continuously in strength such that all the nuclei in the sample are able to come into resonance. Alternatively the magnetic field may be kept fixed and the oscillating radio

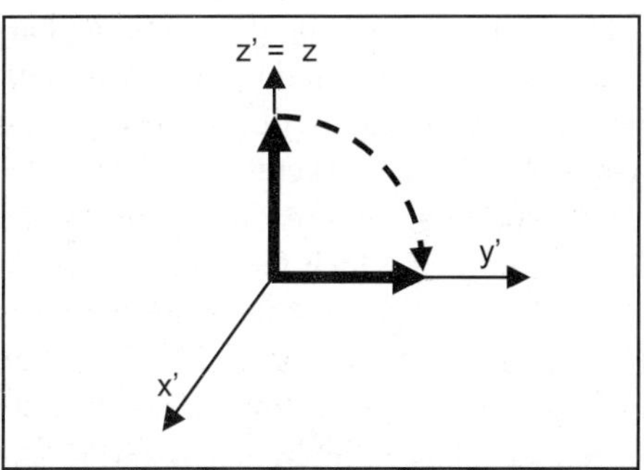

Figure 8.4: The tilting of the total magnetisation on to the x'y' plane.

frequency field may be varied. Either way, the continuous wave method, has been almost entirely replaced by the Fourier Transform NMR (FTNMR).

In FTNMR the radio-frequency stimulus is applied as a pulse, which may last typically for about 200µs. The pulse is sufficiently strong and is constituted of a sufficiently large number of frequencies to excite all the resonances in the sample. As a matter of fact what is usually applied is a square pulse. As explained in a previous chapter (Chapter 7), according to Fourier transform theory, such a square pulse is constituted of all possible frequencies. The response to such an excitation would be an NMR signal from each of the spins in the sample. The composite would be another complicated wave which may be Fourier transformed to extract the NMR spectrum. In order to improve the detection of the signal, or, in other words, the signal to noise ratio, this process of giving the pulse and measuring the response is repeated several times, before the Fourier transform is taken. If the response is measured N times and added, the signal will increase N times while the background noise, which is random will increase only $\sqrt{N}$ times. The expression 'free induction decay' or FID is used to describe the response on the application of the radio-frequency pulse – it describes the decay of the induced signal arising from the free precession of the nuclei in the field B_0

The nuclei in the sample will be present in several different environments, and the resulting FID will consist not only of several different frequencies, but also each different frequency will decay at a different rate. The Fourier transform of the FID therefore contains information not only about the resonance frequency of each nucleus but also about the decay of the resonance and hence the rate and mode of energy transfer from the nucleus to the environment.

8.5: NMR parameters

In general, the following are the measurable NMR parameters: 1) Chemical shift; 2) Intensity; 3) Line-width; 4) Spin-spin coupling constants.

8.5.1: Chemical shift:

The effect of the external applied magnetic field on each of the protons in a molecule will not, in general, be the same. This is because, owing to the electron currents surrounding each nucleus, an additional small field is also created just around the nucleus. The total effective field at each nucleus will thus depend on the neighbouring atoms and chemical groups. Since the magnetic field is different at each chemically distinct nucleus, the frequency at which the resonance occurs will also be different at each nucleus. Consider for example a molecule such as ethanol. A schematic sketch of its ^{1}H NMR spectrum at 100 MHz (i.e. with B_0, the applied field strength, corresponding to a resonance frequency of 100MHz for a free H nucleus) is given in Figure 8.5. The frequency at which the hydroxyl (OH) proton resonates is different from the one at which the methylene (CH_2) protons resonate which is again different from the one at which the methyl (CH_3) protons resonate. Such a shift in the position of the resonance peak in the spectrum due to the chemical environment is called the chemical shift. The chemical shift is always measured with respect to some arbitrary chosen reference point in the spectrum. Usually an inert compound with a well-known resonance position, not much affected by the molecular environment, is used as the standard. The standard may be introduced either internally (i.e. mixed with the sample) or externally (in a small tube placed alongside the sample tube). Chemical

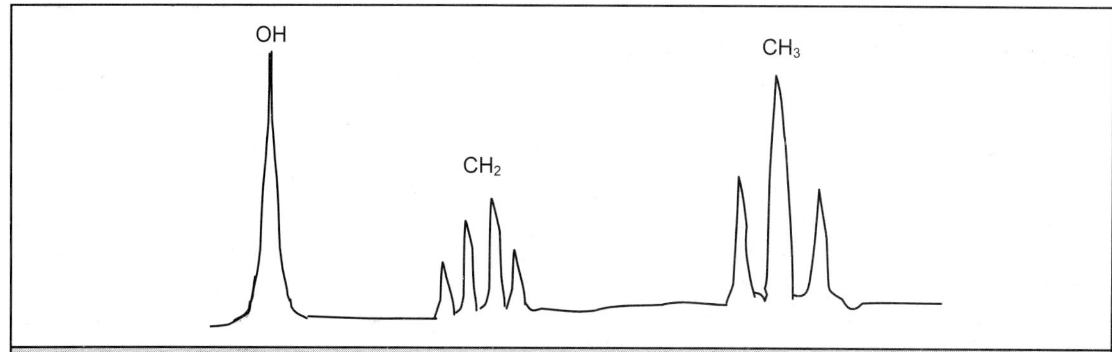

Figure 8.5: The NMR spectrum of ethanol. (This is a schematic diagram. The splitting of the peaks has been exaggerated.)

shifts are expressed in the dimensionless units of parts per million (ppm) and are normalised to be independent of the applied magnetic field. The chemical shift is given by the expression

$$\delta = (\nu_S - \nu_R)/\nu_R \times 10^6$$

where ν_S is the resonance frequency of the sample, and ν_R is the resonance frequency of the reference. By convention, when displaying the NMR spectrum, the resonance frequency of the standard is placed at the right-most end of the scale at 0 ppm, and the scale, in ppm, increases to the left, in decreasing order of frequency. This corresponds to the fact that the chemical shift is actually produced by a shielding, and therefore by a reduction of the applied magnetic field by the electrons in the neighbourhood of the resonating nucleus. Table 8.2 indicates the typical chemical shifts for some important chemical groups.

8.5.2: Intensity:

The signal strength or intensity of the resonance peak is a measure of the number of nuclei that contribute towards it. Signal intensity is obtained by measuring the area under the peak. To be mathematically rigorous, it is necessary to integrate each peak from $-\infty$ to $+\infty$. In practice this is not possible and one has to be content with measuring the area under those parts of the peak that are accessible. This usually accounts for only about 90% of the total area, and hence there is always an error of $\pm 10\%$ in the measurement of the signal strength.

Apart from the number of nuclei resonating at the given frequency, signal intensity is also influenced by the effects of saturation, the nuclear Overhauser effect, the effects of a finite pulse width and other experimental limitations. If these factors are taken into consideration and if appropriate corrections are applied, then accurate measurements of relative concentrations of the different nuclei can be made. If absolute concentrations

Table 8.2: Typical ^{1}H chemical shifts. (in ppm from tetramethysilane or TMS. Where a range is given, the position of the peak is dependent on the attached groups as well as the solvent, pH, etc.)

Chemical group	Chemical shift (δ)
$-CH_3$	0.9
$-CH_2$	1.4
$-CH$	1.5
$-C_6H_6$	7.3
$-CHO$	9.7
$-OH$	$0.5 - 4.0$
$-NH_2$	$2.0 - 5.0$
$-NH$	$2.0 - 5.0$
$-SH$	$1.0 - 2.0$

need to be measured, a known quantity of a standard sample can be used to calibrate the spectrum and measurements can then be made. This technique is especially useful in biomedical applications.

8.5.3: Line width:

The width of each spectral line is related to the relaxation processes in the sample. There are primarily two ways in which the additional energy gained by the nucleus on transition can be released. The energy can either be transferred to a neighbouring spin, or it can be transferred to the neighbouring lattice. A time constant is associated with each and referred to as T_1 and T_2 respectively. Line width is measured at half the signal height in Hz and is represented as $\Delta v_{1/2}$. The relationship to so-called spin-spin relaxation time T_2 is

$$1/T_2 = \Delta v_{1/2}$$

in the ideal situation where only the relaxation contributes to the line width.

8.5.4: Relaxation parameters:

As explained earlier, when a sample is placed in a magnetic field, it becomes magnetised. The direction of the magnetisation M_0 is the same as that of the applied magnetic field, being conventionally taken as the z-axis. The component of magnetisation along the other two directions, i.e. M_{xy} is zero. This is the situation in the equilibrium condition. When a radio frequency pulse is applied, the magnetisation tilts over onto the xy plane and as the perturbing field is removed, the magnetisation returns to the equilibrium situation of $M_z = M_z$ and $M_{xy} = 0$. This happens by the relaxation processes and lasts over a period of time ranging from milliseconds to a few seconds. As mentioned earlier, there are two main relaxation processes, spin-lattice relaxation and spin-spin relaxation.

Spin-lattice relaxation is responsible for the return of the z component M_z to its equilibrium value. It is also called longitudinal relaxation. The lattice referred to in the term is the environment and all the nuclear dipoles surrounding the particular spin, whether they belong to the same molecule or to other molecules. In solid state NMR, the word lattice may refer to the crystal lattice. In solution phase NMR it simply means the neighbourhood of each spin. The return of the nuclear spins to thermal equilibrium, i.e. to the equilibrium of the lattice is characterised by a time constant referred to as T_1. A measurement of T_1 is frequently an important part of the NMR experiment, and is required for the following reasons: (a) To select the most suitable radio-frequency pulse interval, the T_1 values of the various resonances need to be known. (b) Knowledge of T_1 values helps in determining the rates of chemical reactions through NMR. (c) T_1 provides information about the molecular structure. (d) In modern medicine T_1 is the most useful parameter used in NMR imaging. Frequently the image of the tissue seen is nothing but a map of the values of T_1 at different positions within the tissue. T_1 may be measured by FTNMR by applying a sequence of radio-frequency pulses. There are several different pulse sequences that may be used. A commonly used sequence is the inversion recovery sequence in which a 180 ° pulse is first applied. (As already explained, a 180° pulse changes the angle of the bulk magnetisation vector by 180°.) The magnetisation is allowed to relax back to its equilibrium value for

a time interval τ and a 90° pulse is applied and the FID is collected soon after. This signal will contain information about the strength of M_z after time τ. If the process is repeated for different values of τ, the spin-lattice relaxation time constant T_1 may be obtained.

Spin-spin relaxation is responsible for the return of the M_{xy} component of the magnetisation to its equilibrium value of zero. This is also known as transverse relaxation and is characterised by a time constant T_2 known as the spin-spin or transverse relaxation time. Consider Figure 8.6. Immediately after the application of the radio-frequency pulse, the net magnetisation is given by some maximum value (Figure 8.6(a)). If there is only a single spin in the sample resonating at precisely one frequency, the magnetisation (as considered in the rotating frame of reference) remains static, and of constant magnitude. However, if there is spread of frequencies, different nuclei will precess at different frequencies. In the rotating frame of reference the magnetisation vectors for the spins will fan out, resulting in a reduction of the net magnetisation (Figures 8.6(b) and 8.6 (c)). Finally M_{xy} becomes zero, the equilibrium value (Figure 8.6(d)). Since T_2, the time constant characterising this process depends on the spread in frequencies, the width of the resonance peak is a measure of T_2.

Spin-spin relaxation processes are therefore those that cause line broadening. One component of this is the dipole-dipole interaction, which also causes spin-lattice relaxation, and therefore T_2 cannot be longer than T_1. Additional contributions to T_2 come from the z components of the small perturbing fields of the neighbouring spins. T_2 measurements provide information about molecular structure. They can also be used to determine rates of molecular diffusion. Information about T_2 helps to design pulse sequences that may simplify the spectra and enhance the resolution.

8.5.5: Spin-spin coupling:

This is different from the spin-spin relaxation process explained above. Resonances in the NMR spectrum often do not occur as a single peak but are split into doublets, triplets, quadruplets or higher multiplets. The splitting is due to the neighbouring spins that are transmitted via the electrons of the covalent bonds between the atoms. Spin-spin coupling is thus a 'through-bond' coupling and occurs only due to nuclei within the molecule. Splitting of the peak can be caused only by nuclei with spin $I \neq 0$. The magnitude of splitting is designated by the spin-spin coupling constant 'J'. It is measured in Hz and is independent of the applied magnetic field strength. Figure 8.7(a) is the spectrum of acetaldehyde (CH_3CHO) in which the CH_3 signal is split into a doublet due to spin-spin coupling of the protons with the neighbouring CHO proton through the three bonds between them (Figure 8.7(b)). The CHO resonance itself is split into a quadruplet owing to coupling with the three methyl protons. A complete theoretical treatment of spin-spin coupling is beyond the scope of this book. We substitute such a treatment by the following discussion.

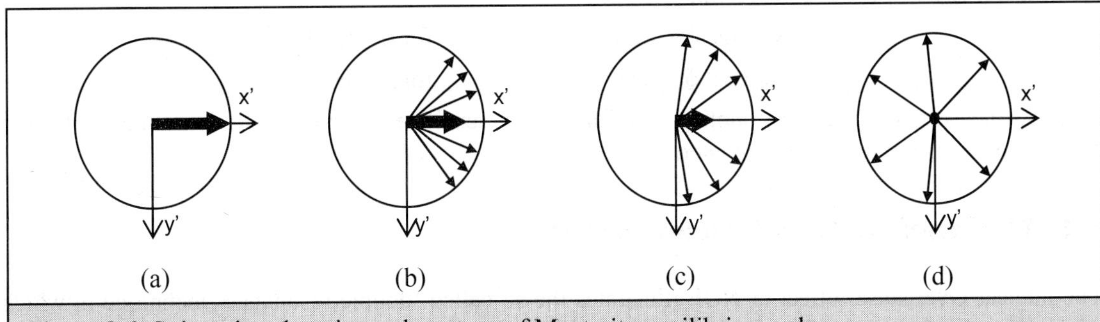

Figure 8.6: Spin-spin relaxation – the return of M_{xy} to its equilibrium value.

of the resonating spin is changed with respect to the neighbouring nucleus. Thus in a molecule such as cyclohexane, protons which are equatorial to the ring will have different chemical shifts from those which are axial. Anisotropy effects are dependent on the order of the neighbouring bonds. Single bonds lead to less anisotropy than double or triple bonds. In aromatic rings, the circulation of π electrons around the ring results in a shielding of nucleus above and below the plane of the ring, leading to an upfield shift, and a deshielding of protons in the plane of the ring, leading to a downfield shift of the resonance. Van der Waals or anisotropy effects are also produced by the solvent in which the sample is dissolved.

Hydrogen bonding produces a downfield shift in the resonance of the protons involved in the hydrogen bonds, since such a bond is between electronegative atoms, which would withdraw the electron away from the proton leading to a deshielding.

8.7.2: Spin-spin coupling:

Spin-spin coupling is dependent on the nature of the neighbouring nuclei and coupling constants carry information about the chemical structure of the molecule. It is however necessary to first of all distinguish between chemical equivalence of atoms and the magnetic equivalence of atoms. In the case of chemical equivalence the spins have identical chemical environments and thus identical chemical shifts. Magnetically equivalent nuclei will not only have the same chemical shift, but will also possess the same coupling constant with every other nucleus in the molecule. Figure 8.8(a) shows tert-butanol in which the methyl hydrogens are both chemically and magnetically equivalent. In Figure 8.8(b) the methylene hydrogens of diflouroethylene are chemically equivalent but magnetically non-equivalent, since each hydrogen is coupled to each of the two flourines with a different coupling constant.

Coupling constants in molecules can be either positive or negative in sign and up to a few tens of Hz in magnitude. Of the chemical factors affecting spin-spin coupling, the dihedral angle, the bond length and bond angle in a H-C-C-H system has already been discussed. Other factors are: (1) the electronegativity of the substituents, (2) valence angles and (3) bond lengths. Vicinal protons, i.e. protons such as those in the system H-C-C-H, show a decrease in the coupling constant if an electronegative substituent is attached to the carbon atoms. If more than one such substituent is present, the decrease is greater. In cyclic systems the steric disposition of the electronegative substituent also affects the coupling constant, being larger when it is equatorial and smaller when it is axial. The effect is the greatest when the substituent is trans-coplanar to one of coupling protons. For

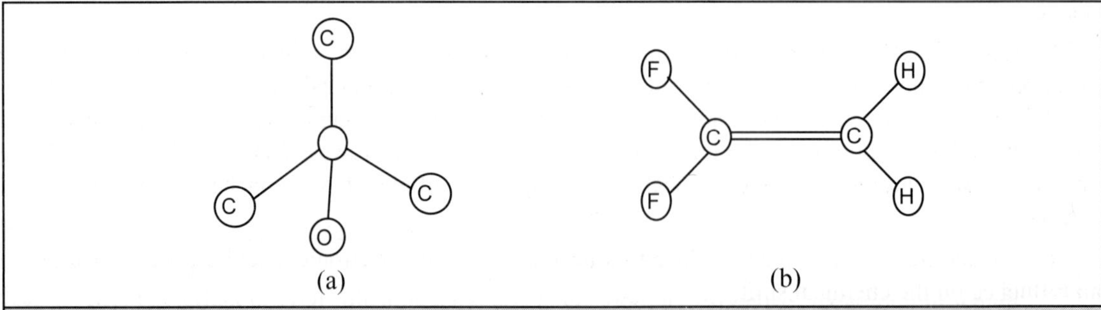

Figure 8.8: Chemical and magnetic equivalence. a) Tert-butanol. b) Difluoroethylene.

geminal protons, i.e. protons attached to the same carbon atom, the coupling is usually negative. An additional electronegative substituent decreases the values further (magnitudes increase, but the sign remains negative). Neighbouring π bonds such as those found in aromatic systems decrease coupling constants still further. Vicinal coupling constants are sensitive to the angles between the C-H and C-C bonds, as well as to the C-C bond lengths. In both cases, the coupling constants decrease as the values of the angles or bond lengths increase. Long range spin-spin coupling extending over four or five bonds is also possible, through the magnitude is usually small, i.e. only of the order of a few Hz.

8.7.3: ^{13}C NMR:

Among the spin ½ nuclei ^{13}C is of interest in chemistry, especially organic chemistry. However the isotope ^{13}C has a natural abundance of only 1.11% as compared, for example to 99.98 % for ^1H. ^{12}C has an abundance of greater than 90% but spin I = 0, and therefore not useful for NMR. In spite of its low abundance ^{13}C NMR is frequently used in chemistry. ^{13}C chemical shifts are in the range 0-230 ppm with reference to TMS. Many of the factors that affect the ^{13}C chemical shifts are the same as for ^1H NMR. The state of hybridisation of the observed ^{13}C nucleus is an important factor. sp^3 carbons resonate between -20 and 100 ppm with respect to TMS; sp^2 carbons between 120 and 240 ppm and sp carbons 70 and 110 ppm. The electronegativity of the substituent is another important factor influencing the chemical shifts, as are steric and van der Waals effects. Mesomeric effects arise from electron-donating groups attached to benzene rings that shield the ortho and para carbon atoms (electron withdrawing groups would deshield these positions). Hydrogen bonding, attachment of heavy atoms, bond conjugation, attachment of deuterium atoms instead of hydrogen, pH and the solvent, all contribute to the ^{13}C chemical shift.

Spin-spin coupling in ^{13}C NMR can be observed between the ^{13}C nucleus and neighbouring protons, as well as between two ^{13}C nuclei. The latter effect is more difficult to observe, since, due to the low natural abundance of ^{13}C isotopes, the chances of two such nuclei occuring as neighbours in the molecule is very small.

^{13}C NMR also lends itself to the study of sample in the solid state. In solutions, local magnetic fields due to the nuclei of neighbouring molecules will be averaged out due to the rapid tumbling motion and will not have any effect on width of the resonance peak. In solids however, each individual molecule would be surrounded by a different static environment. This leads to a widening of the resonance. Until the advent of recent high resolution solid state NMR methods, ^{13}C NMR was only sparsely applied in the solid state, and many interesting aspects of solid state chemistry, polymer chemistry, etc. could not be studied. High-power decoupling cross polarisation (CP) and Magic Angle spinning (MAS) are two of the relatively recently developed techniques which make it possible to obtain high resolution even in solids. So much so, at least for low molecular weight compounds, solid state NMR gives a resolution comparable to the solution state.

8.8: NMR applications in biochemistry and biophysics

In addition to the fact that biologically important molecules are usually of very large molecular weight and size, biological systems also reveal more complex interactions between the molecules or groups of molecules. With the advent of very high-field magnets and techniques such as multi-dimensional

NMR, the study of such complex systems through NMR has in recent times become possible in much greater detail and precision. Even before these developments, however, NMR was being used in biochemistry to determine concentrations, study the pH behaviour of biomolecules and to elucidate the conformation of small biologically important molecules.

8.8.1: Concentration measurement:

The determination of relative concentration by NMR is made simply by measuring the area under the NMR absorption signal. If the spectrum has been calibrated against a signal of known concentration, absolute concentrations can be measured in this way. Precautions however have to be taken against several sources of error that may creep in. One of these is the fact that if, due to different relaxation times, one signal has become saturated, while another is not, the measurements of the relative concentrations may be wrong. Another source of error is the possible uneven intensity of the excitation pulse, which may lead to different frequencies being excited with different intensities. It is however possible to choose signals which do not suffer from such drawbacks, enabling fairly precise measurements of concentrations.

8.8.2: pH titration:

The protonation or deprotonation of functional groups in biological molecules has a detectable influence on the chemical shift. The rates of proton exchange with the solution are much faster than the NMR time scales, and therefore individual exchanges cannot be detected. However the average effect is clearly observed in the NMR spectrum. This effect is not only sufficiently strong to be detected at the protonation/deprotonation site, but even several bonds away in the molecule. The protonation of a particular site depends on its pK. pH titration is the measurement of the chemical shift as a function of pH, and helps to measure the pK of the site. The protonation or deprotonation of the site at its pK value is marked by a sharp change in its chemical shift at that value of the pH. In proteins, histidine residue resonances are especially dependent on the pH of the solution, and together with other such resonances can be used as a monitor of denaturation. This is because protonation sites in the interior of the proteins will be progressively exposed to the solution as denaturation proceeds. This can be detected by the dependence of the spectrum on pH or temperature (or other denaturing conditions). Inorganic phosphates participate in many enzymatic reactions, and exist, during the course of the reaction, in many different protonated states such as H_3PO_4, $H_2PO_4^-$, HPO_4^{2-}, and PO_4^{3-}. By using ^{31}P NMR and pH titration, the course of such a reaction can be followed.

8.8.3: Conformation of biomolecules:

The NMR spectra of biomolecules contain a large amount of structural and conformational information. The problem is to extract this information. Even at relatively high resolution, most of the peaks in a ^1H NMR spectrum of a protein are broad. The causes for such broadening in macromolecules are as follows. (a) A biopolymer usually contains more than one unit of any particular monomer, and each unit would be in a slightly different chemical environment, leading to a widening of the resonance. This cause for the widening is the same as that occurring in solids.   (b) The presence of other nuclei close-by, leads to dipolar relaxation in excess of what would be observed for the monomer and this contributes to the broadening of the resonance. (c) Chemical exchange of protons

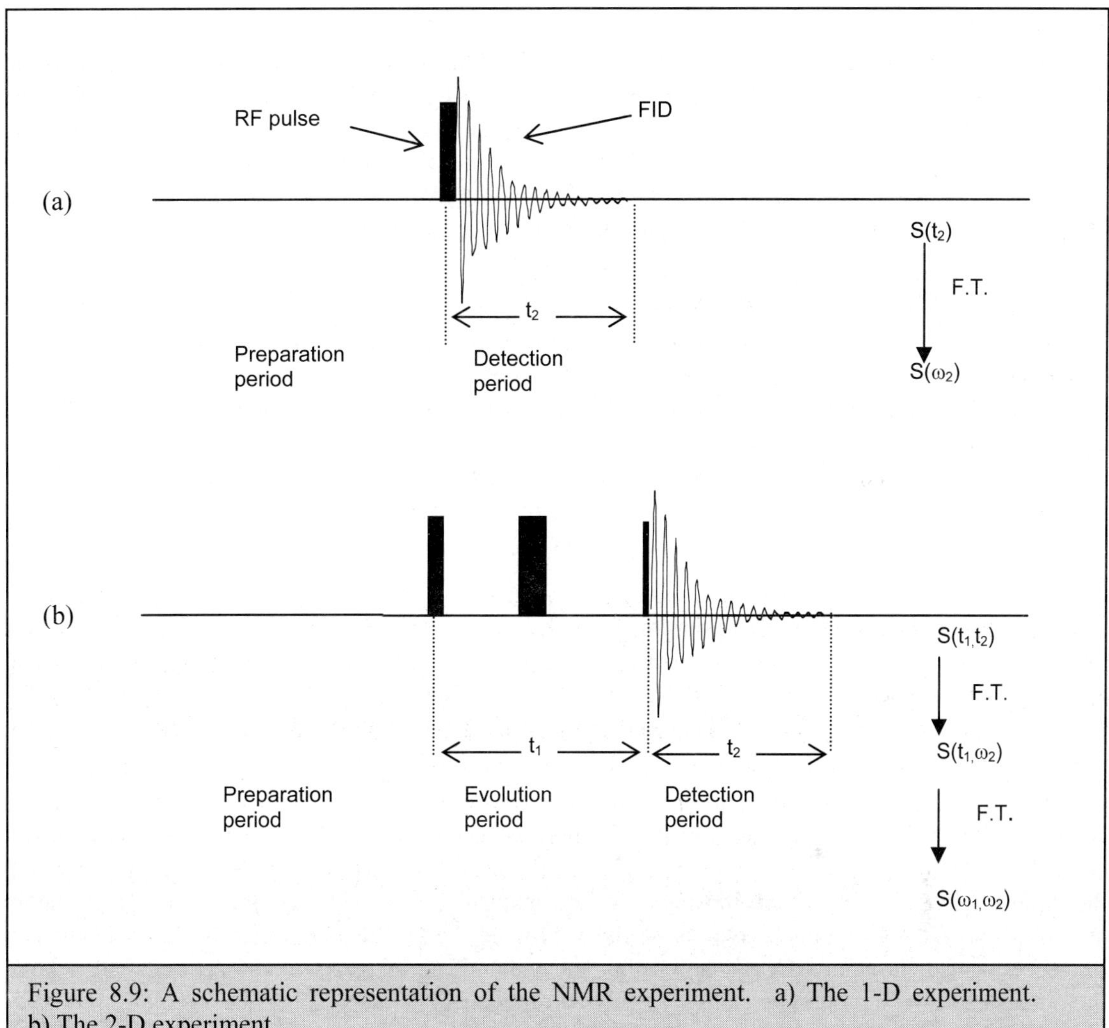

Figure 8.9: A schematic representation of the NMR experiment. a) The 1-D experiment. b) The 2-D experiment.

between different close-by sites can also cause line broadening. (d) Paramagnetic ions are frequently found in biological molecules and are a source of line broadening. The consequence of all this is that detailed conformational studies are possible by the conventional one dimensional and or continuous wave NMR only in the case of small organic molecules (as discussed earlier). However, certain aspects of biomolecular conformation can be studied by conventional NMR. One such is the helix to random-coil transition in polypeptide chains. Similarly one can use NMR to follow the denaturation-renaturation transition in proteins. In either case the NMR spectrum of the random coil would be a composite of the spectra of the constituent amino acids. The spectrum of the ordered structure would deviate markedly from the simple random-coil spectrum. The transition can thus be followed as function of temperature, pH or any other relevant parameter.

8.8.4: Two-dimensional NMR:

As mentioned earlier, detailed conformational studies of macromoleules are possible now with the development of two-dimensional techniques. A one-dimensional FTNMR experiment involves the application of a RF pulse after the sample has been allowed to achieve equilibrium during the 'preparation' period (Figure 8.9(a)). Immediately after comes the detection phase where the FID (i.e. the signal S) is collected as a function of time (here represented by the variable t_2). A Fourier transform of the signal into the frequency domain results in the NMR spectrum.

$$S(t_2) \rightarrow S(\omega_2)$$

In the case of two dimensional NMR spectroscopy an additional time period (denoted t_1) and called the evolution period is introduced, after which the FID is collected (Figure 8.9(b)). During the evolution period another RF pulse may be applied and a further period of time, called the mixing period, may be allowed to elapse before the FID is collected. In each experiment, the time t_1 may be given a different value, ranging from zero to some maximum value depending on the relaxation times. Each time the FID is collected and Fourier transformed to yield $S(t_1,\omega_2)$. The several spectra so obtained are arranged in a data matrix which can be subjected to a second Fourier transform, column wise.

$$S(t_1, \omega_2) \rightarrow S(\omega_1, \omega_2)$$

The spectrum thus will have two dimensions. ω_1 is along one of the dimensions and ω_2 along the other. The ω_2 axis contains information about coupling interactions with other nuclei during the evolution period t_1, and the ω_1 axis contains information about the chemical shifts. The precise pulse sequence applied to the samples, gives rise to different types of coupling information being displayed along the ω_1 axis. There are many different pulse sequences that have been published. The two most commonly used sequences in biological research are abbreviated as COSY and NOESY.

COSY is an abbreviation for correlated spectroscopy. There are different types of COSY pulse sequence. The so-called homonuclear shift correlated spectrum displays the J coupling, or the spin-spin coupling through bonds as a two-dimensional plot. Figure 8.10 is a simple COSY spectrum in which the connectivity of spins in the molecule can be mapped out. COSY spectra are of great use in assigning the peaks to the nuclei that cause them. Solving the chemical structure of the molecule can be accomplished through a COSY spectrum. The assignment of peaks in the spectrum of a large biomolecule such as a protein is relatively easily accomplished.

NOESY is an abbreviation for Nuclear Overhauser Enhancement spectroscopy. The nuclear Overhauser effect is a 'through-space' effect and can be used to determine the spatial separation between nuclear spins. In other words, unlike correlation spectroscopy, or COSY, where the interaction between one resonating nucleus and another is through the bonds connecting them, in NOESY, this interaction takes place directly through space NOESY spectra are similar to COSY spectra except that in this case the off diagonal peaks correspond to 'through-space' connectivity of spins. The strength of the effect varies as $1/r^6$ where r is the distance between the spins, and measurement of the strength of the off diagonal NOE peaks in a 2D NOSEY spectrum can be used to estimate distances between spins. As explained in the next section, such measurements are of great use in arriving at the three-dimensional folding patterns of biopolymers such as proteins in which the polypeptides chain fold back on itself to bring into close proximity nuclei which are widely separated along the chain.

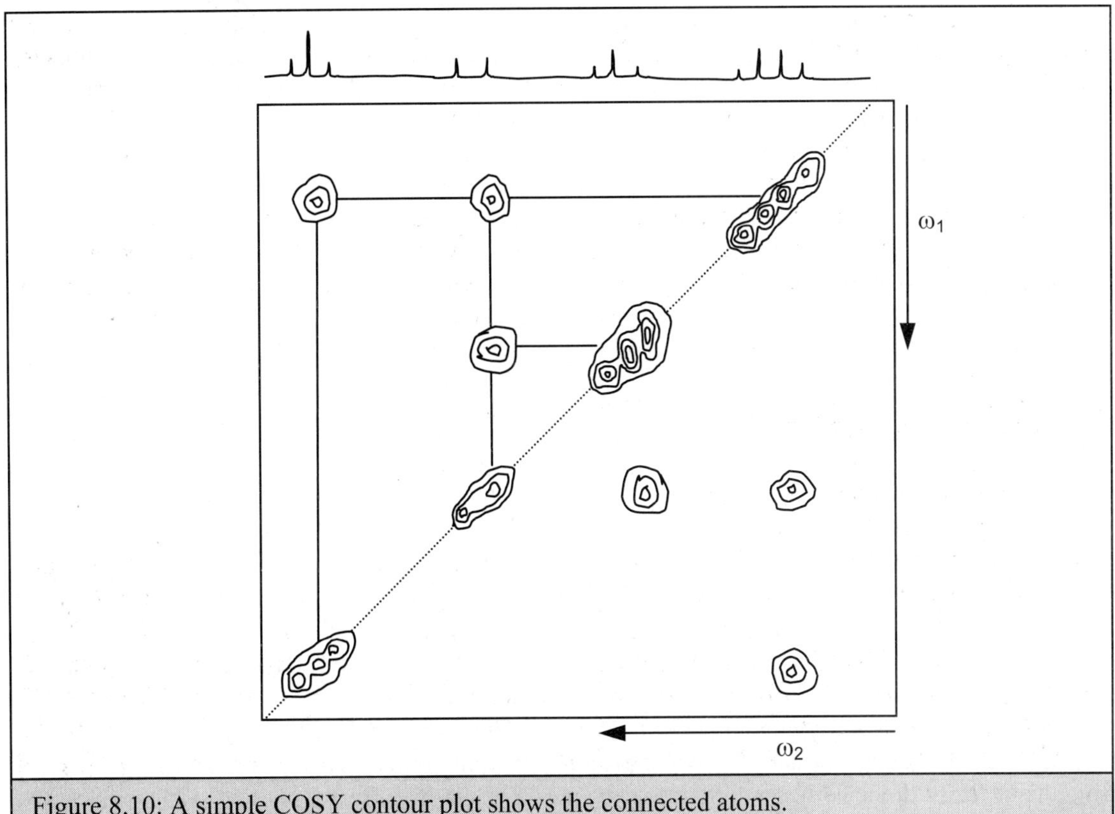

Figure 8.10: A simple COSY contour plot shows the connected atoms.

8.8.5: Determination of macromolecular structure:

The experiment to determine the three dimensional structure of a macromolecule usually consists of two steps. The first step is to assign all the resonance peaks that occur in the spectrum. The high-resolution spectrum of a macromolecule such as a protein is very complex and only through a 2-D spectrum such as COSY is it possible to trace the connectivity in the molecule and assign the peaks. In proteins, especially, but also in nucleic acids, the interpretation of the 2-D COSY spectrum is sometimes made easier by the existence of recognisable patterns of connectivity which can be attributed to specific amino acids or base-sequences. The second step in the structure determination is to identify secondary and tertiary structures from the J-connectivity patterns and from NOE data. A NOESY spectrum will give information about the 'through-space' couplings between the spins in the molecule. The NOE intensities may be divided into classes as strong, medium and weak NOEs corresponding to a distance less than 2.5 Å, 3.0 Å and between 3.0 and 4.0 Å respectively. (Usually no NOEs are observable beyond 4.0 Å). The distance data so obtained may be used to attempt a reconstruction of the three-dimensional folding of the biopolymer chain. This is mathematically a complicated problem and no general algorithm exists which can lead from the distances to the geometry in a reliable way. There are however several approaches to solving this problem, each being successful to a lesser or greater degree. The NOE distances alone are not sufficient to define the structure. Oxygen and nitrogen atoms, for example, are usually not seen in the NMR spectra and their positions must be inferred from the positions of connected protons or ^{13}C nuclei. Distance geometry algorithms therefore use all chemical information available and may be performed in torsional-angle space where the degrees of freedom are the dihedral angles of the main chain and the side chains. Additional constraints in the form of non-bonded van der Waals contacts, hydrophobic interactions and so on, may also be used in arriving at the structure. A commonly used technique is restrained molecular dynamics. As described elsewhere in the book the observed experimental data, in this case the NOE distances, are written as a pseudo potential, along with other semi-empirical potential terms such as bond length energy, bond angle energy etc. This method has been successfully used in many instances in arriving at precise three-dimensional structures of small proteins (Mw ~20,000). Structures of larger proteins (Mw > 50,000) are still somewhat inaccessible to current techniques.

8.9: NMR in medicine

Some of the most spectacular applications of modern NMR techniques have been in the area of medical imaging, namely Magnetic Resonance Imaging or MRI. Unlike NMR spectroscopy, where the NMR signals are obtained as a single spectrum from the entire sample, imaging requires that the signal be resolved in space. The different signals from the different parts of the object are reconstructed on the computer screen as an image of the object. The electromagnetic fields and waves used in NMR pass easily through living bodies and signals from any interior portion can be detected and processed. In this way it is possible to build up clear images which are of great use in medicine. Three NMR measurable parameters are commonly used in imaging. The first is the spin-density or the number of a particular type of nucleus (usually the water protons) per unit volume. This parameter can be deduced from the intensity of the corresponding resonance peaks at all the different portions of the object. The second parameter that can be used in imaging is the spin-lattice relaxation time T_1. Different types of tissue within the object would have different T_1 values and if these values are mapped out, an image is obtained. The third commonly used parameter; viz. the spin-spin relaxation time T_2 is also treated the same way. Thus, in contrast to X-ray imaging (including the CAT scan), where the X-ray absorption is the only detectable parameter, there are at least three parameters which can be detected in NMR imaging and NMR tomography. This leads to not only greater detail in the images, but also allows different types of tissues to be imaged. In addition, NMR does not use any highly ionising radiation to produce its images.

The most basic principle in the imaging technique is the selection of a slice to be imaged from the entire volume of the object. Unlike as in NMR spectroscopy, where the magnetic field is kept steady over the entire sample, in the case of NMR imaging, spatial selection is achieved by applying a magnetic field gradient on top of the static magnetic field. This would mean that particular spins, such as for example protons (^{1}H) come into resonance at different frequencies at different parts of the sample along the axis of the field gradient. If the sample is irradiated by a radio-frequency excitation pulse in a very narrow range of frequencies, then only the spins in a particular spatial slice of the object, perpendicular to the direction of the magnetic field gradient would come into resonance and contribute to the NMR signal. If a 2D image of such a slice is obtained, many such images can be stacked one on top of the other to yield a complete image of the object.

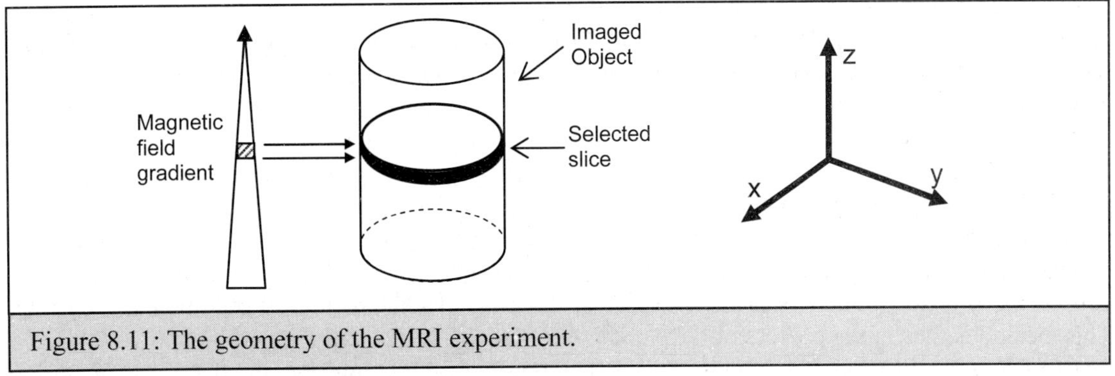

Figure 8.11: The geometry of the MRI experiment.

In order to obtain a 2-dimensional image of the slice, a 2D-Fourier imaging technique is most commonly used. The slice selection gradient is assumed to be applied along the z axis. Thus the slices are perpendicular to this axis and parallel to the xy plane (Figure 8.11). After the first excitation pulse, a gradient G(x) is applied for a time t_x along the x-axis. Immediately after, another gradient G(y) is applied along the y-axis for a time t_y. The NMR signal is then collected. The process is repeated for different values of t_y ranging from 0 to the maximum, each time obtaining the signal for different values of t_y. Thus a matrix of signal of size n^2 is obtained where n is the number of different values of t_x and t_y. It can be shown that the matrix when Fourier transformed will result in 2D image of the slice. The time taken to obtain such an image is usually less than 10 minutes. While this is sufficiently short for imaging live animals, if the process were to be repeated several times in order to obtain the complete three dimensional image, it is difficult to keep the object still and motionless for the few hours that it might take. Other methods have been developed where additional slices are measured almost simultaneously and the entire image reconstructed in one go. This may entail a loss in resolution, contrast, etc.

One of the most important applications of NMR tomography is the imaging of sensitive parts of the human body such as the head. X-rays are highly attenuated by the skull, making X-ray imaging of the brain quite difficult, in addition to the fact that X-rays are ionising radiation and therefore can cause damage. The radio frequency radiation used in NMR passes almost unhindered through the skull and has no as yet known ill effects on the tissues. Certain images of the skull, such as the so-called sagittal view are simply not possible with X-ray tomography. In addition, pathological conditions are easily identifiable especially with the possibility of using different modes of imaging (i.e. spin density, T_1 or T_2). Apart from these three parameters it is also possible to image the organs by means of the difference in the chemical shifts. NMR imaging may also be used to study continuous flow processes within the body such as blood flow. The excitation pulse sequences may be adjusted so that only the flow is visible, leading to a clear image of the blood vessels alone.

NMR imaging at very high spatial resolution is called NMR microscopy. At present the maximum obtainable resolution is about 20 μM, mainly because below this limit there are simply not enough protons in the sample volume to give a signal with sufficiently signal-to-noise ratios. Thus NMR microscopy, while potentially of great use, since the sample is preserved intact even while observing interior portions, has yet to reach the resolution of the best light microscopes.

A related application of NMR in medicine is magnetic resonance spectroscopy (MRS) also known as 'in vivo spectroscopy'. Once again a spatial selection technique is used to obtain a reasonably high resolution NMR spectrum from a small volume of the object, such as typically, the human body. MRS differs from MRI in that in the latter, information is available regarding only one aspect of the sample such as the spin density. MRS looks at the entire information content of the spectrum. The spatial selection techniques are somewhat different in detail from those used in imaging. This is because the portion of the sample volume may be defined with respect to either the laboratory frame of reference or with respect to certain biological structures, i.e. whether the molecule investigated is inside or outside a defined closed structure such as a cell. Typically in vivo spectroscopy is used to monitor the state of various identifiable metabolites as a function of the progress of some biological condition, e.g. disease. This method is thus in the process of developing into a sophisticated early diagnostic tool as well as a way of following the results of therapy. The method however is only now moving out of the laboratory and into the clinics.

Summary

- Nuclear Magnetic Resonance spectroscopy is one of the most important experimental techniques currently in use in biophysics.

- The NMR experiment consists of the placing the sample in a strong external magnetic field, and then irradiating it with radio-frequency electromagnetic radiation.

- The intensity of absorption (or emission) of electromagnetic radiation at each frequency over the radio-frequency range (i.e. from a few MHz to a few hundred MHz) is then measured.

- At a specific frequency for each nucleus, the energy is absorbed or emitted when that nucleus goes from one quantized spin state to another.

- The frequency at which the transition occurs is called the resonance frequency.

- The resonance frequency of a particular type of nucleus, e.g. ^{1}H, will be different when it is placed in different chemical environments.

- By changing the frequency of the exciting radio-frequency field and measuring the intensity of the absorption at each frequency, it is possible to detect the nuclei and in each case identify their respective environments.

- This is useful in identifying various chemical groups in molecules.

- The value of spin for any nucleus can either be an integer or a half-integer ($^1/_2$, $^3/_2$, $^5/_2$...).

- In chemistry and biology, the most useful nuclei are those with spin ½, such as ^{1}H, ^{13}C, etc.

- The resonance condition is given by the basic NMR equation $\omega_0 = \gamma B_0$, in which ω_0 is the resonance frequency, B_0 is the applied magnetic field, and γ is a constant for each nucleus called the magnetogyric (or gyromagnetic) ratio.

- There are two main methods by which the NMR experiment is performed, viz. the continuous wave method and the Fourier transform method.

- The NMR condition may also be derived from considerations of classical mechanics. Such a derivation helps understand Fourier Transform NMR

- In FTNMR the radio-frequency stimulus is applied as a square pulse, which may last typically for about 200μs.

- According to Fourier Transform theory, such a square pulse is constituted of all possible frequencies.

- The composite response from all the nuclei in the sample would be another complicated wave which may be back-Fourier transformed to extract the NMR spectrum.

- The Fourier transform contains information not only about the resonance frequency of each nucleus but also about the decay of the resonance and hence the rate and mode of energy transfer from the nucleus to the environment.

- In general, the following are the measurable NMR parameters: 1) Chemical shift; 2) Intensity; 3) Line-width; 4) Spin-spin coupling constants.

- The chemical shift is given by the expression $\delta = (\nu_S - \nu_R)/\nu_R \times 10^6$ where ν_S is the resonance frequency of the sample, and ν_R is the resonance frequency of a suitable reference compound placed in the NMR machine along with the sample.

- The return of the nuclear spins to thermal equilibrium, i.e. to the equilibrium of the lattice is characterised by a time constant referred to as T_1.

- In modern medicine T_1 is the most useful parameter used in NMR imaging.

- Transverse relaxation is characterised by a time constant T_2 known as the spin-spin or transverse relaxation time.

- Spin-spin coupling or J coupling is a 'through-bond' coupling and occurs only due to nuclei within the molecule.

- The nuclear Overhauser effect or NOE is 'double resonance' technique in which a second radio frequency field is applied to the sample in addition to the observing radio frequency field.

- This results in a transfer of magnetisation through space that may be observed as a 'through-space' coupling between atomic nuclei not bonded to each other but proximate in space.

- The proton and ^{13}C chemical shifts in an NMR spectrum are a sensitive measure of the chemical environment of the nucleus. Their measurements form the mainstay of chemical applications of NMR.

- Spin-spin coupling is dependent on the nature of the neighbouring nuclei and coupling constants carry information about the chemical structure of the molecule.

- ^{13}C NMR lends itself to the study of sample in the solid state.

- The protonation or deprotonation of functional groups in biological molecules has a detectable influence on the chemical shift, and therefore may be assayed by NMR.

- Conventional NMR can be used to study certain aspects of biomolecular conformation.

- Multidimensional NMR has been used to determine the structures of small proteins.

- Some of the most spectacular applications of modern NMR techniques have been in the area of medical imaging, namely Magnetic Resonance Imaging or MRI.

- Three NMR measurable parameters are commonly used in imaging – spin-density or the number of a particular type of nucleus (usually the water protons) per unit volume, the spin-lattice relaxation time T_1, and the spin-spin relaxation time T_2.

- Different types of tissue within the object have different values for these parameters, and if these values are mapped out an image is obtained.

Chapter 9

Molecular Modelling

9.1: Introduction

X-ray analysis or NMR or other methods of structural characterisation of the molecules give, as output, a set of numbers that specify the relative positions of each of the atoms of the molecules in space. In order to use the information present in that list, human capabilities almost always demand that it be converted to a form in which it can easily be visualised. Before the advent of powerful computer graphics, the co-ordinates were used to build models in wood or plastic. These suffered from the following disadvantages: 1) They were tedious to build. 2) They were static; they could not be easily altered to try out new ideas about the structure. 3) They were not accurate. 4) They were fragile. 5) They were heavy, prone to collect dust and were difficult to maintain. 6) They took up a lot of space. 7) They were expensive.

Despite these disadvantages they were widely used and several new ideas in biology can be credited to such models, the most notable being the double-helical structural model of DNA, constructed by J.D. Watson and F.H.C. Crick with metal parts.

Computers have made all the difference to molecular modelling. They possess none of the disadvantages listed above. The chief problem with computer graphics is that they are two-dimensional pictures of 3-dimensional objects. Nevertheless one can get over this problem to a great extent by using techniques such as stereopsis. Most programs also have a facility which allows the user to 'rotate' the picture on the display screen so that different parts of the molecule come into view continuously giving the illusion of watching a real three-dimensional model. Thus the advantages of using computer graphics far, far, outweigh the single disadvantage.

9.2: Generating the model

If the co-ordinates of the molecule are already available from, say, an X-ray crystallographic analysis, then they are just fed into the computer program which will interpret each data point as an atom and

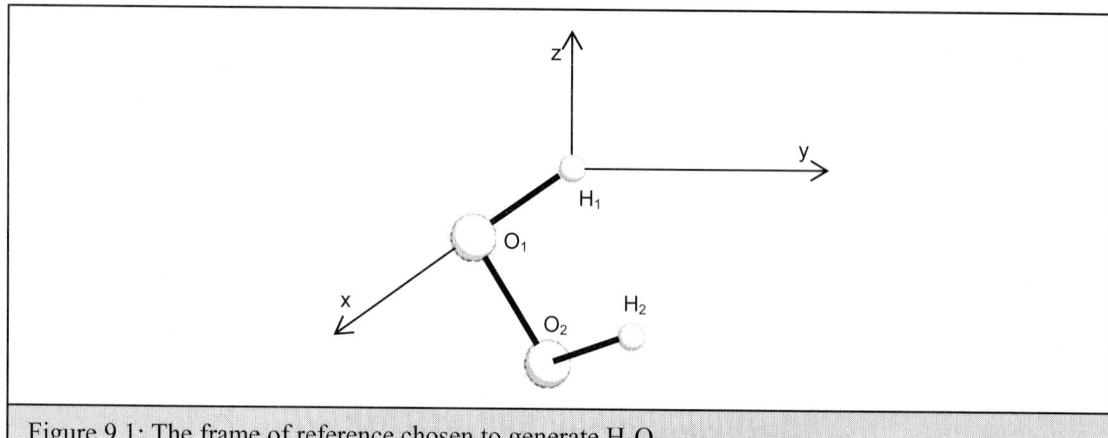

Figure 9.1: The frame of reference chosen to generate H_2O_2.

display it according to the mode and orientation chosen by the user. If the co-ordinates are not available, but if the chemistry of the molecule is known (i.e. the chemical formula for a small molecule; and the amino acid sequence for a protein), it is possible to generate the co-ordinates for the molecule, *ab initio*, on the computer. As an example we demonstrate the generation of the co-ordinates for a very simple molecule, namely, hydrogen peroxide

9.2.1: Building the structure of H_2O_2:

Hydrogen peroxide, H_2O_2 has the following chemical structure.

Chemical studies have established that the molecular dimensions are as given below:

Distances	Angles
O - H = 1.1 Å	H - O - O = 120°
O - O = 1.5 Å	

The one variable in this molecule, which is left to the choice of the modeller, is the dihedral angle[9] or the torsion angle H-O-O-H. If this is changed, it implies a rotation of one hydrogen atom about the O-O bond, with the other hydrogen atom kept fixed. In order to generate the co-ordinates for such a

---

[9] A dihedral angle is the angle between two planes, measured as the angle between the normal vectors to the two planes. A torsion angle is the angle made between the two outermost bonds when three consecutive bonds are considered. For example, consider the atoms P, Q, R and S bonded in a chain, P to Q, Q to R and R to S. The torsion angle between P-Q and R-S is measured by first viewing the four atoms along the central Q-R bond. In this view, atoms Q and R will superpose on one another. The angle made by the far bond, say R-S, with respect to the near bond, P-Q, measured in the clockwise direction, is the torsion angle. This is the same as the angle between the plane PQR and the plane QRS, and hence is sometimes called the dihedral ('two planes') angle.

molecule, we first define the frame of reference as in Figure 9.1. Note that it is conventional to choose a right-handed Cartesian co-ordinate system. We first place one of the hydrogen atoms, which we call H1, at the origin. Co-ordinates of H1 are therefore

H1 0.0 0.0 0.0

Next we place the O1 atom along the x-axis, at a distance of 1.1 Å from origin. The co-ordinates of O1 are therefore

O1 1.1 0.0 0.0

The third atom O2 is placed in the xy plane, at a distance of 1.5 Å from O1 and such that the O2-O1 bond makes 120° with the H1-O1 bond. An elementary geometrical analysis shows that the co-ordinates of O2 are

O2 1.5 1.3 0.0

Finally in order to determine H2, we place it at a O2-H2 distance of 1.0 Å and O1-O2-H2 angle of 120°. This still leaves us the freedom to place the atom anywhere on a circle. If we choose the torsion angle to be + 60°, the co-ordinates of H2 are the following.

H2 1.7 2.0 0.8

Thus the spatial positions (co-ordinates) of all the atoms in hydrogen peroxide molecule can be obtained by simple geometrical considerations. The lengths between the bonded atoms, the angles and the torsion angles, are called the internal parameters or internal co-ordinates of the molecule. A complete set of internal co-ordinates is exactly equivalent to the set of Cartesian co-ordinates and one can be derived from the other in the complete and non-redundant way.

9.2.2: Building protein structure:

The geometry of each amino acid in the protein is known, i.e. the relevant bond lengths and angles are known. Also known is the geometry of the linking peptide bond. Despite this, building a protein chain is not a trivial task since the torsion angles are not known. In the amino acids with the longer side chains, especially, the orientations of the various chemical groups depend on the side chain torsion angles. Apart from this, in the main chain, or the peptide backbone, the ϕ and the ψ angles are not fixed by chemistry. Hence, unless the fold of the protein is known or guessed at, a three-dimensional model of the entire protein molecule cannot be built.

It is known that information regarding the final three-dimensional structure of the protein is inherent in its amino acid sequence. Hence knowledge of the latter should, in principle, lead us to the former. However, the rules of protein folding are still unknown. It is not yet possible to go from the sequence to the structure without additional experimental information. This is called the 'Protein Structure Prediction Problem' and is one of the great, unsolved problems in molecular biology. It may be succinctly stated as follows: Given the amino acid sequence of a protein, how does one predict its three-dimensional structure. Several methods of structure prediction are available, though none of them have been completely successful. One of the most popular prediction methods was developed by P.Y.Chou and G.D.Fasman. The Chou and Fasman method has been quite successful in predicting the secondary structure of proteins. The method is based on a statistical analysis of the known protein structures that have been collected in a computer database.

Extensive collections of published three-dimensional structures are now available in computer readable format. A collection of all the published protein structures is available at the Research Collaboratory for Structural Biology in USA and is called the Protein Database or PDB for short. A typical entry in this database would consist of the name of protein, the authors of its structure, the methods used to determine the structure, and a list of the co-ordinates and thermal parameters of each atom in the structure including solvent atoms, ligands, etc. There are now over 50,000 entries in the database. Most of the structures deposited in the database have been solved by X-ray crystallography. There are also about 1500 structures determined by NMR techniques. The number of structures in the database is increasing rapidly and about 1000 to 2000 new structures are added to the database every year. The database can be accessed through computer networks or can be obtained on CD-ROMS. A similar high quality database is the crystallographic structural database or the CSD at the University of Cambridge, UK. This contains more than 4,50,000 entries of the co-ordinates of the crystal structures of small organic molecules, i.e. those with a molecular weight less than about 1000.

Table 9.1: Chou-Fasman propensity values of the 20 amino acid residues for each of three secondary structures.			
Residue	α-helix	β-strand	reverse-turn
Ala	1.45	0.97	0.15
Cys	0.77	1.30	0.31
Asp	0.98	0.80	0.33
Gln	1.53	0.26	0.12
Phe	1.12	1.28	0.19
Gly	0.53	0.81	0.45
His	1.24	0.71	0.18
Ile	1.00	1.60	0.15
Lys	1.07	0.74	0.27
Leu	1.37	1.22	0.14
Met	1.20	1.67	0.18
Asn	0.73	0.65	0.45
Pro	0.59	0.62	0.41
Glu	1.17	1.23	0.15
Arg	0.79	0.90	0.27
Ser	0.79	0.72	0.41
Thr	0.82	1.20	0.27
Val	1.14	1.65	0.08
Trp	1.14	1.19	0.30
Tyr	0.61	1.29	0.33

Chou and Fasman used the protein structural database to estimate the frequency with which each amino acid occurred in one of the three main secondary structural motifs, i.e. α-helix, β-strand and reverse-turn. The normalised values that they obtained are given in Table 9.1 The numbers represent the propensity for each amino acid to take up the respective conformation. Thus glutamine has a very high propensity to occur in a helix, i.e. to be a helix-former. Similarly isoleucine strongly prefers to exist in a β-strand. The propensities for the reverse turn are not as well expressed. However, it may be noticed that some residues like glycine or asparagine have a fair degree of propensity for the reverse turn, without a strong preference either for the α-helical or β-strand conformation. The way this table is used to predict the secondary structure of a polypeptide chain is illustrated by means of an example.

Consider a 16-residue sequence such as shown in Figure 9.2. The α-helical, β-strand and reverse turn propensities for each residue is also indicated as taken from Table 9.1. Also indicated in the figure are the average values of the propensities taken six at a time. (Hexa-peptides are the normally used probe length, though other lengths also may be used. For the β-strand especially the probe length can be as short as five peptides). An examination of the averages shows that the average strand propensity is the highest for the hexa-peptide that starts at residue 1. A β-strand can be considered to nucleate at

Sequence	1 Lys	2 Ile	3 Thr	4 Val	5 Val	6 Gly	7 Ala	8 Gly	9 Ala	10 Asn	11 Gly	12 Leu	13 Ala	14 Gln	15 Ala	16 His
Propensities																
α-helix	1.07	1.00	0.82	1.14	1.14	0.53	1.45	0.53	1.45	0.73	0.53	1.37	1.45	1.53	1.45	1.24
β-strand	0.74	1.60	1.20	1.65	1.65	C.81	0.97	0.81	0.97	0.65	0.81	1.22	0.97	0.26	0.97	0.71
β-turn	0.27	0.15	0.27	0.08	0.08	C.45	0.15	0.45	0.15	0.45	0.45	0.14	0.15	0.12	0.15	0.18
Average Propensity (window = 6)																
α-helix	1.01	1.03	0.95	0.96	0.91	0.97	0.92	0.82	0.96	1.01	1.18	1.18	1.26	1.41	1.42	1.41
β-strand	1.30	1.37	1.28	1.43	1.30	1.26	1.09	0.95	1.02	0.91	0.81	0.81	0.82	0.83	0.65	0.65
β-turn	0.19	0.17	0.22	0.19	0.24	0.22	0.28	0.34	0.29	0.30	0.24	0.24	0.20	0.15	0.15	0.15

Figure 9.2: Secondary structure prediction of a 16-mer oligopeptide using the Chou-Fasman method.

this point, since the average propensity for an α-helix is rather low at the same point. The strand can be extended in both directions until a region is reached where the helix is a more probable structure. Thus on comparing the propensities for both types of structure it can be determined that the residues 1 to 6 are most likely to be β-strand, while residues 11 to 18 are most likely to be α-helical. The determination of possible reverse turns is more complicated since each residue has a propensity value for its occurrence at each position of the reverse turn. (A reverse turn has four residues i.e. four positions). The principle however is the same, and, after a thorough analysis, the entire polypeptide chain can be classified as α-helix, β-strand, reverse turn or none of the above, i.e. random coil. Such methods of secondary structure prediction are now available as easy-to-use computer programs. Given the sequence of a protein one can use the program to quickly predict its secondary structure.

The problem of arriving at the structure of the protein from knowledge of its sequence is nevertheless far from solved. Firstly the secondary structure prediction methods, such as the Chou and Fasman method are still quite approximate, and all the regular structural elements of the protein chain cannot be identified. Predictions made have often turned out to be wrong and the overall rate of success does not exceed 80% in the best test cases. Methods are still evolving and combinations of methods are expected to have much higher success rates.

The other major obstacle in protein structure prediction is that the arrangement of secondary structural elements in space follows rules that are not completely understood. This comes in the way of arriving at the tertiary fold of the chain from knowledge only of the sequence. The prediction of the three dimensional structure of a series of proteins with homologous sequences is much easier if the structure of one of them is known. Thus almost the first task that is done when the sequence of a protein is known to compare it with all other known protein sequences and identify those which are similar. If the structure of one of the similar proteins is known through other experimental or theoretical techniques, it can be used as a guide in generating the structure of the new protein. The principle which is followed is that stretches of similar sequences must have similar or the same secondary structure. If the two sequences show a high degree of overall similarity (homology), then the tertiary fold of the two sequences can also be taken to be similar, and the co-ordinates of the unknown protein can be generated by appropriately modifying the co-ordinates of the known protein. This method of generating protein structures is especially used to build the initial trial model in the molecular replacement method in X-ray crystal structure analysis.

9.2.3: Building nucleic acid structure:

To build a DNA double helix is relatively simple as compared to building a protein chain. All one needs are the co-ordinates of one nucleotide unit and the helical transform that generates the next repeat along the strand. A rotation about the two-fold axis perpendicular to the helix axis generates the other strand. Monomer co-ordinates for A, B and Z type DNA are available in the literature. Other unusual, regular DNA structures such as tetraplexes, triplexes, etc., can also be generated by following the same principle. "Real" DNA molecules are not perfectly regular and show a sequence dependent variation from the standard structure at each base-pair step. Since no data on the correlation between the sequence and the microstructure exists, it is difficult to build models that show this variation. Thus most DNA models are regular helices.

Models of RNA structures are much more difficult to build. Only a few structures of large single stranded RNA molecules exist and the principles of RNA structure are not well known. An analysis of the sequence of the RNA molecule, if known, enables prediction of secondary structure, specifically regions of the molecule which base pair with each other and therefore can be thought to exist as short double-helical stretches. The way in which these double helices are put together to form the three-dimensional RNA structure is an unsolved problem.

9.2.4: Displaying and altering the generated model:

There are many ways of representing the three dimensional molecule on the two dimensional computer display. Choosing the appropriate mode of representation depends on the particular information that is required to be emphasised.

Small molecules are typically represented by the so-called ball and stick model (Figure 9.3), where each atom is shown as a sphere and each covalent bond as a stick. The radius of the sphere is commonly taken to be proportional to the van der Waals radius of the respective atom. In colour representations, a commonly used code is hydrogen – white, carbon – black, nitrogen – blue, oxygen – red, phosphorous and sulphur – yellow. If the anisotropic thermal vibrations of the atom also need to be represented, then ellipsoids are used instead of spheres to depict the atoms. The radius of the atoms may be increased to an extent where the sticks representing the bonds are hidden. Such a model is called a space-filling representation. A chemically accurate version of a space-filling model was first designed in plastic by Corey, Pauling and Koltun and is called the CPK model and may be thought to represent the electron cloud of the molecule. A computerised version of this draws the CPK model on the screen. Large molecules, such as most proteins, are effectively seen when drawn as line diagrams rather than ball and stick models. Even then, there is too much detail to be fully appreciated (Figure 9.4). A further abstraction is to plot and join only the C^α atoms (Figure 9.5). It may be sufficient to show only the atoms of backbone, if the interest is only in the protein fold. Another way of getting over the problem of information overload in the diagram may be to show the picture in stereo (Figure 9.5). Cartoon diagrams (Figure 9.6) are useful to show the disposition of the secondary structural elements. It is conventional to depict α-helices as cylindrical rods and β-strands as flat arrows. If the interest is only in the surface of the protein, diagrams such as Figure 9.7 may be used.

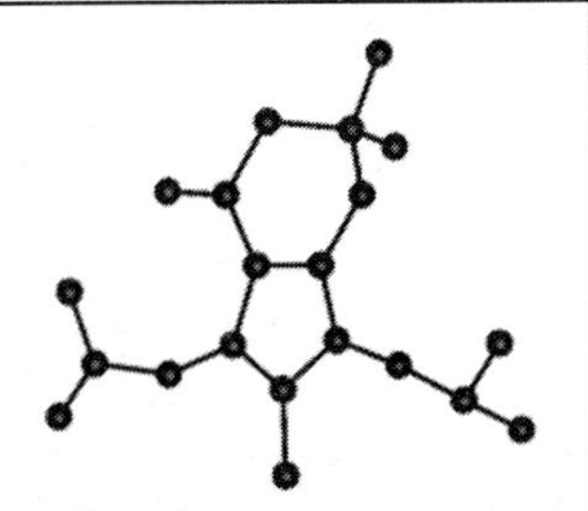

Figure 9.3: A ball-and-stick representation of an indole molecule. Here, all the atoms are represented by spheres of equal radius.

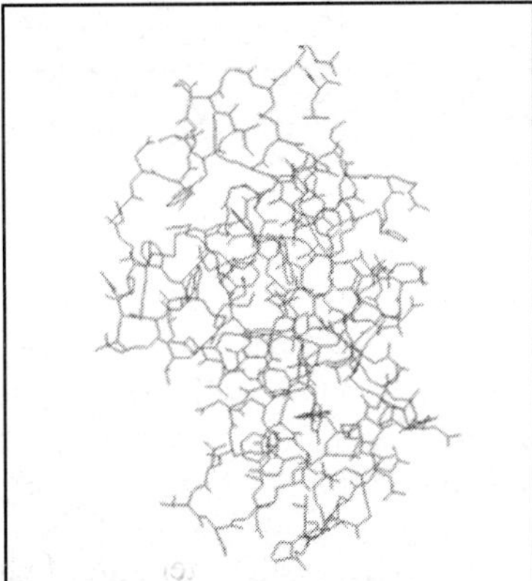

Figure 9.4: Line diagram of lysozyme.

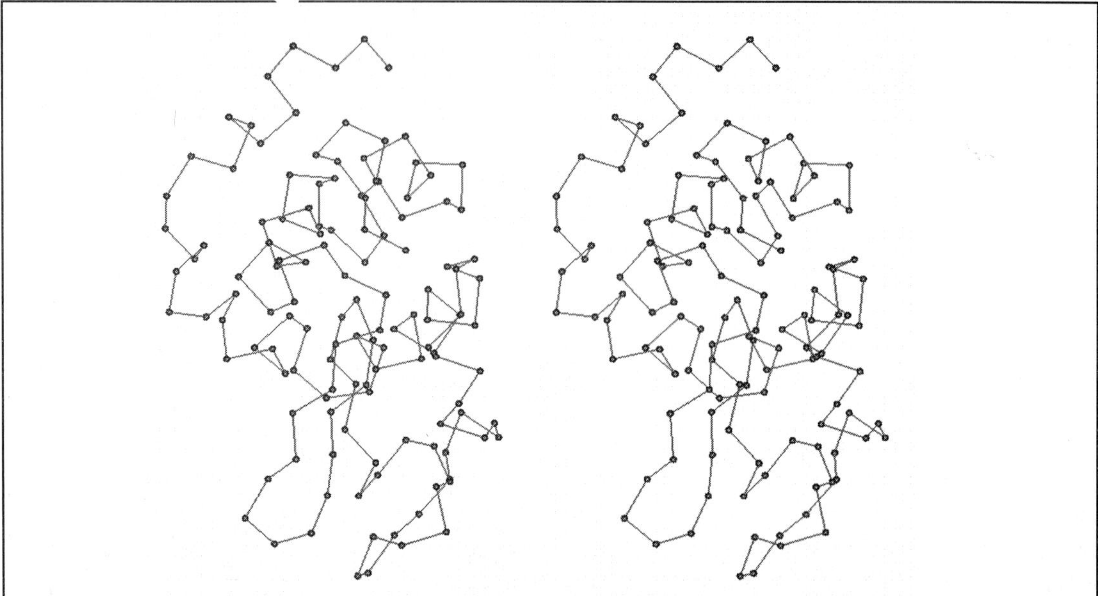

Figure 9.5: Stereo picture of the C^α trace of a protein. [Though only a two-dimensional image of the object falls on each retina, the brain is able to synthesize the three-dimensional picture from the two images since they are produced by views at slightly different angles. Stereo pictures mimic this effect by providing views of the molecule as seen from two different angles. In order to obtain the 3-D effect it is necessary to allow the two eyes to simultaneously focus only on one picture each, to the exclusion of the other. This may be done by placing a card between the eyes and blocking out the second image. Many people are able to obtain the 3-D effect without any such aids. The reader may try to stare at the example given above, focusing and defocusing his/her eyes until he/she obtains the 3-D effect. The first attempt may take a couple of hours before it is successful.]

9.3: Optimising the model

The generated molecule needs to be checked for compliance with the laws of physics and chemistry before it can be accepted for further processing. The stereo-chemical criteria used during the construction of the model are usually insufficient to decide upon the final correct structure. The dynamics of the molecule, i.e. the various forces acting on various parts of it have to be explicitly calculated and the structure has to be modified to optimise the action of these forces. A simple first step in the structure optimisation is to ensure that the bond lengths and bond angles fall within the ranges measured in accurate spectroscopic or X-ray diffraction experiments. Atomic sizes are also known from experiment and the fold of the generated molecule should not be such that the distance between two non-bonded atoms is less than the sum of the atomic radii, i.e. there should be no inter-penetration of the atomic spheres. These chemical and physical data can be checked and optimised using computer programs. It is usually not possible to exactly match the requirements.   However the

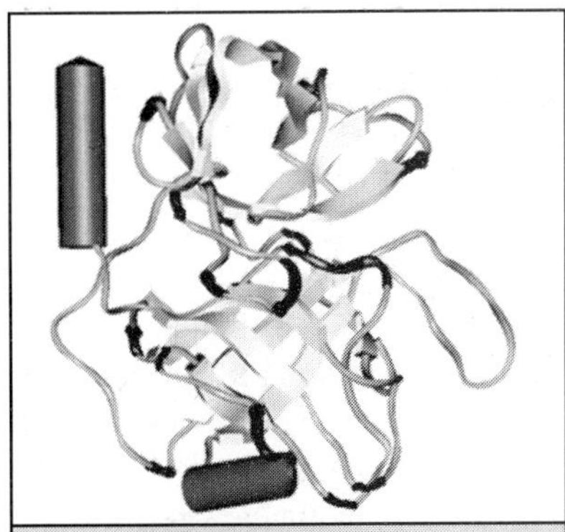

Figure 9.6: A cartoon diagram of a protein.

difference between the expected and observed geometrical values is minimised using least squares techniques.

Such geometrical optimisation is sufficient to give a plausible model. It is however not sufficient to serve as a basis for new interpretation regarding the structure-function relationships in the molecule, i.e., no new 'science' can be expected from it. Energy minimisation and structural optimisation using molecular dynamics are at the next higher level of modelling. The free energy of the molecule is an experimentally measurable thermodynamic quantity. This quantity can also be computed. A very important physical principle requires that at the equilibrium state of any system, the free energy be the least. In other words, the minimum energy state is the equilibrium state and all systems tend to move towards this state. The process of structure optimisation is used to determine the minimum energy configuration. In order to calculate the free energy for various states or various conformations of the molecule it is necessary to set up the quantum mechanical equations for the whole molecule and to solve them. This task is beyond the scope of currently available mathematical and computational techniques. However the value of the energy can also be calculated from a semi-empirical expression such as the following.

$$E_{tot} = E_{bond\text{-}length} + E_{bond\text{-}angle} + E_{tor} + E_{non\text{-}bonded} + E_{elec} + E_{hydrogen\text{-}bond}$$

where E_{tot} is the total energy. $E_{bond\text{-}length}$ is the contribution to the energy due to the deviation of the bond lengths from their expected values. $E_{bond\text{-}angle}$ is the contribution due to the deviation of the angles from their standard values. E_{tor} is the contribution due to the torsion angles and is dependent on the disposition of the four atoms that make up the torsion angle (e.g. an eclipsed configuration has a very high energy). $E_{non\text{-}bonded}$ is the contribution due to the van der Waals contacts between the atoms. Interpenetration will raise the energy enormously. E_{elec} is the contribution due to the electrostatic interaction of attraction and repulsion between the atoms. Even though the molecule as a whole may be electrically neutral, quantum mechanics tells us that each atom will posses a 'partial' charge, positive or negative, and it is these partial charges which take part in the electrostatic interactions. This is in addition to interactions between charged residues such as Glu or Asp. Finally $E_{hydrogen\text{-}bond}$ is the energy contribution due to hydrogen bonds. Other terms, such as one for the 'anomeric effect' are sometimes added to this expression for the energy. We note that this is the expression for the potetial energy and not the free energy. However for all computational purpose, we may use this expression, only bearing in mind that in comparing the values obtained with experimental values, the appropriate conversions are required.

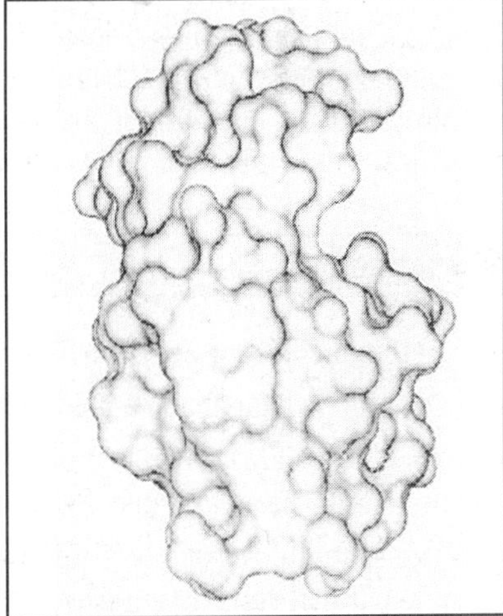

Figure 9.7: Surface representation of lysozyme.

The conformation of the molecule is optimised with respect

to the energy. In other words, the conformation is changed so that the value of the energy is a minimum. The minimisation is carried out by some well-known mathematical techniques such as least squares or gradient search methods. Such a procedure will give the final 'correct' conformation of the molecule. In addition new interactions and structures not built into the original model may appear.

The molecule may also be subjected to the molecular dynamics calculations to study its dynamical behaviour. In this procedure each atom is thought to be moving randomly at a high velocity, depending on the temperature used in the calculation. Each atom is influenced by forces such as bonding forces, angle distortion forces, etc. Since the force on each atom depends on the positions of all the other atoms, and since the positions are continuously changing, the technique is carried out in iterative cycles. The first step is to assign the velocities to all the atoms and calculate the initial force felt by each atom in the initial configuration of the molecule. The next step is to calculate the new positions of the atoms after a very short time interval, usually of the order of 0.01 picoseconds (1 picosecond = 10^{-12} second). The new positions of the atoms are then used to calculate the new force field. Thus the motion of the various parts of the molecule can be followed. The results are usually checked every 100 steps or so and give 'snapshots' of the molecule at intervals of about 1 picosecond. The simulation usually cannot be carried on for longer than about 5000 picoseconds for two main reasons: a) Enormous amounts of computer time are necessary even on the fastest computers. b) The force fields are not realistic enough and the molecule tends to assume unrealistic configurations if the simulation is carried on for longer than about 5000 picoseconds. However, large advances have been made in the past five years both in computer technology and in modelling the force-field accurately. This has allowed several interesting simulations of small proteins and DNA fragments to be carried out for as long as 1 millisecond or 10^6 picoseconds. The great advantage of the molecular dynamics technique is that it allows a much more natural picture of the molecule. Molecules are always in motion, but X-ray crystallography and many other techniques give only a static picture.

Apart from building models of molecules, molecular modelling, optimisation and molecular dynamics are used to model interactions between molecules. Successful applications include studying the docking of proteins to substrates, virus assembly, etc. Modelling studies on drug-DNA interactions and drug-protein interactions are used extensively in the rational design of new drugs. Many of the drugs now in the market are the result of such studies.

Summary

- Before the advent of powerful computer graphics, molecular models were built of wood or plastic.

- Computers have made a great deal of difference to molecular modelling.

- The co-ordinates of the molecule are fed into a computer program, which interprets each data point as an atom and displays it according to the mode and orientation chosen by the user.

- The spatial positions (co-ordinates) of all the atoms in small molecule can be obtained by simple geometrical considerations.

- This requires knowledge of the internal parameters of the molecule, i.e. the bond lengths, the angles and the torsion angles.

- However building the structure of a protein chain *ab initio* is not a trivial task since the torsion angles are not known.

- In principle, knowledge of the amino acid sequence of a protein should lead to its final three-dimensional structure.

- However, the rules of protein folding are still unknown and it is not yet possible to go from the sequence to the structure computationally.

- The Chou and Fasman method has been quite successful in predicting the secondary structure of proteins.

- Chou and Fasman used the protein structural database to estimate the frequency with which each amino acid occurred in one of the three main secondary structural motifs.

- This table of the propensity values is used to predict the secondary structure of a polypeptide chain.

- However these secondary structure prediction methods are still quite approximate.

- The major obstacle in protein structure prediction is that the arrangement of secondary structural elements in space follows rules that are not completely understood.

- The prediction of the three dimensional structure of a series of proteins with homologous sequences is much easier if the structure of one of them is known. The procedure is termed 'homology modelling'.

- To build a DNA double helix one needs the co-ordinates of one nucleotide unit and the helical transform that generates the next repeat along the strand.

- Models of RNA structures are much more difficult to build, since the principles of RNA structure are not well known.

- There are many ways of representing the three dimensional molecule on the two- dimensional computer display.

- These include stick models, ball-and-stick models, CPK space-filling models, etc.

- The generated molecule needs to be checked for compliance with the stereochemistry and geometry before it can be accepted for further processing.

- The difference between the expected and observed geometrical values is minimised using least squares techniques.

- Energy minimisation and structural optimisation using molecular dynamics are at the next higher level of modelling.

- The value of the energy can be calculated from a semi-empirical expression such as
$$E_{tot} = E_{bond-length} + E_{bond-angle} + E_{tor} + E_{non-bonded} + E_{elec} + E_{hydrogen-bond}$$

- The conformation of the molecule is optimised with respect to this energy.

- Molecular modelling, optimisation and molecular dynamics are also used to model interactions between molecules.

Chapter 10

Macromolecular Structure

10.1: Introduction

The central paradigm of biophysics, indeed of physics in general, is the following.

Structure ➜ Function

Thus in order to understand and perhaps modify the behaviour of biomolecules, it is necessary to know and understand their structure. One of the most dramatic illustrations of this principle came with the discovery of the double-helical structure of DNA. It was immediately clear how DNA could function as the storage and carrier of genetic information. The vast advances in biotechnology since then stem almost entirely from this single discovery. Protein structure and virus structures also illustrate the truth of this principle. In this chapter we shall consider in detail the structural principles of some the important classes of biological molecules.

10.2: Nucleic acid structure

The storage, transmission and expression of genetic information in the cell is carried out by nucleic acids. Deoxyribonucleic acid or DNA is responsible for storing genetic information and passing it on to the progeny cells. DNA is therefore involved in replication and transcription. The other type of nucleic acid present in the cell is ribonucleic acid or RNA. It is involved in many different functions. The genetic information in the DNA is transcribed into messenger RNA or mRNA, which is then translated into the protein in the ribosome. Transfer RNA (tRNA) molecules mediate this process. RNA is also a part of the ribosome and is then called ribosomal RNA or rRNA. Small RNA molecules (~100 nucleotides) can fold into structures called ribozymes, which have catalytic activity. During each of these different functions the nucleic acid molecules have different three-dimensional structures.

10.2.1: The chemical structure of nucleic acids:

Both DNA and RNA are polymers. The repeating unit, i.e. the monomer in the case of DNA is a deoxyribonucleotide, and a ribonucleotide in the case of RNA. A mononucleotide can be thought as made up of three parts, a) a phosphate group, b) a sugar moiety and c) a nitrogenous base.

The phosphate group is simply PO_4^-, as illustrated in Figure 10.1. When it is a part of the polynucleotide, the single extra negative charge is distributed between the two so-called 'pendant' oxygen atoms. It is this negative charge that is responsible for the acidity of nucleic acids. The phosphate group is a rigid unit and has no internal flexibility. The four oxygen atoms occupy the corners of a tetrahedron, with the phosphorous atom at the centre. When the group becomes a part of a polynucleotide, the geometry changes only slightly, with the lengths of the P-O bonds increasing by about 0.2 Å.

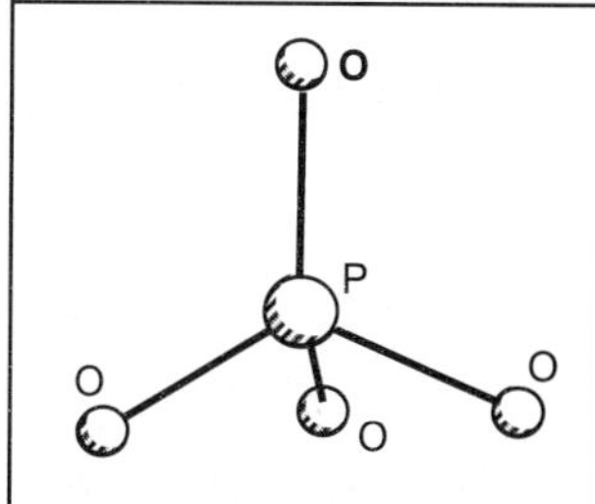

Figure 10.1: The chemical structure of the phosphate group.

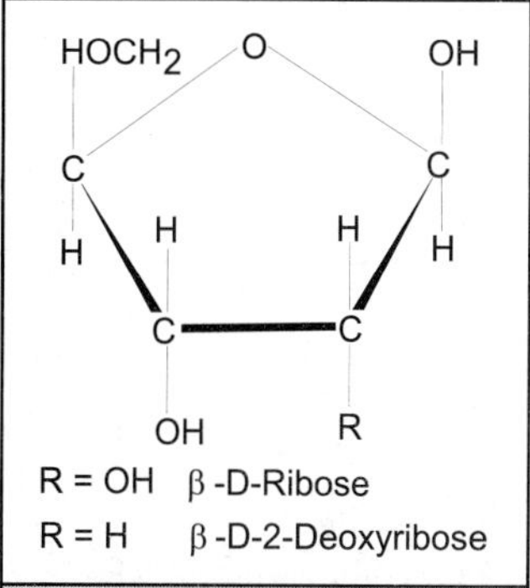

Figure 10.2: The chemical structure of Furanose.

The sugar moiety consists of a five-membered furanose ring with exocyclic alcohol and hydroxyl groups (Figure 10.2). If both the 2' and 3' position have a hydroxyl attached, the sugar is ribose (RNA). If there is a hydroxyl group only at the 3' position it is deoxyribose (DNA). Though it is cyclised, a substantial amount of conformational variation can still take place. Rotation about the C4'-C5' bond produces the three preferred conformations shown in Figure 10.3. In addition, the five-membered ring is not planar, but puckered.

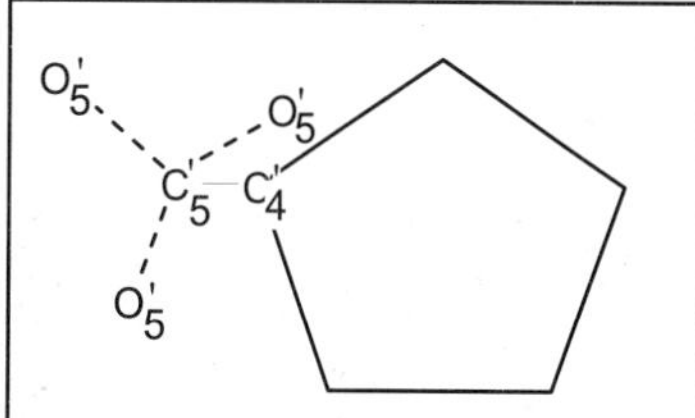

Figure 10.3: The preferred conformations about the C4'-C5' bond.

Due to the sp³ hybridised bonds at the carbon atoms of the ring, the energetically favourable conformation is one in which only three (or four) of the five atoms lie in the same plane. Further, in order to avoid short contacts between the pendant hydrogen atoms, the best way to relieve such energy strains is to push the C2' atom or the C3' atom out of the plane (Figure 10.4). Thus, in nucleic acids, one comes across C2'-endo or C3'-endo sugar puckers, indicating that the displacement of the atom out of the plane of the other four atoms is in the same direction as the C5' atom. If the displacement is in the direction opposite to the C5' atom, the conformation is termed C2'-exo or C3'-exo.

The planar bases constitute the chemically variable portion of nucleic acids. Bases can be either pyrimidines or purines (Figure 10.5). The bases guanine, cytosine, adenine and thymine are

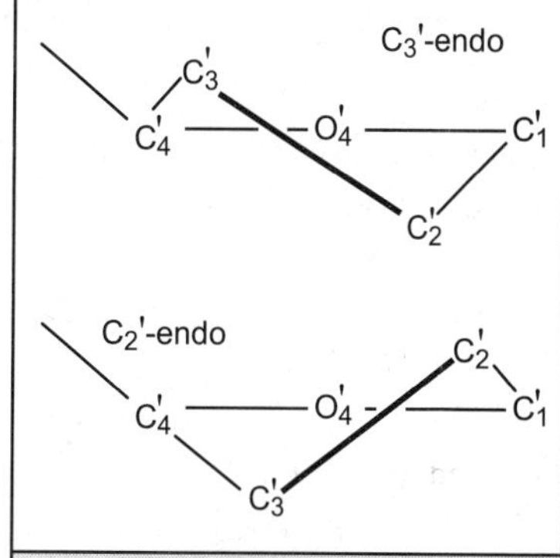

Figure 10.4: The puckering of the sugar ring.

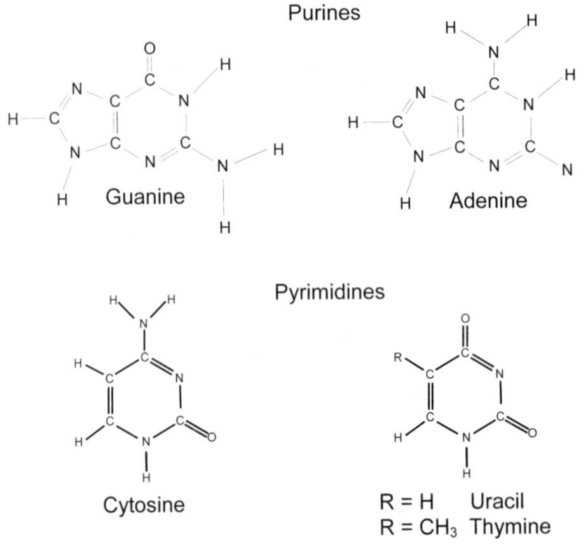

Figure 10.5: The chemical structure of the nucleic acid bases.

found in DNA. In RNA the pyrimidine thymine is replaced by uracil. Like the phosphate group, the bases have no internal conformational flexibility.

10.2.2: Conformational possibilities of monomers and polymers:

A nucleoside is formed by the covalent attachment of one of the bases to a sugar moiety. The bond is called a glycosidic bond and occurs between the C1' atom of the sugar, and either the N9 atom of a purine or the N1 atom of a pyrimidine. The plane of the base is almost perpendicular to the approximate plane of the sugar ring. As compared to its components considered separately, a nucleoside has additional conformational possibilities by way of rotation about the glycosidic bond leading to the *anti* or *syn* orientations of the base with respect to the furanose (Figure 10.6).

The formation of a nucleotide by the addition of a phosphate group to the sugar, at the 5' end or the

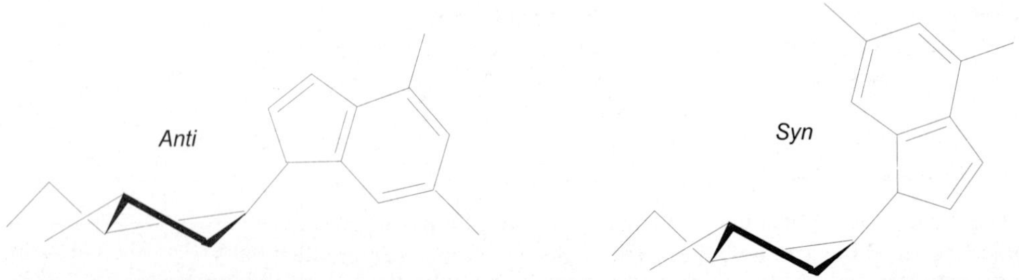

Figure 10.6: The conformation of the base about the glycosidic bond.

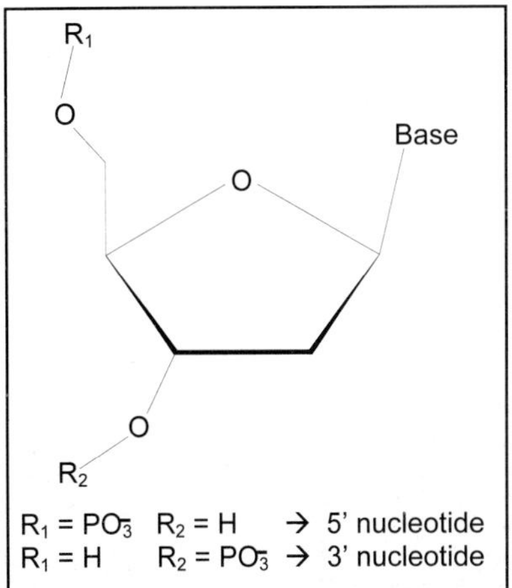

3' end (Figure 10.7), further increases the conformational possibilities. However, model building, theoretical calculations and experimental evidence from NMR and X-ray crystallography have shown that all conformations are not equally likely.

Mononucleotides link to each other through phosphodiester bonds to form the polynucleotide chain (Figure 10.8). In a dinucleotide there are a total of seven bonds about which rotation is possible. In addition, the sugar puckering is also a variable. However, only a few combinations of these angles lead to regular structures when repeated along the chain. Upon formation of the polynucleotides, certain new chemical interactions are introduced which do not exist in the mononucleotide. The most powerful of these are the stacking interactions between the planar bases. These interactions occur between the π-electron clouds, and lead to the bases being arranged in parallel planes, one over the other like a stack of coins. The inclusion of such interactions in the calculation of the optimum three-dimensional structure results in a further decrease in the number of possibilities. Some of these structures are discussed below.

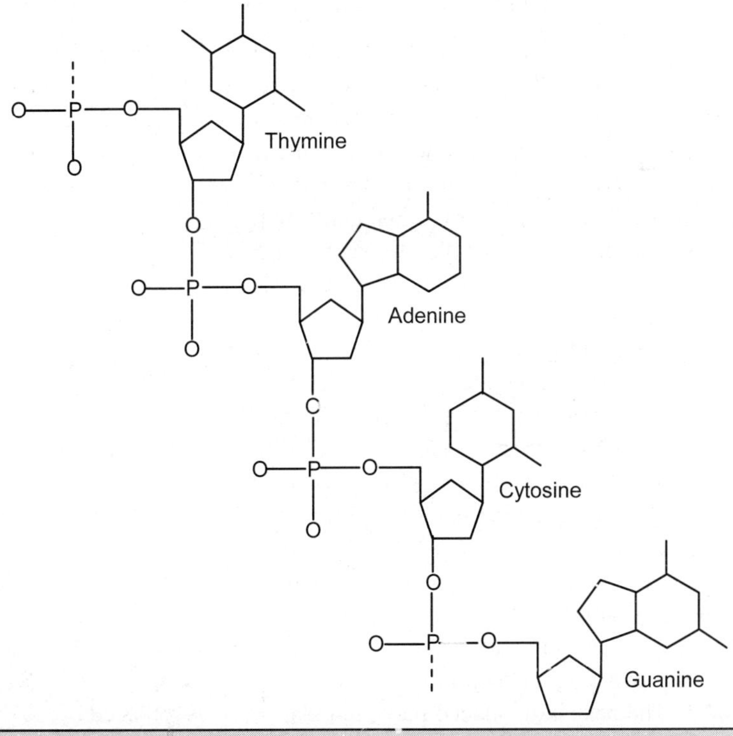

$R_1 = PO_3^-$ $R_2 = H$ → 5' nucleotide
$R_1 = H$ $R_2 = PO_3^-$ → 3' nucleotide

Figure 10.7: A nucleotide unit.

Figure 10.8: The chemical structure of a polynucleotide chain.

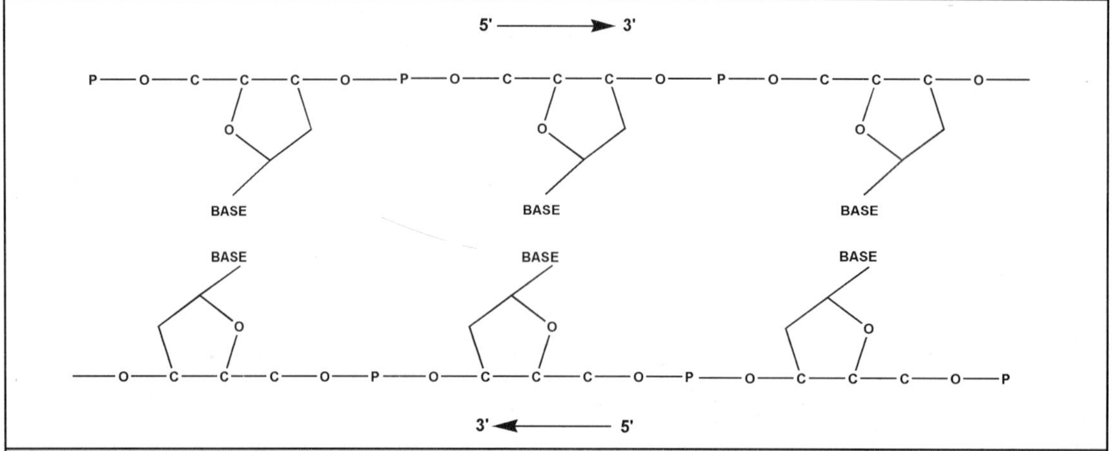

Figure 10.9: Antiparallel arrangement of the two nucleic acid strands.

10.2.3: The double helical structure of DNA:

DNA in the cell is one of the largest molecules found in nature. It is a polymer whose length can range from a few thousand nucleotides in bacteria to more than a billion (10^9) nucleotides in cells of higher mammals. The discovery of the double helical structure of DNA in 1953 by J.D. Watson and F.H.C. Crick was one that marked an epoch and is the basis for all of modern biotechnology.

It had been noticed that in the DNA molecules, the number of adenine nucleotides was exactly the same as the number of thymine nucleotides and the number of guanine nucleotides equalled the number of cytosine nucleotides. Furthermore, X-ray diffraction experiments on DNA fibres had suggested that each DNA molecule consisted of two strands which wrapped around each other to form a helix and that these two strands were arranged antiparallel to each other (Figure 10.9). (Note that the attachment of the phosphate group at the 5' position of the sugar ring is different from its attachment at the 3' position. This gives rise to directionality in the polynucleotide chain). The concept of base pairing explained the base ratio by suggesting that adenine formed specific hydrogen bonds with thymine, and guanine with cytosine (Figure 10.10). Other hydrogen bonding schemes between the bases were much less favourable. Additionally, both the A:T and the G:C base pairs were of the same size. These facts led to the proposal of the structure shown in Figure 10.11. One may visualise the DNA double helix as a flexible rope ladder wrapped around a central pole. The sugar phosphate chains of the two polynucleotide backbones form the sides of the ladder. The base pairs placed perpendicular to the central helix axis form the rungs. The geometrical details

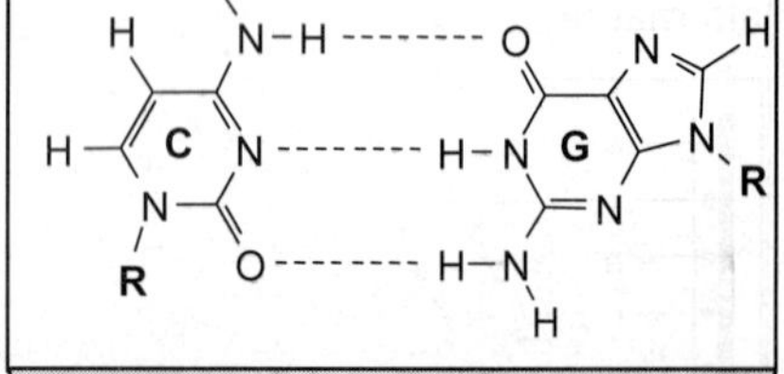

Figure 10.10: Base pairing in nucleic acids (DNA). In RNA, T is replaced by U.

of the structure are given in Table 10.1.

The complementary base-pairing scheme immediately suggests how genetic information is stored and passed on from generation to generation. If the information is 'written' as the sequence of bases, then it may be seen that when the two strands of the DNA molecule separate, if a complementary strand is built on each of the two strands, it results in two identical copies of the original double helical molecule. This can be repeated again and again and each time the cell divides, and in each daughter cell the DNA would be identical to that in the parent cell (Figure 10.12).

10.2.4: Polymorphism of DNA:

The structure of DNA proposed first by Watson and Crick explained the X-ray diffraction patterns produced by 'wet' DNA fibres or the 'B' type patterns. This model has thus come to be called B-DNA. Fibres at much lower humidity produced another pattern, with sharper spots and therefore indicating a higher degree of order. They were called the A type patterns and Watson and Crick proposed a somewhat altered model of the double helix to explain them (Figure 10.13). The differences in the helical parameters are quite substantial, as may be seen from Table 10.1. The crucial features that remain unchanged in the two models of DNA are their double helical nature, the antiparallel orientation of the strands and the base-pairing scheme. The A type structure also may be imagined as a rope

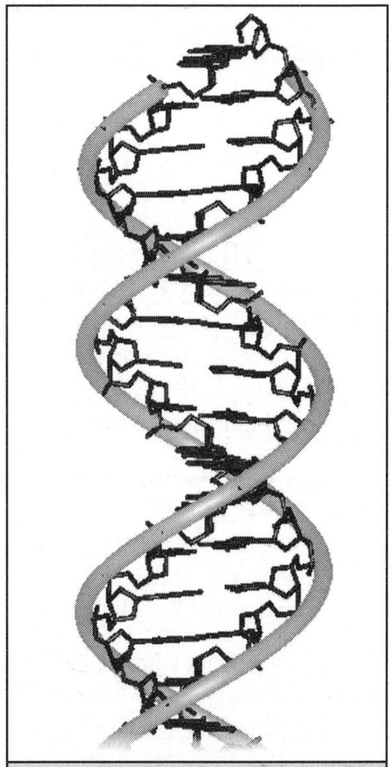

Figure 10.11: The double-helical, base-paired structure of DNA.

ladder wrapped around a central pole, but the imaginary pole is far thicker in this case. A way of recognising this is to observe the amount of displacement of the base pairs from the helix axis (Figure 10.14). In B DNA, the axis passes through the base pairs, while in A DNA, the axis is displaced by about 2 Å into the major groove. Also the B-type model is slimmer than the A-type model.

Both A and B type models are right-handed helices in which the helix proceeds as in a right hand screw. A dramatic example of the polymorphism of DNA was discovered in the left-handed Z type DNA (Figure 10.15). It was first observed in sequences of the type 5'...CGCGCGCG...3', i.e. alternating pyrimidine-purine. Circular dichroism and UV absorption studies indicated a change in the

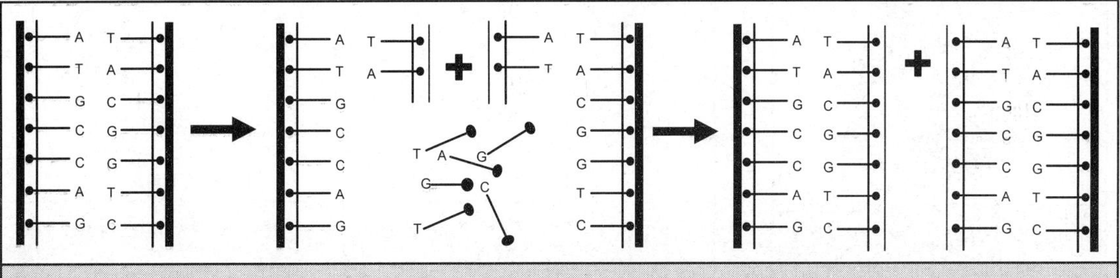

Figure 10.12: A schematic representation of how the double helix explains DNA replication.

Table 10.1: The structural parameters of three polymorphic forms of the DNA double helix.

	A	B	Z
Axial rise per residue (Å)	2.56	3.38	3.7
Twist angle per residue (°)	32.7	36.0	-30.0[*]
Pitch (Å)	28.2	33.8	45.0
Number of base pairs per turn	11	10	12
Handedness of the helix	right	right	Left
Number of nucleotides in the repeat unit	one	one	Two
Tilt of base pairs from helix axis (°)	20.0	-6.0	-7.0
Dislocation of base pairs into groove (Å)	4.4 minor groove	-0.14 major groove	-5.5[#] major groove
Major groove width, depth (Å)	2.7, 13.5	11.7, 8.5	8.8, 3.7[@]
Minor groove width, depth (Å)	11.0, 2.8	5.7, 7.5	2.0, 13.8

[*] Average of -60° for a dinucleotide repeat. The twist is actually -10 and -50° for the first and second steps respectively of the dinucleotide.

[#] This is an approximate value.

[@] This groove is filled up with exocyclic base atoms.

helical sense from right handed to left handed in sequences of this type. Later a single crystal structural study of the hexanucleotide 5'-CGCGCG-3' showed that, not only is the helical sense left-handed but also the phosphate groups follow a zigzag pattern. Hence it is called by the name Z DNA. The helical parameters of Z type DNA are also given in Table 10.1. It may be noted that left-handed Z DNA has been conclusively and positively identified only in vitro. Though antibodies raised against such sequences have been shown to cross-react with rat chromosomes, definitive evidence for a biological role for Z DNA is still lacking.

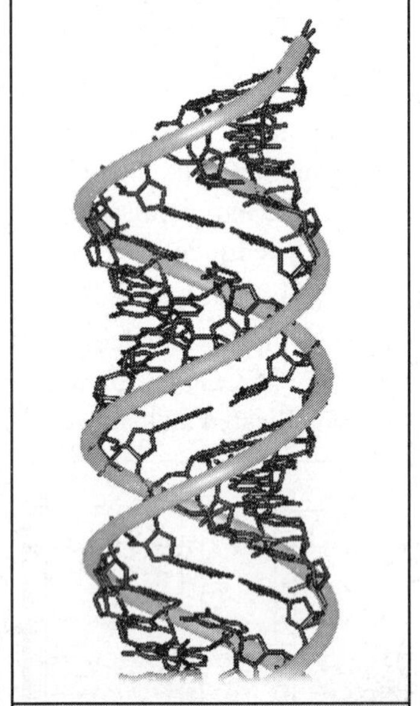

Figure 10.13: The A type double-helical structure.

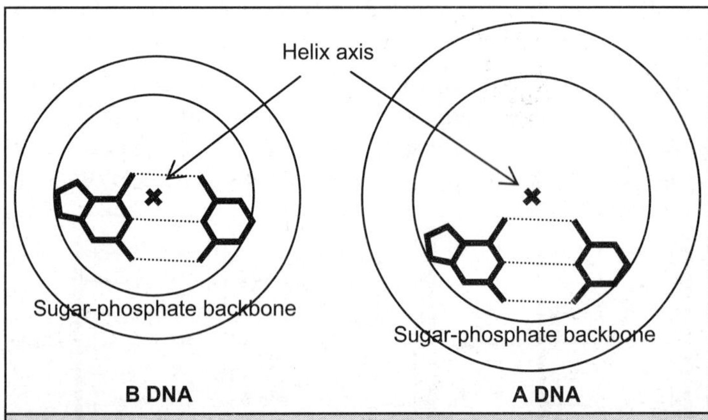

Figure 10.14: The displacement of the base pairs into the minor groove in A type and B type DNA.

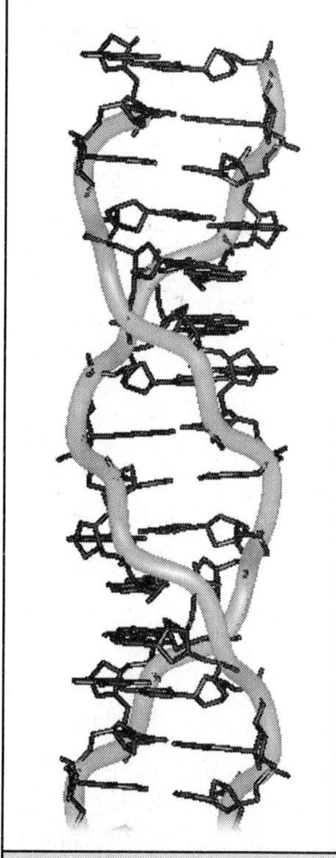

Figure 10.15:
Left-handed Z DNA.

Since 1978, single crystal studies of small DNA fragments have yielded high-resolution data on DNA structure. One of the most important findings is that, the structure of DNA deviates from the standard models described above in a sequence dependent manner. The regular models of A, B, and Z DNA, which are described above, and in Table 10.1, have been constructed in such a manner that, except for the change in the base, the structure of the molecule, especially the sugar-phosphate backbone, is exactly the same in one part of the molecule as in another. Given the co-ordinates of one monomer and helical transform it is possible to generate the co-ordinates for any length of the double helix.

The crystal structure results go against this and show a micro-heterogeneity, as illustrated in Figure 10.16. In this figure, the dark lines represent the regular helical model obtained from fibre X-ray diffraction. The light lines represent the structure of the same sequence of DNA determined by single-crystal X-ray crystallography. The micro level structure in each portion of the helix appears to depend on the sequence in that part, though the correlation between the sequence and structure is unclear. Sequence specific structure has implications for biology in that it could be another way of transferring information from the DNA molecule (Figure 10.17).

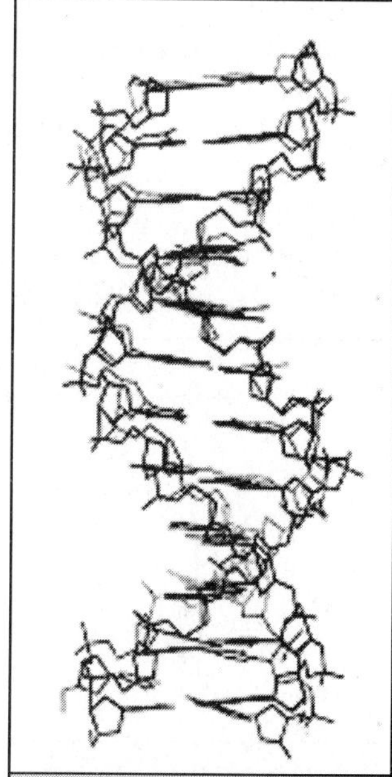

Figure 10.16: Micro-heterogeneity in DNA structure.

10.2.5: DNA supercoiling and unusual DNA structures:

As we have seen, owing to the nature of its stereochemistry, the energetically most favourable state of the DNA molecule is two strands coiled around each other. If, for some reason, the number of times

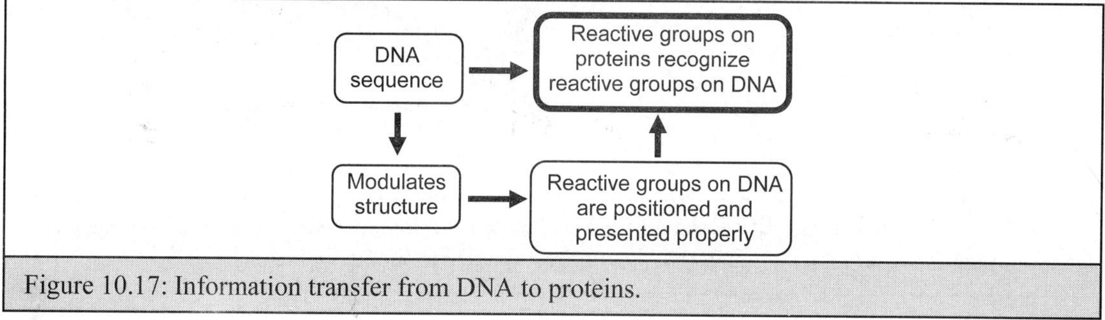

Figure 10.17: Information transfer from DNA to proteins.

one strand goes around the other is changed from the optimum value, supercoils (Figure 10.18) relieve the extra strain. Exactly the same effect may be seen, for example, in household electrical wires made of two or three twisted strands. In supercoiled DNA, two double helices further wind around each other, like two strands of a rope, which are themselves made of coiled fibres. Another way of relieving the strain is by increasing the so-called 'writhe', by which the DNA molecule wraps round and round an imaginary or real object. In these terms, for example, A-type DNA has a higher degree of writhe than B-type DNA. The structure of chromatin consists of DNA coils wrapped around histones (Figure 10.19). Each such histone-DNA unit is a nucleosome and in the chromatin state, the nucleosomes are arranged like beads on a string. In its more condensed state, the nucleosome 'necklace' is further wound around to form a solenoid-like structure.

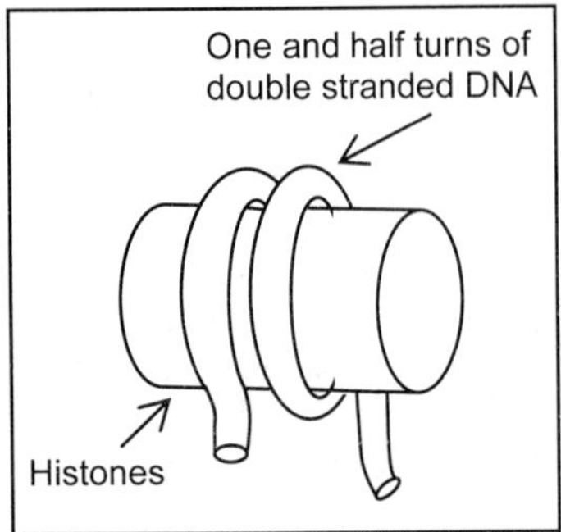

Figure 10.19: The nucleo-some core particle.

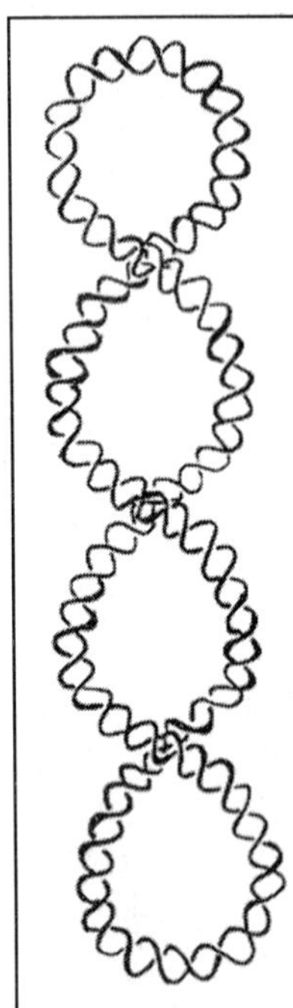

Figure 10.18: Supercoiling in DNA.

Apart from the double helix, certain DNA sequences have been shown to assume other types of structures. One of these is when stretches of guanine bases come together to form a G-quartet structure (Figure 10.20). Here the guanine bases form hydrogen-bonded tetrads that are stacked one over the other. The 'i'-motif is another unusual DNA structure (Figure 10.21a). It is formed by cytosines and has two distinctive features: (i) The cytosine-cytosine base pair. (ii) The intercalated structure of the two duplexes that make up the tetraplex. Other unusual DNA structures, observed using various biophysical techniques, include triplexes (Figure 10.21b), loops, cruciforms, slipped DNA, etc.

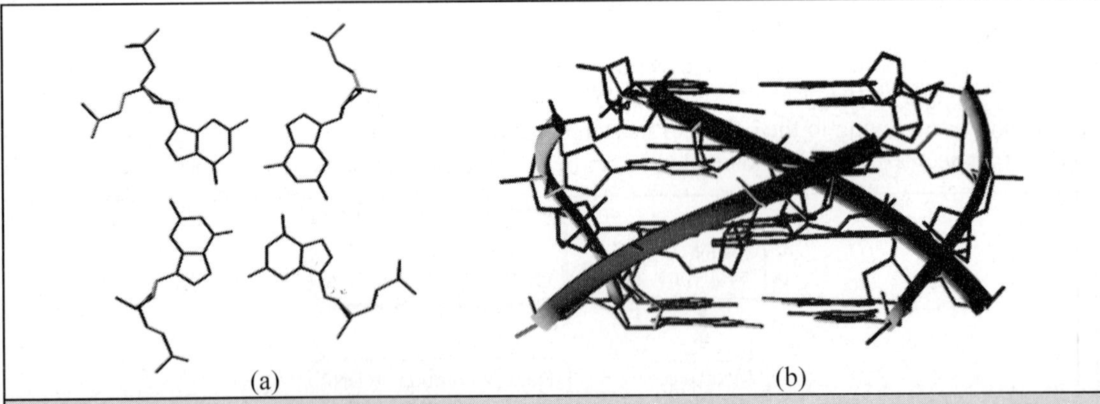

Figure 10.20: The structure of G-plates. (a) View down the helical axis showing the arrangement of four guanine residues in a plane. (b) View perpendicular to the helical axis.

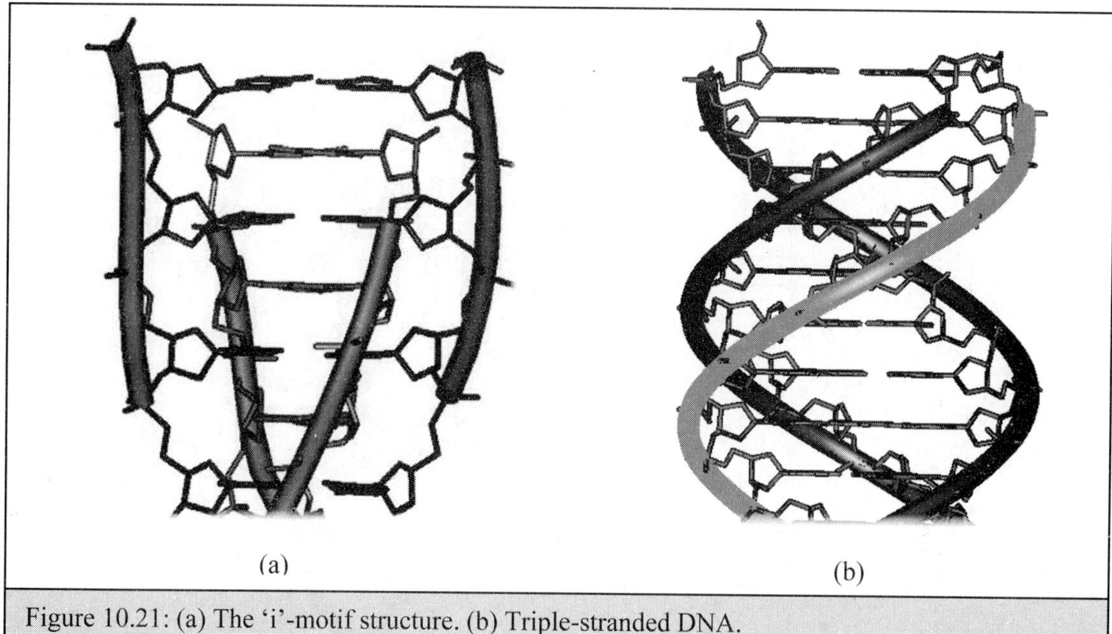

Figure 10.21: (a) The 'i'-motif structure. (b) Triple-stranded DNA.

10.2.6: DNA nanostructures

In recent years, DNA has been made to play a role that is quite different from its biological function of storing and transmitting genetic information. The rise of nanotechnology has lead to the possibility of using DNA as a structural molecule, to create and sustain useful structures and shapes at nanometre length scales. The DNA double helix is fairly rigid to a length of about 10 nanometres (or 100 Å – about 30 base pairs). Furthermore, the fact that it consists of two self-complementary strands makes it possible to create junctions, at which one stiff, double-helical rod may be linked up with another, through appropriately designed sequences (Figure 10.22). This basic structure may then be repeated, in two or three dimensions to create lattices and scaffolds. Such nanostructures may in the future be used

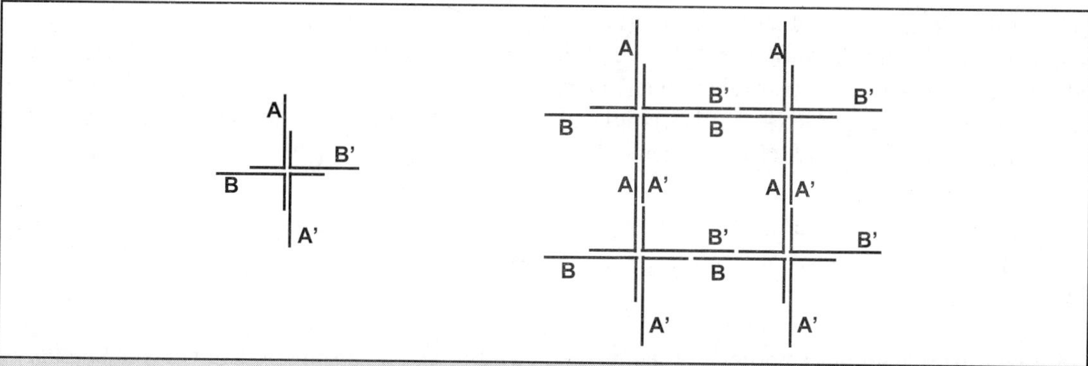

Figure 10.22: A square DNA nano-lattice. Four strands of DNA with appropriate sequences and 'sticky ends' (A, A', B and B') can join together by Watson-Crick base-pairing to build the lattice.

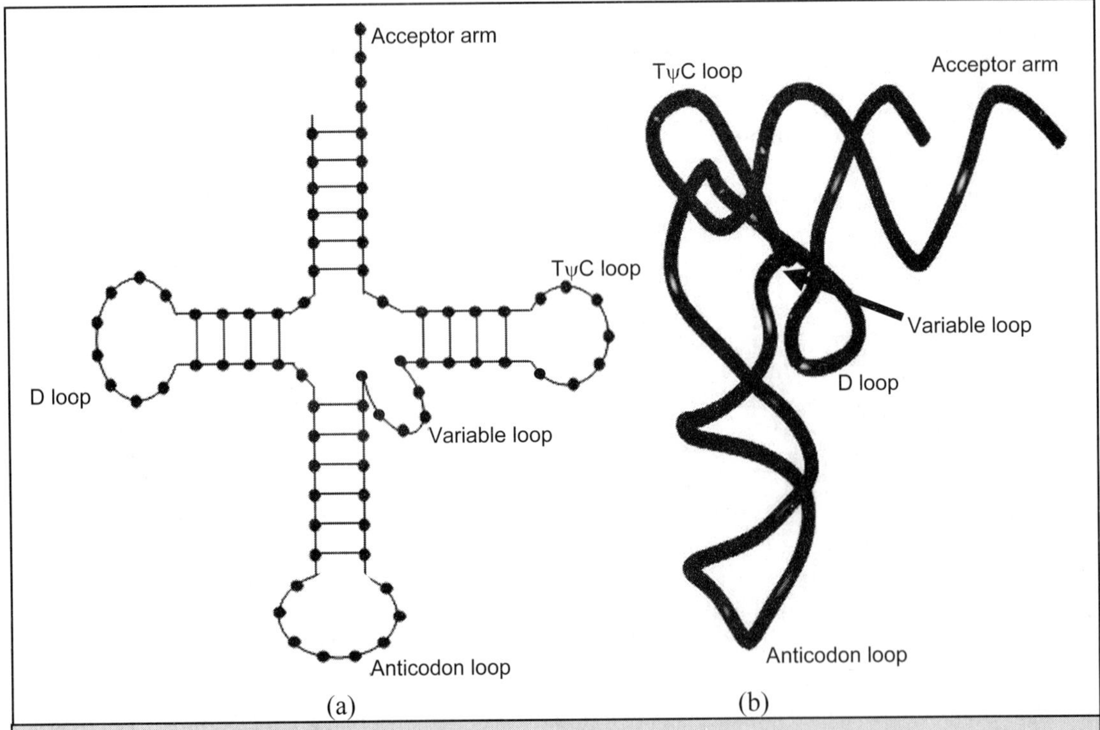

Figure 10.23: (a) The clover leaf secondary structure of tRNA. (b) The three-dimensional structure of tRNA. Only the trace of the sugar-phosphate backbone is shown.

to build precise molecular sieves. By attaching proteins and other molecules at appropriate places on the scaffolds, this may lead to nanometre scale 'factories'.

10.2.7: The structure of transfer RNA:

Most of the ribonucleic acid in the cell exists as rather short single strands, compared to the very long double stranded form assumed by DNA. For example, transfer RNA (tRNA) molecules are only about 70-80 nucleotides long. Some of these nucleotides consist of unusual bases such as psuedouridine and dihydrouracil. Messenger RNA (mRNA) molecules have varying lengths, depending on the code they carry. Ribosomal RNA (rRNA) molecules, which form a part of the ribosome, are also long nucleotides.

Until recently, only the structure of tRNA was known in detail. In the last decade detailed structures of a variety of other RNA molecules, including that of the entire ribosomal particle, have been determined through X-ray crystallography or NMR. Here we will describe tRNA structure. Being a single strand, the structure is not a double helix. However portions of the sequence can form Watson-Crick base pairs with other portions of the molecule and these regions are double helical. An analysis of the sequences of tRNAs suggests that they can be arranged to form four base-paired regions so that the general shape is that of a clover leaf (as seen in playing cards) (Figure 10.23a). Earlier theoretical and modelling studies, supported by diffraction and other physico-chemical studies, had shown that RNA double helices must necessarily be of the A-type. This is because steric hindrances due to the extra oxygen atom at the 2' position of the sugar which would prevent the other structural types of helices. In tRNA too, the base paired stems form short A-type helices. The three dimensional arrangement of these short helices and the extra loops is such that the structure looks like the letter 'L' (Figure 10.23b). The anticodon region, i.e. the portion that carries the code is at one end and the amino acid is attached at the other end.

10.3: Protein structure

Proteins are an important class of biological polymers. They are the molecules that actually do almost all the work in the cell and are largely responsible for expressing the information present in the DNA molecule. They range in size from a few kilodaltons to hundreds of kilodaltons. All proteins are built up from 20 amino acids, which constitute the monomers. The sequence in which these amino acids are arranged differs from protein to protein and is referred to as the primary structure of the protein. Some sequences of amino acids are known to arrange themselves into regular three-dimensional structures, that usually do not involve amino acid residues very far apart in sequence, and that are referred to as the secondary structure of the proteins. At the next level of structural organisation, the secondary structural units arrange themselves into globular shapes. This is called the tertiary structure and is stabilised mainly by interactions between the amino acid side chains. Thus the side chain conformations of the amino acids also form part of the tertiary structure. Also discernible at this level are domains corresponding to compactly arranged groups of secondary structural features. In the next higher level of structure, globular folded polypeptide chains come together in specific arrangements called the quatenary structure. Thermodynamic experiments by Anfinsen have shown that the functional three-dimensional structure of the protein is encoded in its primary amino acid sequence. Thus a polypeptide chain in solution will automatically fold into its three dimensional structure given only that the conditions of temperature, ionic strength, pH, etc., are within a fairly wide range. Ultimately the structure and therefore the function of proteins is directly derived from the chemical nature of the amino acids.

10.3.1: Amino acids and the primary structure of proteins:

Alpha amino acids can be considered as being chemically put together around the central carbon atom called C^α. Carbon atoms have a valency of four and in amino acids this valency is satisfied by an amino group, a carbonyl group, a hydrogen atom and a side chain. The only exception relevant to proteins is the amino acid proline. Figure 10.24 gives details of the side chain of the twenty amino acids that are found in naturally occurring proteins. It is not known why only these twenty amino acids were chosen. Presumably the answer lies in the way in which life originated on earth.

The 20 amino acids have two major features in common: a) They are all α- amino acids. b) They are all L-amino acids. They can be separated into groups with other common features. Glycine, alanine, valine, isoleucine, leucine and phenylalanine have hydrophobic, non-polar side chains. Phenylalanine, tryptophan and tyrosine have aromatic planar rings. Aspartic acid and glutamic acid carry a negative charge at neutral pH, while lysine and arginine each carry a positive charge. Histidine has a highly

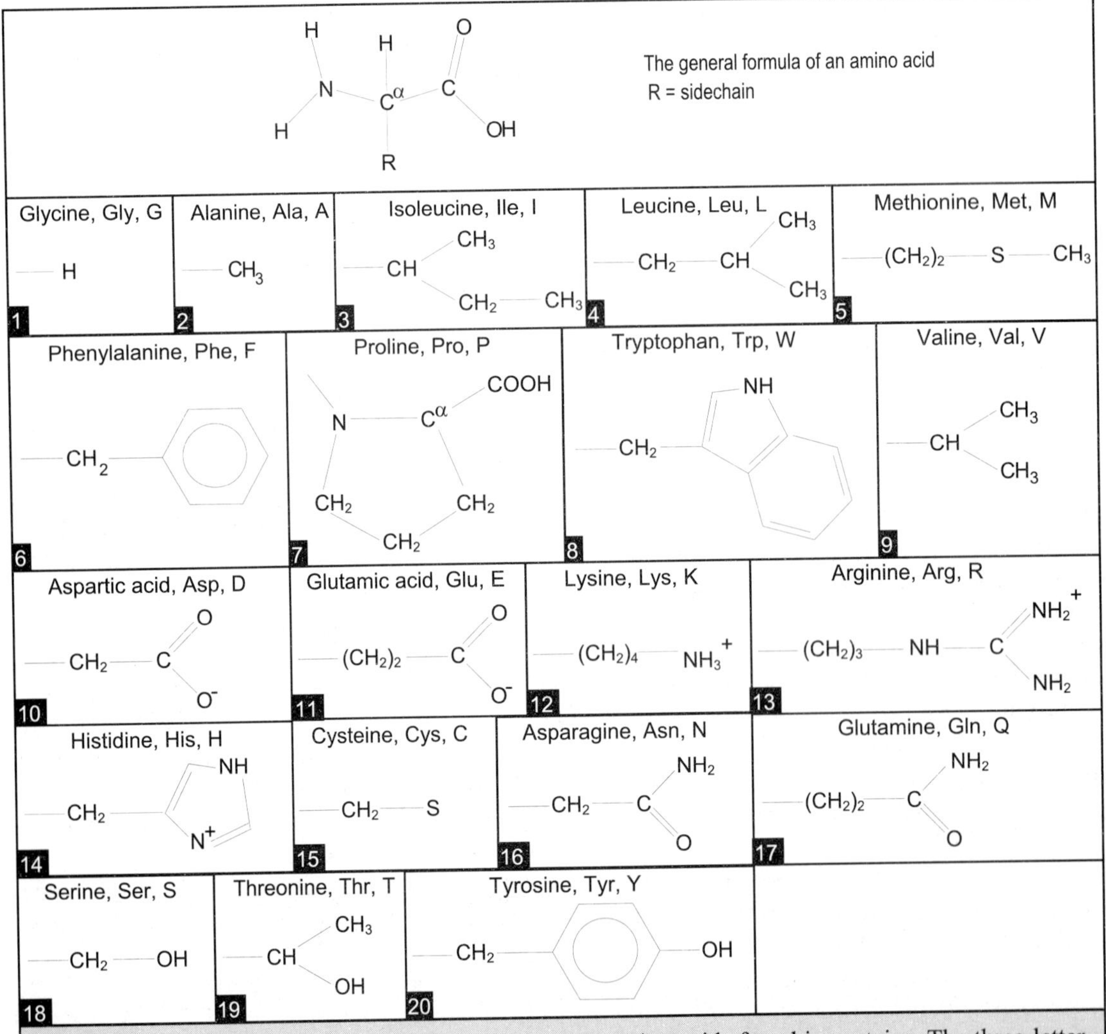

Figure 10.24: The side chains (R groups) of the twenty amino acids found in proteins. The three-letter codes and the single-letter codes for each amino acid are also given. Numbers 1 to 9 are the hydrophobic groups; numbers 10 to 14 are the polar charged groups; and numbers 15 to 20 are the uncharged polar hydrophilic groups.

reactive 5 membered planar ring that also carries a positive charge at neutral pH. Cysteine, serine, threonine, asparagine, glutamine and tyrosine have neutral, polar (and therefore hydrophilic) side chains.

The chemical nature and size of the side chain determines its activity and also its position in the protein. Histidine frequently occurs at the active site of enzymes. Hydrophobic side chains usually occur in the interior and hydrophilic groups on the surface of the proteins that are active in aqueous media. The situation is the reverse in the case of membrane proteins. The sequence in which the amino

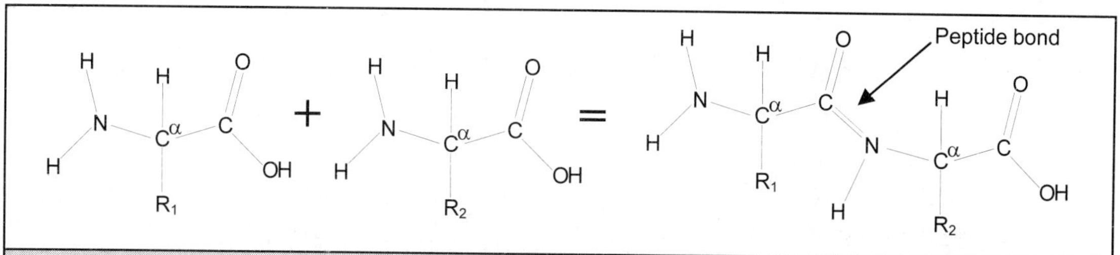

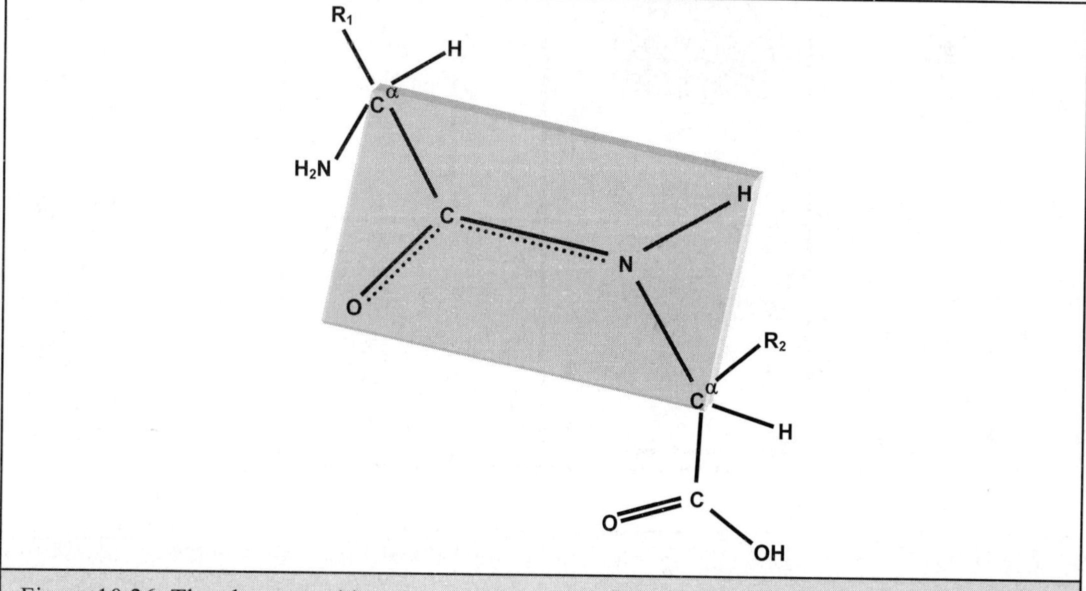

Figure 10.25: The peptide bond. The dotted line indicates that it is a partial double bond.

acids occur along the polypeptide chain therefore closely controls how the chain will fold into the three-dimensional structure. In other words the primary structure of the protein determines its secondary, tertiary and quaternary structures and ultimately the function of the protein. The polypeptide chain is built up when the amino acids join to each other by means of a peptide bond.

10.3.2: The peptide bond and secondary structure of proteins:

The peptide bond is formed as shown in Figure 10.25. The C-N bond has a partial double bond character, which means that the molecule cannot rotate about it. The six atoms that form the peptide group are constrained to lie on a plane thereby forming the planar peptide group (Figure 10.26). Ramachandran and his colleagues at the University of Madras realised that this property simplified the geometrical analysis of the polypeptide chain. If the rotational possibilities of the longer side chains are ignored and the backbone of the protein chain alone is considered, then each residue is described by two torsion angles ϕ and ψ (Figure 10.27). The freedom of rotation about even these bonds is not absolute, but is restricted by possible steric hindrances. For example, if the $C'_{i-1} - N_i$ bond is *cis* to the

Figure 10.26: The planar peptide unit.

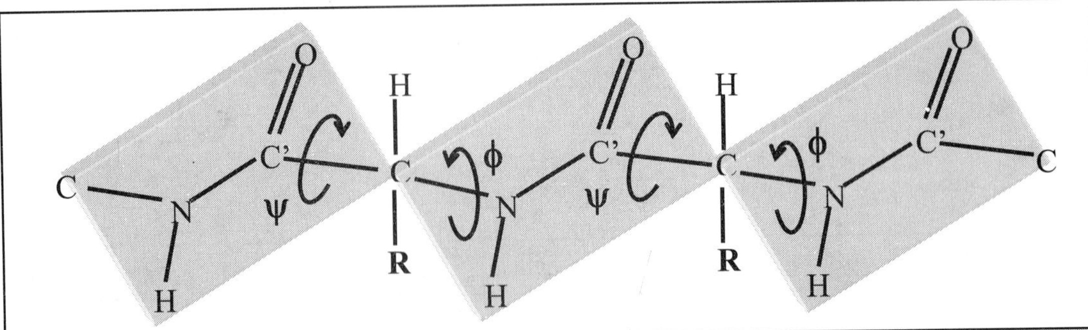

Figure 10.27: The torsion angles ϕ and ψ used to describe the conformation of a polypeptide.

$C^{\alpha}_i - C'_i$ bond when the $C_i - N_{i+1}$ bond is *cis* to the $N_i - C^{\alpha}_i$ bond, i.e. when both ϕ and ψ are $0°$, then a severe clash between H_{i+1} and O_{i-1} will occur, making this particular pair of values disallowed. A systematic search of all pairs of values of ϕ and ψ reveals that, only about 18% of the values are allowed. This is represented as the Ramachandran map (Figure 10.28). This map indicates the situation when the atoms are modelled as hard spheres allowing no interpenetration. The radii of the spheres were derived from the interatomic distances seen in the crystal structures of small molecules. Apart from these normal values, somewhat smaller values for the contact distances were also obtained. These were called the extreme limits, and when they were used, the allowed area on the Ramachandran Plot

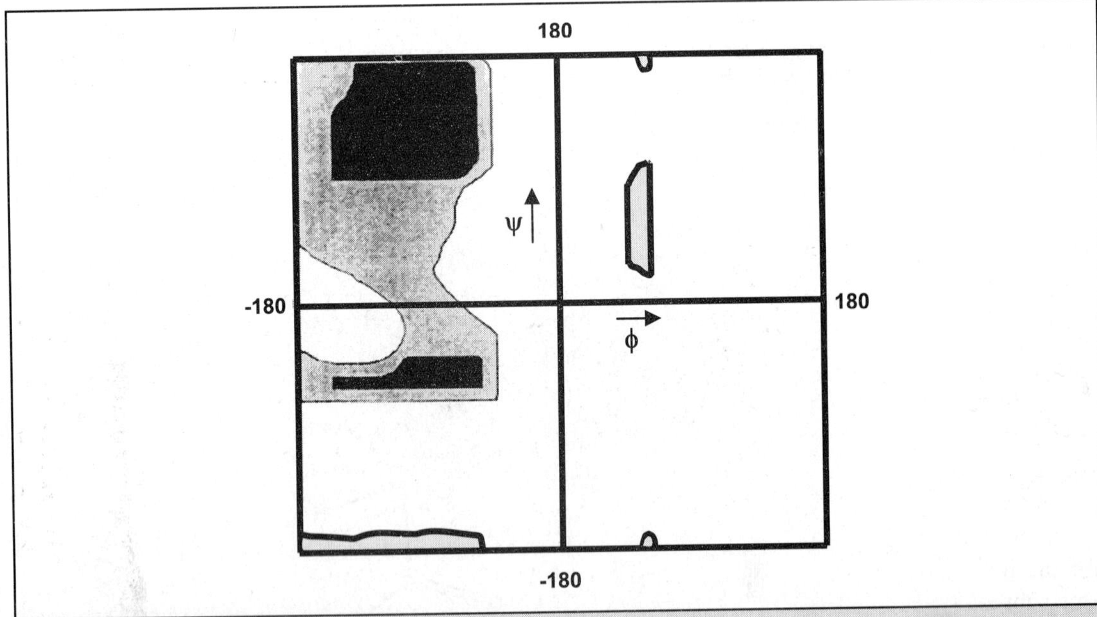

Figure 10.28: The Ramachandran map. The strictly allowed regions are black, and the generously allowed regions are grey.

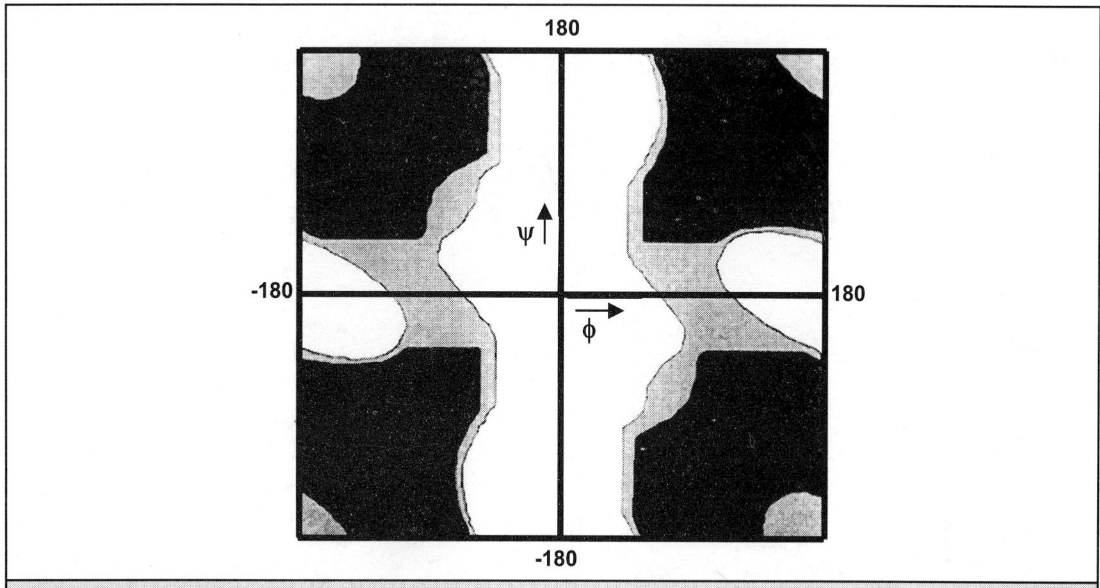

Figure 10.29: The Ramachandran map for glycine. The strictly allowed regions are black, and the generously allowed regions are light grey.

expanded to 22%. The map is correct for eighteen of the twenty side chains. The exceptions are glycine and proline. Glycine has a much larger allowed region owing to the absence of the C^β atom (Figure 10.29). The side chain of proline binds to the peptide nitrogen atom and this limits the value of ϕ to -60 ± 20°.

The peptide units are not absolutely planar, as assumed in constructing the above maps. The torsion angle ω about the peptide bonds can deviate by 10 to 20° from planarity. This changes the ϕ-ψ map somewhat. The map also changes when the 'rigid sphere' model of the atoms is replaced by a semiempirical energy function involving interatomic distances. The changes are however quite small and over 98% of the experimentally determined values of ϕ and ψ fall within the allowed regions of the Ramachandran plot.

If some of the pairs of values of ϕ and ψ which fall within the allowed region are repeated for many residues along the polypeptide chain, it leads to regular recognisable secondary structures, such as α-helices and β-sheets.

<u>10.3.2.1: α-helix:</u> This is one of the most commonly occurring secondary structures in proteins, with (ϕ,ψ) values of (-57°, -47°) (Figure 10.30). The helix has 3.6 residues per turn. The occurrence of such a non-integral value for the helical repeat was a surprise when Linus Pauling first proposed it. But subsequent experimental and theoretical studies have overwhelmingly confirmed the structure. The stability of the α-helix is enhanced by the hydrogen bonds between the C=O group of residue 'i' and the NH group of

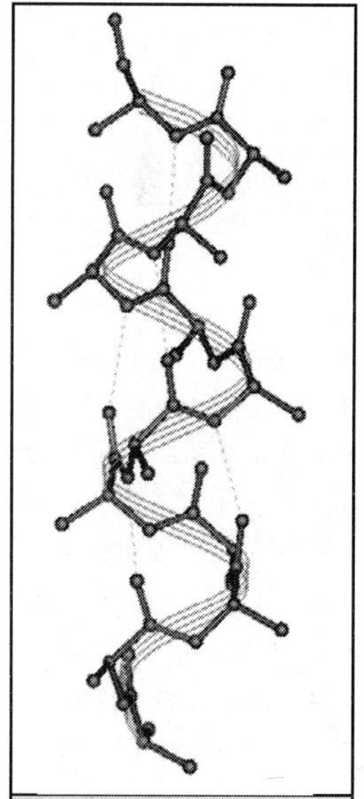

Figure 10.30: The alpha helix.

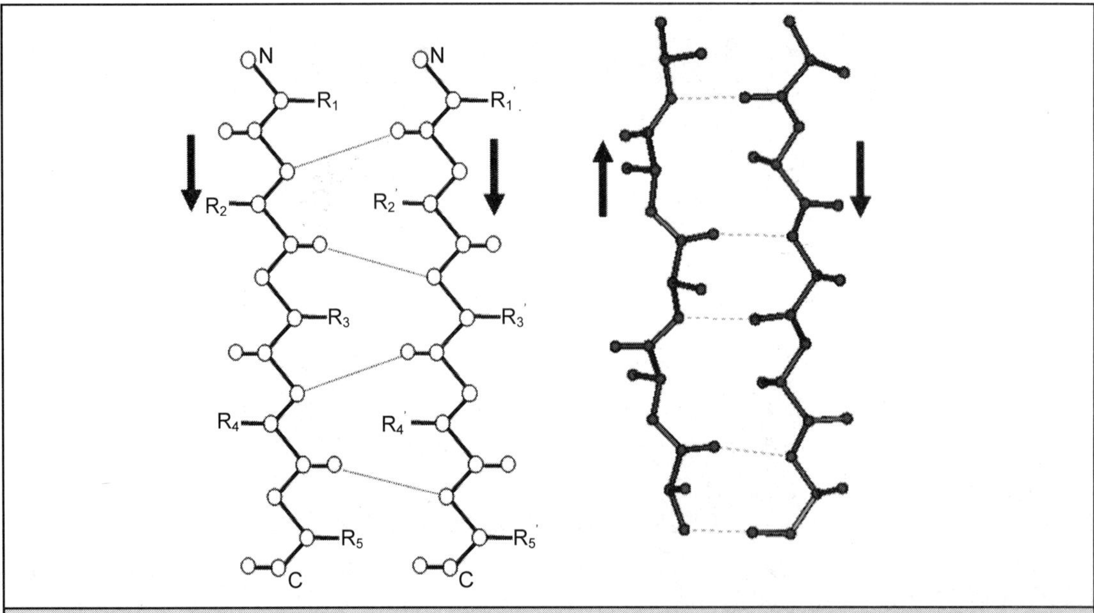

Figure 10.31: Beta sheets. (a) Parallel beta sheet (schematic). (b) Anti-parallel beta sheet (3-D representation).

residue 'i+4'. Each turn of the 'α-helix is 5.4 Å long, indicating a rise per residue of 1.5 Å. The width of the helix is about 4 Å. For L-amino acids only right handed α-helices are allowed and therefore almost all α-helices observed in protein structures have a right handed twist. Very short (3-5 residues) left-handed α-helical regions have also been observed in proteins, though very rarely. Owing to the fact that all the hydrogen bonds in the α-helix point in the same direction, the helix as a whole has a dipole moment such that the amino end or the N terminal has a net partial positive charge and the carboxy end or the C terminal has a net partial negative charge. This induces negatively charged ions such as PO_4^- to bind to the amino end. However, positively charged ions rarely bind to the carboxyl end. Alpha helices are preferentially made up of certain of the 20 amino acid residues. Among the strongest helix-formers are alanine, glutamine, leucine and methionine. Conversely proline, glycine, tyrosine and serine occur in helices only rarely. An α-helix often occurs on the surface of a protein. Then it is usually made up of hydrophobic residues on the side facing the interior of the protein and hydrophilic residues on the other.

<u>10.3.2.2: β-strands and β-sheets:</u> β-strands (Figure 10.31) are highly extended polypeptide chains and the value of (ϕ,ψ) are in the large allowed region in the top left quadrant of the Ramachandran Map. β-strands interact with other β-strands to form β-sheets. This interaction can occur in two ways. Firstly, all the interacting β-strands could run in the same direction, i.e. the amino terminii of all strands are on the same side. In this case, a parallel a β-sheet is formed (Figure 10.31(a)) in which somewhat distorted H-bonds are formed between the amino groups of one strand and the carboxy groups of the other. In the other arrangement, the strands run in opposite directions, resulting in an antiparallel sheet (Figure 10.31(b)). Here again N-H...O hydrogen bonds stabilise the structure. In both cases, the amino

acid side chains project out of the sheet on either side. The sheets are usually not flat but have a twist (Figure 10.32) and always in a right handed sense. Mixed parallel and antiparallel β-sheets also occur, but only rarely.

<u>10.3.2.3: β-turns:</u> Whenever the peptide chain has to reverse its direction, as frequently happens in globular proteins, it can sharply and efficiently do so through a structure known as the β-turn or a reverse turn. Three consecutive peptide units are involved in such a turn and the most prominent stabilising factor is the hydrogen bonds that are formed between O_i and N_{i+3} (Figure 10.33). The (ϕ,ψ) values of all the residues involved in this secondary structural unit do not fall in the same region of the Ramachandran plot. At the residue i+1, the values are (-60, -30°) and at the residue i+2, the values are (-90,0°), assuming that the turn is made up of the residues i, i+1, i+2 and i+3. This is called the Type I β-turn. In the Type II turn, the values are (-60,120°)

Figure 10.32: The twist in the beta sheet.

and (80,0°) at residues i+1 and i+2 respectively. However, in this turn there is a steric clash between side chain R_{i+2} and O_{i+1}, so that R_{i+2} can only be H, i.e. glycine. A type III reverse turn has regular structure with (ϕ,ψ) values of (-60, -30°) at both central residues. A repetition of these angles over a long stretch leads to a type of helix called a 3_{10} helix, which is found in proteins, though rarely.

<u>10.3.2.4: Collagen helix:</u> This is a special type of helix discovered by Ramachandran and his colleagues. It occurs in the fibrous protein collagen, which is the most abundant protein in animals. It is a left-handed helix, with (ϕ,ψ) values about (-60,140°). It is not a completely regular helix and the (ϕ,ψ) values are repeated exactly only at every 3rd residue, i.e. at the glycine residue (the sequence of collagen is always (gly-x-y) where x and y are frequently proline and hydroxyproline respectively). The protein collagen itself consists of three of these single strand helices wound around each other to form a triple helical fibre (Figure 10.34). The structure was able to explain why such a repetitive

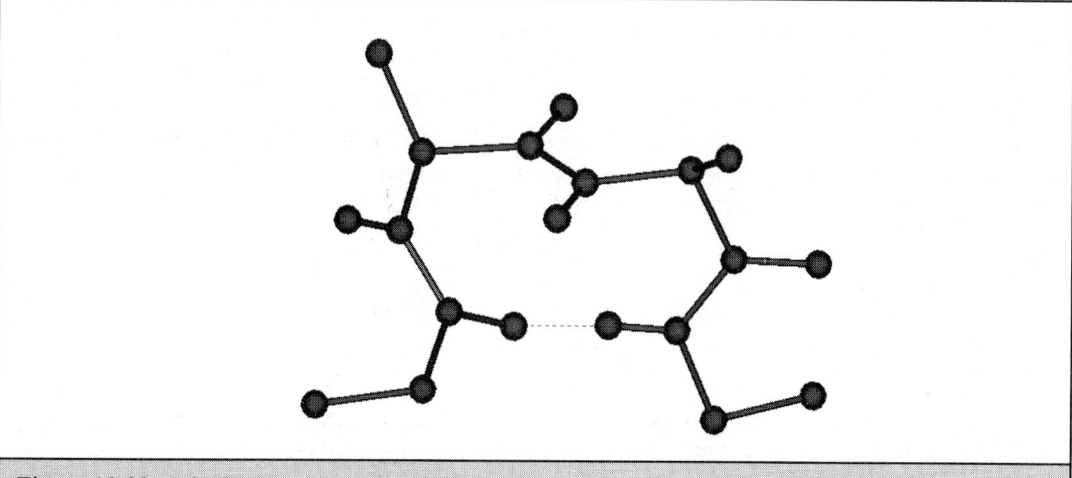

Figure 10.33: A beta turn.

sequence was required, since every third residue came close to the superhelix axis. Side chains larger than that of glycine would introduce a large degree of instability.

10.3.3: Tertiary structure - supersecondary and domain structure:

Various segments of a protein chain fold into various secondary structural units such as α-helices or β-sheets. These units further fold up into compact supersecondary structural units or domains. This is called the tertiary structure of the proteins and involves interactions between amino acid residues that are not necessarily close together in the primary sequence. One of the most important of such tertiary interactions is the disulphide bond.

The disulphide bond is formed between the sulphur atoms of two cysteine residues (Figure 10.35). Such bonds could occur within a single polypeptide chain or between two different chains. The chief function of the disulphide bridge is to provide mechanical stability to the protein structure. They may also determine chemical properties by stabilising the correct active conformation. In some proteins the disulphide bond may play a catalytic role.

Another major stabilising factor in the tertiary structure of proteins is the so-called entropic or hydrophobic forces. Since the protein is made up of both polar and non-polar side chains, in an aqueous solution the protein molecule behaves like a drop of oil, with polar groups on the surface, in contact with the solution and the non-polar groups in the interior of the molecule. The solvent molecules in the near vicinity of the protein also assume a certain amount of ordering around polar groups. For non-polar groups, however, the solvent molecules make a large negative contribution to the entropic energy, leading to the folded state being favoured over the denatured state. Other factors which stabilise the tertiary structure of proteins include salt bridges or strong Coulombic interactions between oppositely charged amino acids, such as glutamic acid and aspartic acid. Hydrogen bonds in the interior of the protein are another important stabilising factor not only of the secondary structural motifs but also of the tertiary structure.

The secondary structural units often form clearly identifiable domains or super-secondary structures. The same domain structure may occur in proteins with widely different functions, though

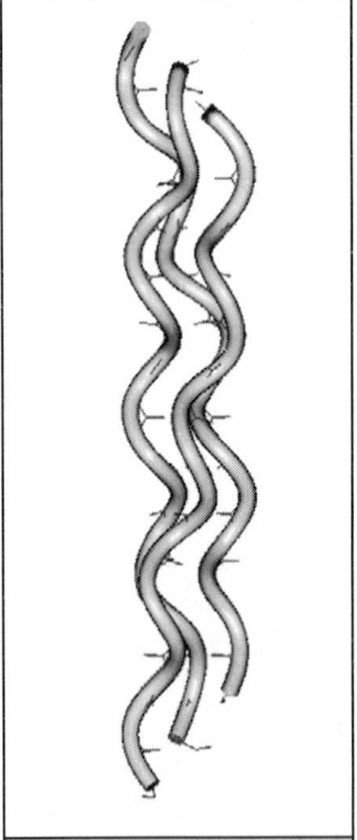

Figure 10.34: The triple-helical structure of collagen.

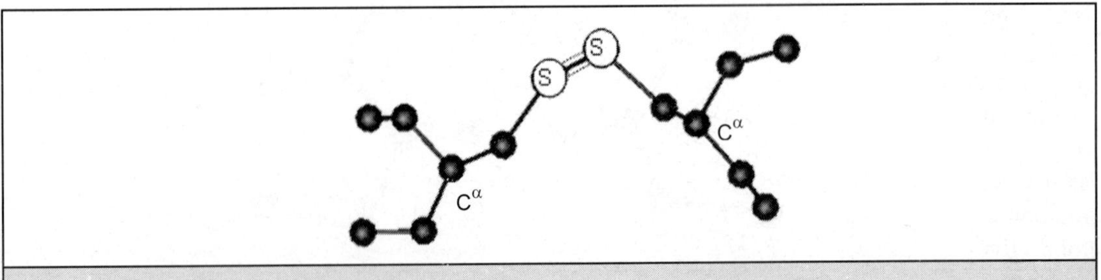

Figure 10.35: The disulphide bond.

the function of the domain itself within the protein may be the same. In proteins like cytochrome b_{562}, four helices come together to form the so-called 4 helix bundle (Figure 10.36). This is a widely occurring super-secondary structure. Motifs containing both β-sheets and α-helices also occur in many proteins. An example is α/β barrel structure. It has been noticed that the core of such barrels is usually packed with hydrophobic side chains. Other super-secondary structural motifs include β-barrels such as the one seen in retinol binding protein. Domains in proteins can be loosely defined as an almost self-contained globular portion of the protein, loosely connected to other domains to form the complete molecule. A clear example of a domain structure is the protein troponin C (Figure 10.37). In other cases the domains may not be so clearly demarcated.

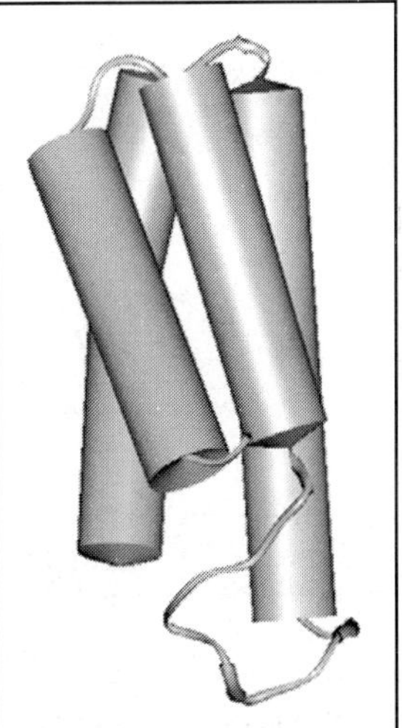

Figure 10.36: A four-helix bundle.

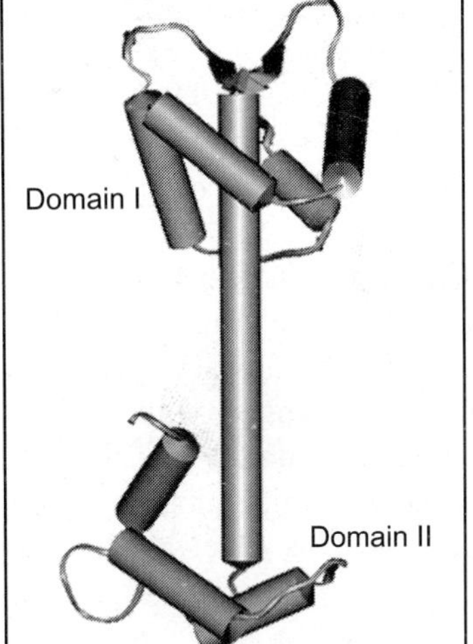

Figure 10.37: The structure of troponin C showing the two domains.

10.3.4: Quaternary structure:

Many proteins in their active or functional forms exist as aggregates of more than one folded polypeptide chain. The macromolecular structure so built up is called the quaternary structure of proteins. Usually the subunits of quaternary structure associate by non-covalent interactions, though disulphide linkages may exist between different subunits. The monomers, which form the quaternary structure, may be identical or may be different. For example, haemoglobin is a tetrameric protein consisting of two chains each of two different polypeptides called the α-chain and the β-chain. In most multimeric proteins, the number of subunits ranges from 2 to about 12, with a majority possessing 2 or 4 subunits. The exceptions are large enzyme complexes and viruses, which may have even 100 or more monomers. In proteins made up of only a few monomers or a small number of subunits, all the interactions that are possible by subunits must be completely fulfilled. Otherwise larger aggregates would occur. This implies that the subunits must be regularly packed around a central point. In other words most protein multimers possess point group symmetry consisting of one or more intersecting rotation axes. (Mirror inversion is also a part of point group symmetry but is not possible in protein assemblies, as it would require the presence of D-amino acids.) Some of the rotation axes found in proteins include two-fold, three-fold, four-fold, five-fold, six-fold and eight-fold axes. A tetrahedral symmetry is also seen where the monomers are placed at the

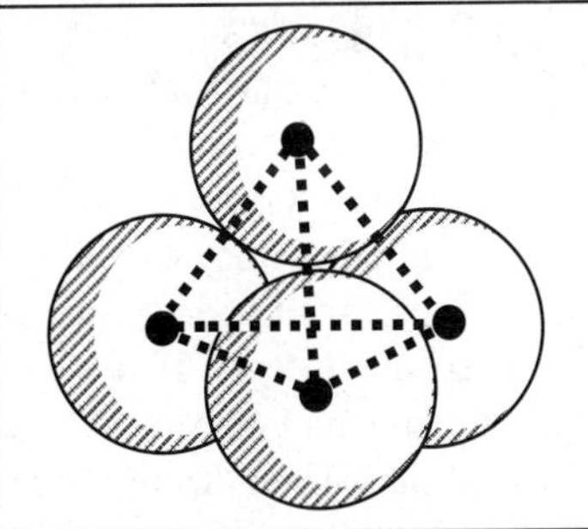

Figure 10.38: Tetrahedral symmetry in the quaternary structure of proteins. Each sphere represents one sub unit of a tetrameric protein in this example.

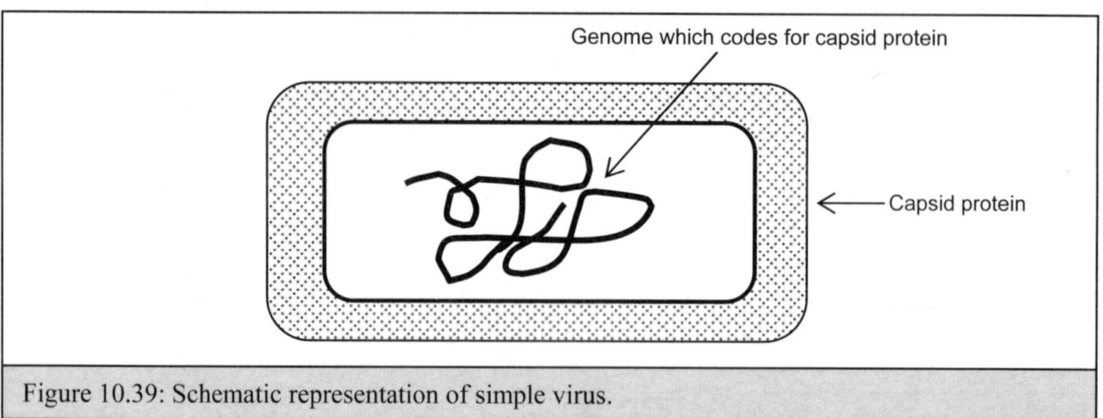

Figure 10.39: Schematic representation of simple virus.

vertices of an imaginary tetrahedron (Figure 10.38). Hexameric proteins may form trigonal prisms, hexagons or octahedra. Octameric proteins may form cubes or square prisms.

In order to attain a definite function, there are several reasons why an assembly of subunits is evolutionarily favourable, rather than giant covalently linked molecules. One of the reasons is that less DNA is needed to code such a protein, since the monomers also contain the information to self-assemble into the multimer. Again, it is easier to make smaller units in an error free fashion than large units. Similarly, defective units can be more economically discarded if they are small. A very important reason is probably that subunits with a particular function may be combined with different subunits to make up different proteins. This modular approach to building up a protein exists at the level of domains too. A particularly striking example is the structure of the nucleotide-binding region, which is almost identical in several proteins with different functions, such as glyceraldehyde-3-phosphate dehydrogenase, phophoglycerate kinase and phosphorylase.

10.3.5: Virus structure:

Viruses are large macromolecular assemblies on the border between living and non-living things. The simplest virus conceptually consists of a genome (DNA or RNA) which codes for a protein that will build up a cover to protect the genome (Figure 10.39). However, the viral genome usually codes for a few other proteins too, which are necessary to allow the virus to infect its host. Even with a genome which codes for the coat protein alone, the nucleic acid molecule is too large to allow the protein to completely enclose it. If the protein were to be larger, the genome would also have to be larger, requiring an even larger protein, and so on. The evolutionary answer to this problem is to construct the viral coat or 'capsid' from several copies of a single, relatively small protein. The capsid is roughly spherical in shape as for example in the common cold virus or in HIV. Other viruses are long cylinders or rod shaped, e.g. Tobacco Mosaic virus. More complex viruses such as bacteriophage T4 have more complex shapes.

Spherical viruses have the shape and symmetry of an icosahedron (Figure 10.40). Each of the twenty faces of the icosahedron is a triangle with a 3-fold symmetry and can accommodate 3 identical protein molecules. The viral capsid as a whole thus has 60 subunits. If the requirement for exact symmetry is somewhat relaxed, a larger number of subunits can be accommodated. However, only certain multiples

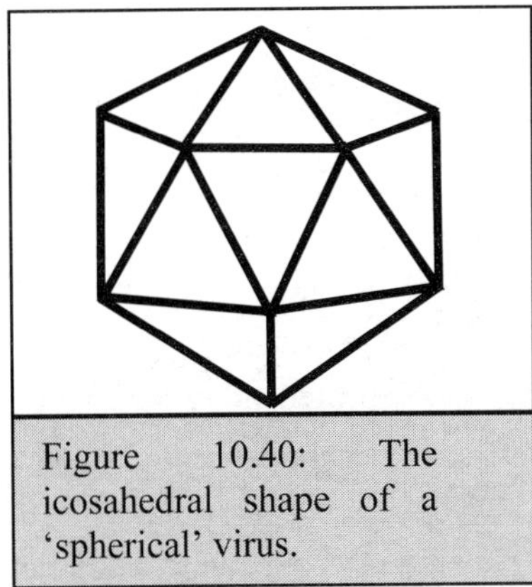

Figure 10.40: The icosahedral shape of a 'spherical' virus.

(1,3,4,7...) of 60 subunits are likely to occur, under the assumption that identical subunits may have quasi-equivalent enviro-nments, rather than exact equivalence. More elaborate structures are also known for spherical viruses. The protein capsid of a class of viruses known as picorna viruses (which includes polio virus and common cold virus or rhino virus) contains 60 copies of each of four different polypeptide chains.

Rod-like viruses such as the tobacco mosaic virus (TMV) follow different structural principles. Here the main motif is a helix. In TMV a total of about 2130 subunits are arranged as a helix 3000 Å long and 180 Å in diameter. The pitch of the helix is 23 Å and there are 16.3 subunits per turn (i.e. 49 subunits in 3 turns) (Figure 10.41). The arrangement is reminiscent of a bunch of bananas with a long RNA molecule winding round the central stem of the bunch. Electron microscopy and electron diffraction studies indicate that the rod-like tail of bacteriophage T4 also consists of identical subunits arranged as a helix.

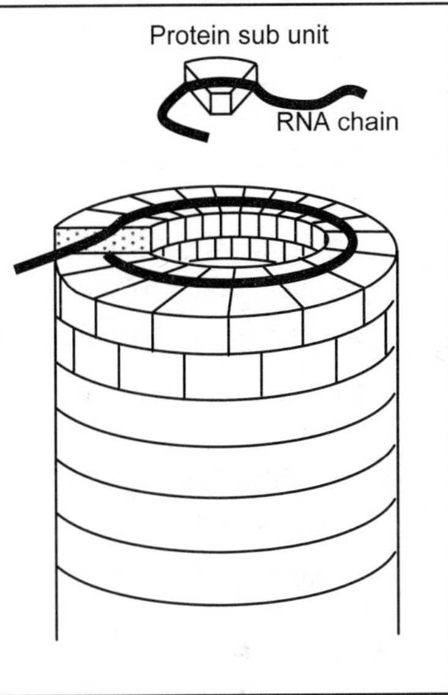

Figure 10.41: The structure of a rod-like virus – the Tobacco Mosaic virus.

Summary

- In order to understand the behaviour of biomolecules, it is necessary to know and understand their structure.

- The storage, transmission and expression of genetic information in the cell is carried out by nucleic acids.

- DNA is involved in replication and transcription.

- RNA has many different forms, such as mRNA, tRNA, rRNA, etc.

- Both DNA and RNA are polymers. The monomer repeating unit is a nucleotide.

- A mononucleotide can be thought as made up of three parts, a) a phosphate group, b) a sugar moiety and c) a nitrogenous base.

- The phosphate group is simply PO_4^-.

- The sugar is a five-membered furanose ring with exocyclic alcohol and hydroxyl groups.

- Though it is cyclised, a substantial amount of conformational variation can still take place, by change in puckering, as well as rotation about the C4'-C5' bond.

- In nucleic acids, the sugar puckering is preferentially C2'-endo or C3'-endo.

- A nucleoside is formed when a glycosidic bond attaches the sugar to the base. A nucleotide is formed when the phosphate group is attached to the sugar.

- Mononucleotides link to each other through phosphodiester bonds to form the polynucleotide chain.

- In a dinucleotide there are a total of seven bonds about which rotation is possible. In addition, the sugar puckering is also a variable.

- Only a few combinations of these angles lead to regular structures when repeated along the chain.

- In solution, the bases are preferentially arranged in parallel planes, one over the other like a stack of coins.

- X-ray diffraction experiments showed that each DNA molecule consists of two strands that wrap around each other to form a helix and these two strands are arranged antiparallel to each other.

- The concept of base pairing says that adenine on one strand forms specific hydrogen bonds only with thymine on the other, and vice versa. Similarly guanine pairs only with cytosine and vice versa.

- One may visualise the DNA double helix as a flexible rope ladder wrapped around a central pole. The sugar phosphate chains of the two polynucleotide backbones form the sides of the ladder. The base pairs placed perpendicular to the central helix axis form the rungs.

- The structure of DNA proposed first by Watson and Crick is called B-DNA.

- Watson and Crick also proposed a somewhat altered model of the double helix called A-DNA.

- A dramatic example of the polymorphism of DNA was discovered in the left-handed Z type DNA.

- In a DNA double helix, if the number of times one strand goes around the other is changed from the optimum value, supercoils relieve the extra strain.

- Apart from the double helix, certain DNA sequences have been shown to assume other unusual types of structures, such as G-plates and G-quartets, i-motif structures, and junction structures.

- tRNA is a single strand RNA molecule about 75 nucleotides long that folds into a clover-leaf shaped secondary structure and an L-shaped three-dimensional structure.

- Proteins are built up from 20 amino acids, which constitute the monomers. Proteins have four levels of structure – primary, secondary, tertiary and quaternary.

- The amino acid sequence is the primary sequence of a protein.

- Some sequences of amino acids arrange themselves into regular three-dimensional structures, such as such as α-helices and β-sheets. These are the secondary structures.

- The secondary structural units arrange themselves into globular shapes. This is called the tertiary structure and is stabilised mainly by interactions between the amino acid side chains.

- Globular folded polypeptide chains come together in specific arrangements called the quaternary structure.

- Ramachandran and his colleagues at the University of Madras mapped all the possible conformations of a dipeptide unit on a so-called Ramachandran map.

- Viruses are large macromolecular assemblies on the border between living and non-living things.
- The viral coat or 'capsid' is usually constructed from several copies of one, or a few, relatively small proteins.

Chapter 11

Energy Pathways in Biology

11.1: Introduction

We have seen from Chapter 1 that a process can occur spontaneously only if the sum of the entropies of the system and the surroundings increases. Therefore, for a natural process to take place, energy must be dissipated. Living systems continuously do work within themselves or on the environment. When work is being done by a living system no temperature changes take place and hence these systems must spend their energy in some other form either internally or externally. The ultimate source of energy for all living systems is solar energy, which is harnessed through green plants, algae and certain types of bacteria. This process by which solar energy is converted into chemical energy that can be used by the plants and other living systems is known as

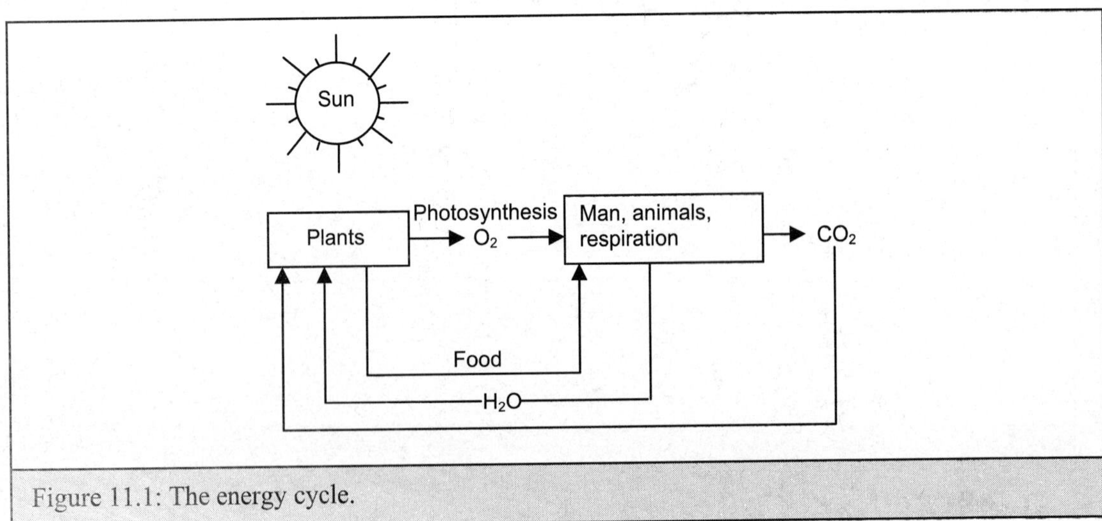

Figure 11.1: The energy cycle.

photosynthesis. The various processes that are involved in the transfer of energy in the biosphere may be represented in a diagram called the energy cycle (Figure 11.1).

Living organisms do not undergo an increase in internal disorder (i.e. entropy), when they metabolise their nutrients. However this does not violate the second law of thermodynamics because the entropy of the surroundings increases. In other words living systems retain their internal order at the cost of the environment. Sometimes the entropy of a living system is said to be negative because it is rich in information content. According to information theory, information is a form of energy and is called negative entropy.

11.2: Free energy

A continuous supply of free energy is required by living things for carrying out the active transport of ions and molecules, for muscle contraction and other mechanical work, and for the synthesis of macromolecules from their constituents. This free energy requirement is derived from the environment. Phototrops derive it from light while chemotrops obtain it in the form of foodstuff. The free energy derived from oxidation of foodstuff and from light is partly converted into a carrier molecule called adenosine triphosphate or ATP (Figure 11.2). ATP is also known as the universal currency of free energy in biological systems. ATP consists of an adenine base, ribose sugar and a triphosphate unit. The energy rich part is the triphosphate unit. Free energy is liberated when ATP is hydrolysed to adenosine diphosphate (ADP) and orthophosphate P_i or when it becomes adenosine monophosphate (AMP) and a pyrophosphate (Figure11.3).

$$ATP + H_2O \rightarrow ADP + P_i$$

$$ADP \rightarrow AMP + P_i$$

ATP, ADP and AMP are inter-convertible. ATP is formed from ADP when fuel molecules are oxidized in chemotrophs or when light is absorbed by phototrophs. The ATP-ADP cycle is the

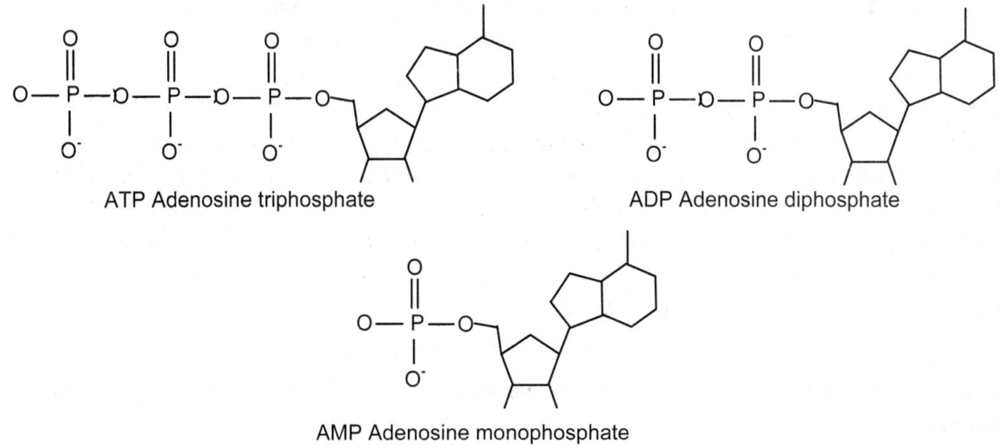

Figure 11.2: Adenosine phosphates.

fundamental energy exchange in biological systems. The hydrolysis of ATP occurs as given by the following reaction.

$$ATP^{4-} + H_2O \rightarrow ADP^{3-} + HPO_4^{2-} + H^+$$

At equilibrium and under standard conditions the amount of energy released in this reaction is -7.3 Kcal/mole. The function of ATP as an energy carrier molecule is dependent on the fact that ΔG^0 (change of free energy at pH 7.0 and under standard conditions) of the above reaction is negative. It is important to note that ATP is not the only phosphate compound with these features. There are others like phosphoenolpyruvate (ΔG^0 = -14.8 Kcal/mole), 1,3-diphosphoglycerate (ΔG^0 = -11.8 Kcal/mole), glucose-1-phosphate (ΔG^0 = -3.3 Kcal/mole), and glycerol-1-phosphate (ΔG^0 = -2.2 Kcal/mole). The free energy of hydrolysis of ATP falls in the middle of the range of values for the other phosphate compounds occurring in biological systems. Hence it is an ideal candidate for energy transfer by coupled reactions with common intermediates. The hydrolysis of ATP releases free energy, which becomes useful only when it is coupled with other cellular processes that require free energy. In a typical cell, ATP molecules are consumed within a minute of their formation. Thus, ATP acts as an immediate donor of free energy rather than as an energy storage molecule. Different processes produce ATP in different biological systems. The two main processes are (1) by the oxidation-reduction reaction in the inner membranes of mitochondria. (2) By photosynthesis which takes place on the membranes of the grana in chloroplasts of green plants. The processes can be divided into those which depend on the oxygen availability i.e. respiration and those which are light driven i.e. photosynthesis.

11.3: Coupled reactions

A thermodynamically unfavourable reaction can be driven by a favourable reaction, both reactions together having a net negative (and therefore favourable) difference in the free energy. These reactions are coupled by a common intermediate. Consider the following example.

$$A <=> B + C \qquad \Delta G^0 = +6 \text{ Kcal/mole} \quad (1)$$
$$B <=> D \qquad \Delta G^0 = -9 \text{ Kcal/mole} \quad (2)$$
$$\overline{\qquad\qquad\qquad\qquad\qquad\qquad\qquad\qquad\qquad}$$
$$A <=> C + D \qquad \Delta G^0 = -3 \text{ Kcal/mole}$$

Under standard conditions reaction (1) cannot proceed, as the ΔG^0 value is positive. But B can be converted to D as reaction (2) is thermodynamically favorable (ΔG^0 is negative). As free energy changes are additive, the conversion of A to C and D is possible as the net ΔG^0 is negative when reactions (1) and (2) are coupled through B, the intermediate. The hydrolysis of ATP shifts the equilibria of coupled reactions. Therefore any thermodynamically unfavorable reaction can be converted into a favorable one by coupling it with the hydrolysis of a sufficient number of ATP molecules. For example the hypothetical reaction

$$A <=> B \ (\Delta G^0 = 3 \text{ Kcal/mole})$$

can be made to proceed when it is coupled to the hydrolysis reaction of ATP to ADP. The coupled reaction may be written as follows.

$$A + ATP + H_2O <=> B + ADP + P_i + H^+$$

With ΔG^0 = -4.3 Kcal/mole. This final ΔG^0 value is obtained by adding the ΔG^0 of the A => B conversion (+3 Kcal/mole) with that of the ATP → ADP reaction (-7.3 Kcal/mole).

11.4: Group transfer potential

Compare the following reactions

$$ATP + H_2O <=> ADP + P_i + H^+ \qquad \Delta G^0 = -7.3 \text{ Kcal/mole}$$

$$\text{Glucose-6-phosphate} + H_2O <=> \text{Glucose} + P_i \qquad \Delta G^0 = -3.3 \text{ Kcal/mole}$$

The ΔG^0 for the hydrolysis of glucose-6-phosphate is much smaller than that of ATP. This means that ATP can transfer its terminal phosphoryl group to water more readily than glucose-6-phosphate. ATP is said to have a higher phosphate group transfer potential than glucose-6-phosphate. This higher potential has a structural basis. At biological pH 7.0, ATP is almost completely ionized and exists as ATP^{4-}. On hydrolysis we get

$$ATP^{4-} + H_2O \rightarrow ADP^{3-} + HPO_4^{2-} + H^+$$

Under standard conditions, ATP^{4-}, ADP^{3-} and HPO_4^{2-} will be present at concentrations of the order of 1.0 M, while the H^+ ion concentration at pH 7.0 is only 10^{-7} M. On the basis of the law of mass action, the low concentration of H^+ pulls the reaction towards the right. Further, at neutral pH there are four closely spaced negative charges in the ATP molecule. These charges repel each other and the dissociation of the terminal phosphoryl group is required, in order to release at least some of the stress. This repulsive force also prevents the products of the reaction coming together again to reform ATP. A third factor contributing to the high group transfer potential of ATP is resonance stabilisation. Orthophosphate exists in several significant resonance forms (Figure 11.3) while ATP has fewer resonance forms. In the hydrolysis of ATP the products HPO_4^{2-} and ADP^{3-} are in more stable resonance forms and contain less free energy than when they are still combined as ATP^{4-}. Thus once again the forward reaction is favoured over the reverse reaction.

In the case of Glucose-6-phosphate, the reaction is

$$\text{Glucose-6-phosphate} + H_2O \rightarrow \text{glucose} + HPO_4^{2-}$$

In this case there is no extra H^+ ion formed and chemical pressure on the reaction to proceed towards the right does not exist. Also, no repulsive force exists between glucose and HPO_4^{2-} and hence these two compounds can recombine to form Glucose-6-phosphate readily. This molecule thus has a much lower transfer potential as compared to ATP.

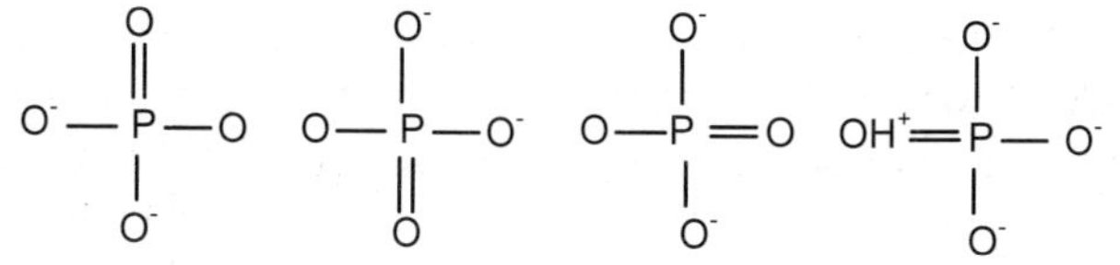

Figure 11.3: Resonance forms of orthophosphate.

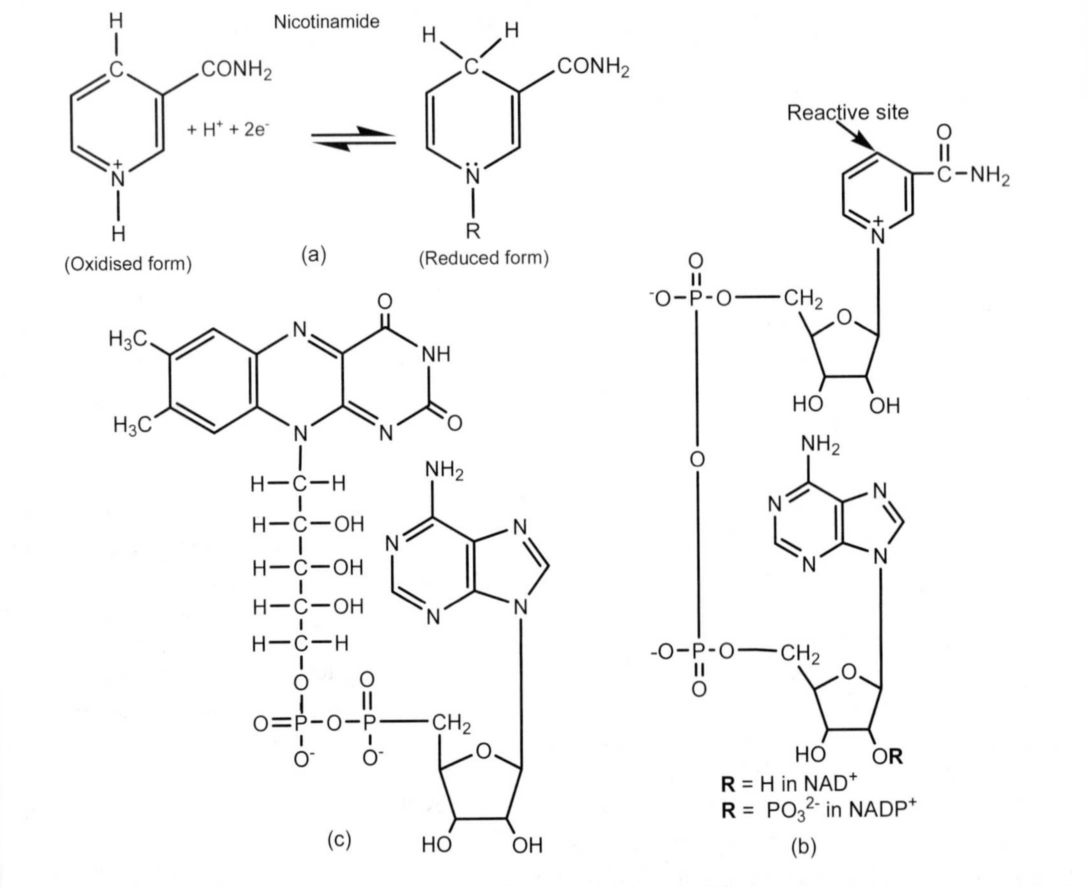

Figure 11.4: Pyridine nucleotides. (a) Oxidised and reduced forms of nicotinamide. b) NAD and NADP. (c) FAD.

11.5: Role of pyridine nucleotides

Chemotrophs get their energy from oxidation of fuel molecules like fatty acids, glucose etc. These reactions are catalysed by enzymes, which in turn require cofactors. The most common cofactors are nicotinamide adenine dinucleotide (NAD), nicotinamide adenine dinucleotide phosphate (NADP) and flavin adenine dinucleotide (FAD) (Figure 11.4). The cofactors act as carriers of electrons in the oxidation reduction reactions. In aerobic organisms electrons from the fuel molecules are transferred first to these carriers. The reduced form of these carrier molecules transfers the electrons to O through an electron transport chain located in the inner membrane of the mitochondria. ATP is formed from ADP and orthophosphate during this process, which is known as oxidative phosphorylation. NAD^+ accepts a hydrogen ion and two electrons from the substrate in the oxidation of the substrate molecule. In the process, NAD^+ is converted from the oxidised to the reduced state.

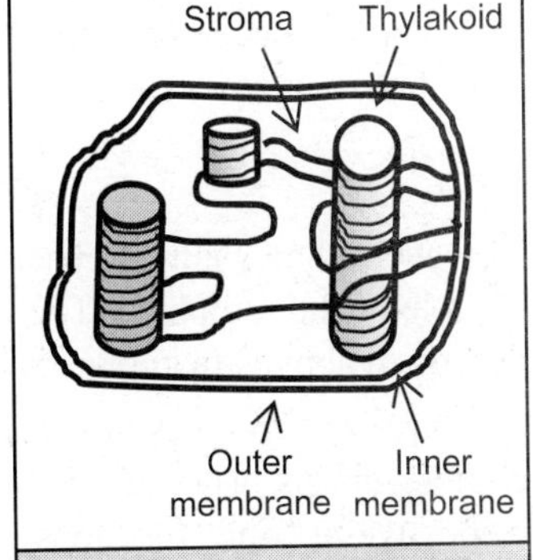

Figure 11.5: Oxidised (a) and reduced (b) forms of FAD.

$$NAD^+ + R_1 {-}{-}{-} \overset{\overset{\textstyle H}{|}}{\underset{\underset{\textstyle OH}{|}}{C}} {-}{-}{-} R_2 \quad \Leftrightarrow \quad R_1 {-}{-}{-} \overset{\overset{\textstyle H}{|}}{\underset{\underset{\textstyle O}{\|}}{C}} {-}{-}{-} R_2 + NADH + H^+$$

Similarly, the oxidised and reduced forms of FAD are shown in Figure 11.5. NADP and NAD have structural similarity but their biological functions differ. NADP is used in reductive biosynthesis while NAD is used primarily in the energy metabolism. The cell has a mechanism by which NADPH can be converted to NADH and vice versa. Enzymes involved in this reaction are known as trans-hydrogenases. In the absence of catalysts, NADH, NADPH and FADH react very slowly with O_2. Similarly the hydrolysis of ATP in the absence of a catalyst is a slow process. The stability of these molecules is essential for their biological function as it provides a way for enzymes to control the reduction mechanism and hence the flow of free energy.

11.6: Photosynthesis

Photosynthesis is the process by which the energy of sunlight is captured and used to sustain life on earth. Green plants, algae and some bacteria are capable of carrying out this process which takes place in the thylakoid membranes of the organelle called chloroplast (Figure 11.6). Light is absorbed by the pigment molecules (carotenoids, chlorophylls and phycobillins) (Figure 11.7) in these membranes and is transferred to the so-called photosynthetic reaction centres. The light energy is converted into chemical energy at these centres. The aggregate of the reaction centre and the photosynthetic pigments is called a photosynthetic unit. Each photosynthetic unit may consist of 50 to 400 chlorophyll molecules. The absorbed light excites the pigment aggregate and this in turn transfers the energy to the reaction centre. The photosynthetic reaction takes place in two phases (Figure 11.8). In the first phase a reductant is produced in the presence of light. In the second stage this reductant is utilised to form sugar in the absence of light. The

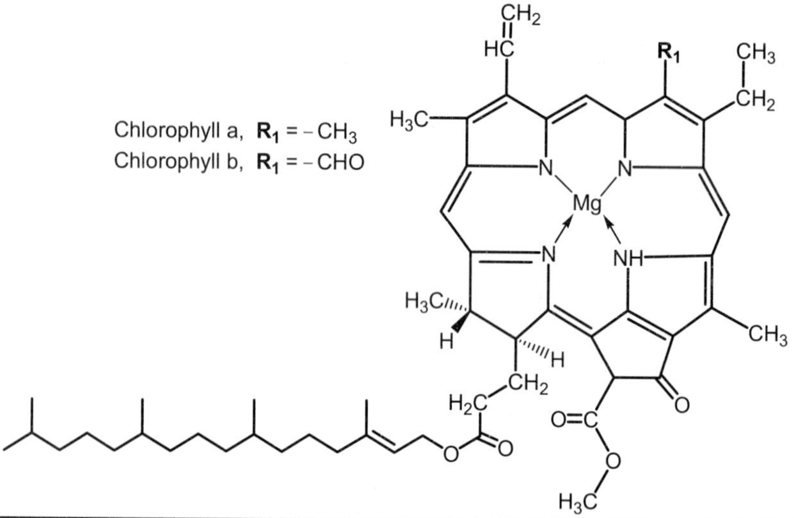

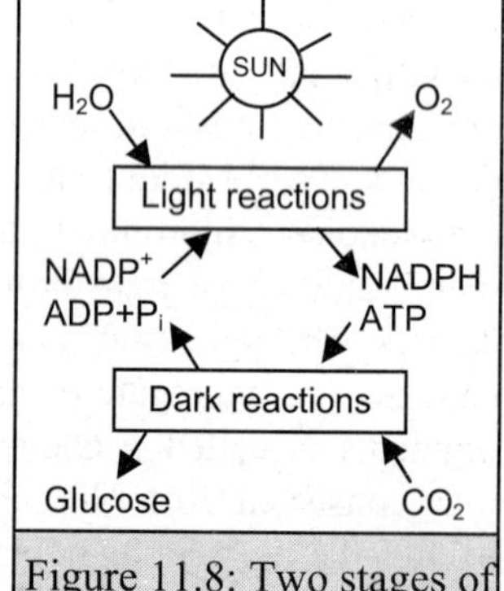

Figure 11.7: The chemical structures of the pigment molecules in chloroplasts, chlorophyll a and b.

excitation energy transfer from chloroplast to the next member in the chlorophyll cluster (light harvesting or antenna molecules) takes place through inductive resonance transfer till it reaches a specialised chlorophyll molecule at the reaction centre. The reaction centre is a protein-chlorophyll complex system. These reactions are described by the following equations.

$$S + h\nu_a \rightarrow S^*$$
$$S^* + A \rightarrow S + A^*$$
$$A^* \rightarrow A + h\nu_e$$

Here $h\nu_a$ is the absorbed light, $h\nu_e$ is the emitted light, S is the sensitiser molecule, i.e. the reductant, and A is the acceptor molecule, and S^* and A^* are the excited stages. The photosynthetic reaction can be written as

$$nH_2O + nCO_2 \xrightarrow{\text{light}} [CH_2O]_n + nO_2$$

However, photosynthetic bacteria do not produce oxygen but they give off elemental sulphur. The two phases of photosynthetic reaction described above are known as light reactions and dark reactions.

There are two types of reactions involving light that take place during photosynthesis. They are known as photosystem I and photosystem II respectively. The reactions that occur in photosystem I produce NADPH. In photosystem II, O_2 is evolved and a reductant is generated. Photosystem I has a greater number of a particular type of chlorophyll molecule called chlorophyll 'a' and is maximally excited at longer

Figure 11.8: Two stages of photosynthetic reaction.

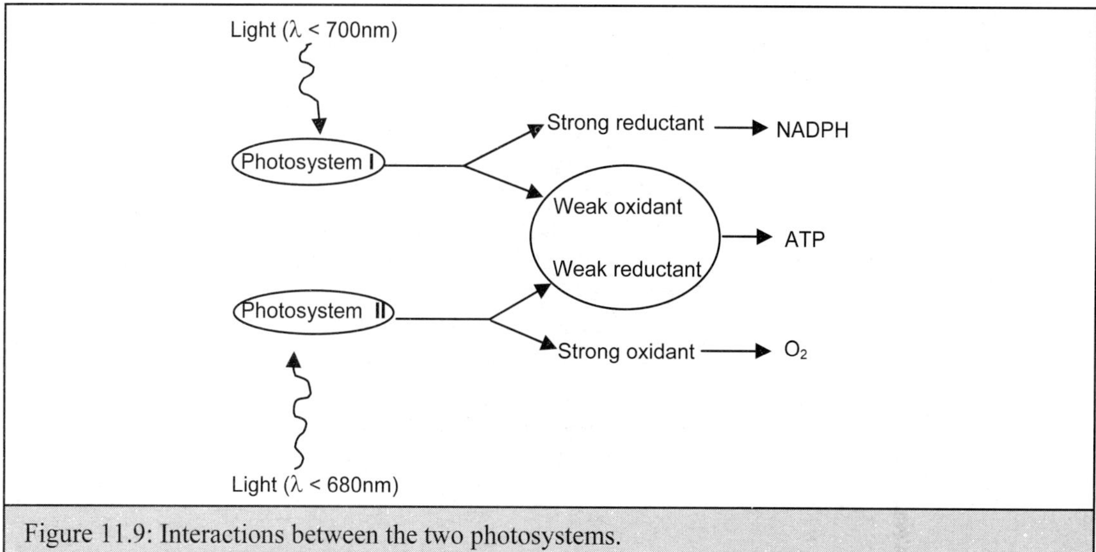

Figure 11.9: Interactions between the two photosystems.

wavelengths. Photosystem II is maximally activated at wavelengths shorter than 680 nm and has more of chlorophyll 'b'. All photosynthetic cells of higher plants and cyanobacteria that evolve oxygen contain both photosystems I and II. Those bacteria that do not evolve oxygen contain only photosystem I. Since photosystem I is maximally excited by light of wavelengths 700 nm, the reaction centre of this photosystem is called as P700. Similarly, the reaction centre of the photosystem II is called P680. Figure 11.9 shows the interactions between and within the two photosystems. When a photon is absorbed by one of the chlorophyll molecules, the molecule goes from the ground state to an excited state. Generally when the molecule returns to the ground state it emits radiation known as fluorescence. There exists a coupled event wherein the de-excitation of one chlorophyll molecule is accompanied by the excitation of another molecule. This process is also known as sensitised fluorescence. As shown in Figure 11.10 the energy levels of the excited state of the reaction centre are lower than that of other chlorophylls. Hence it acts as an energy trap. This transfer of energy takes place in less than 10^{-10} sec.

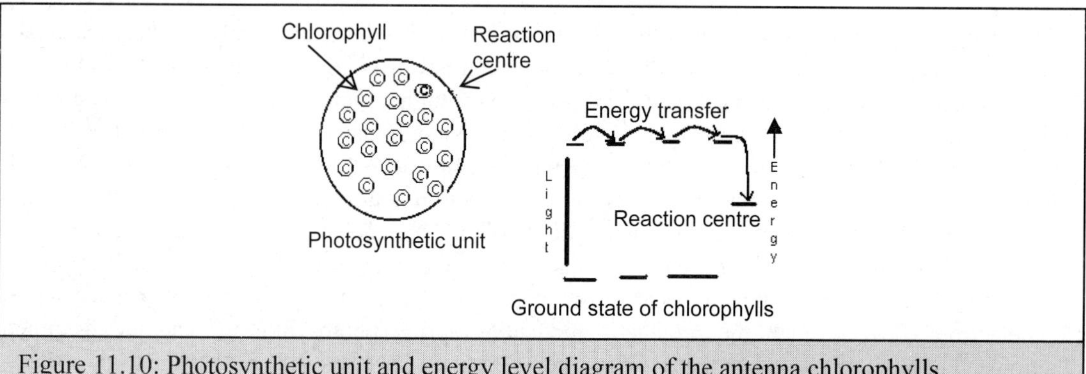

Figure 11.10: Photosynthetic unit and energy level diagram of the antenna chlorophylls.

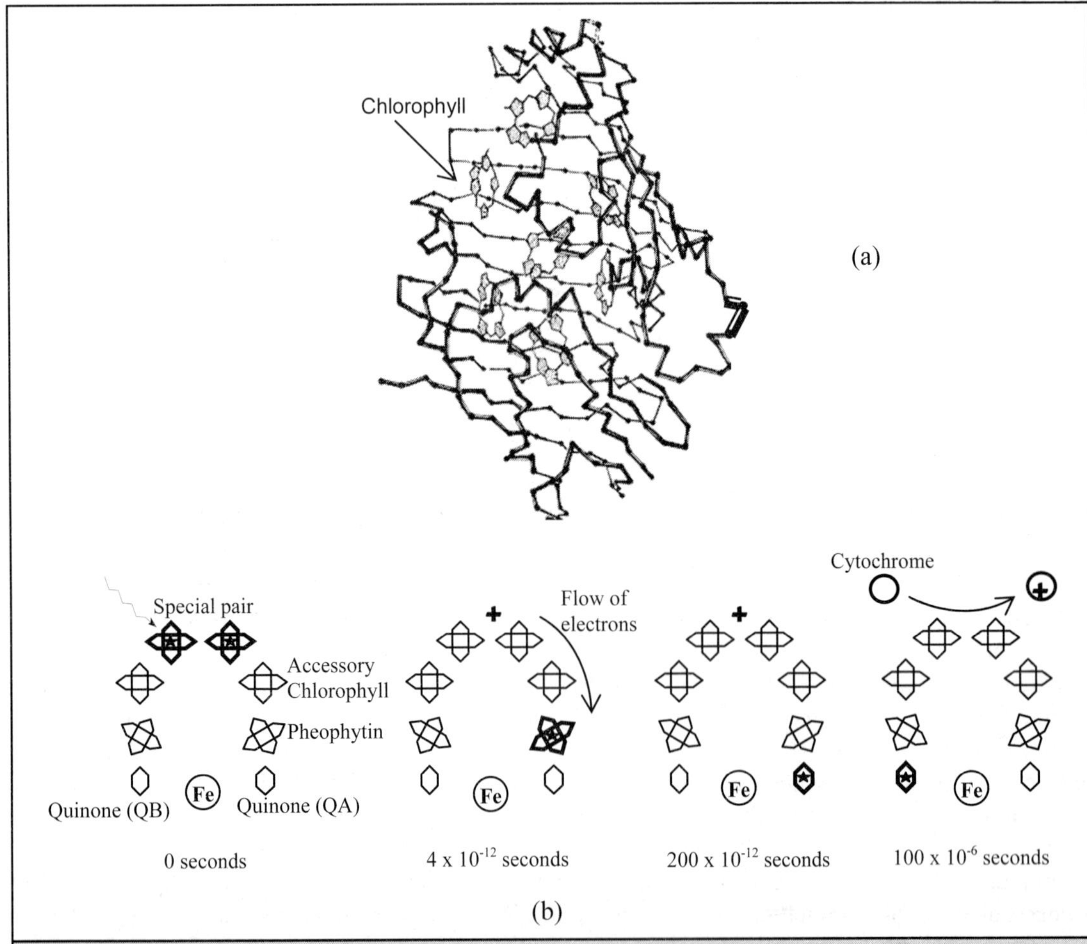

Figure 11.11: (a) Structure of a bacteriochlorophyll-protein complex. (b) Mechanism of electron transport at photosynthetic reaction centre. Excited groups are marked in bold with a star.

The mechanism of photosynthesis in the bacteria of Rhodopseudomonas family has been elucidated by the combined effort of X-ray crystallographers, spectroscopists and molecular geneticists (Figure 11.11). Photosynthesis in these bacteria differs from that of higher plants in that no oxygen is given off during the process. However, the results obtained can be extrapolated to higher plants due to many other similarities. The study throws light on the mechanism of electron transport from one end of the reaction centre to the other end. The photosynthetic reaction centre primarily consists of a protein, embedded in the membrane of the photosynthetic vesicles. One end of the reaction centre is near the inner surface of the membrane and the other end is near the outer surface. Several small molecules known as prosthetic or helper molecules are also partially embedded in the protein complex. Photoelectrons flow through the prosthetic molecules during photosynthesis. The photosynthetic reaction centre has a two-fold symmetry axis running from one surface of the membrane to the other.

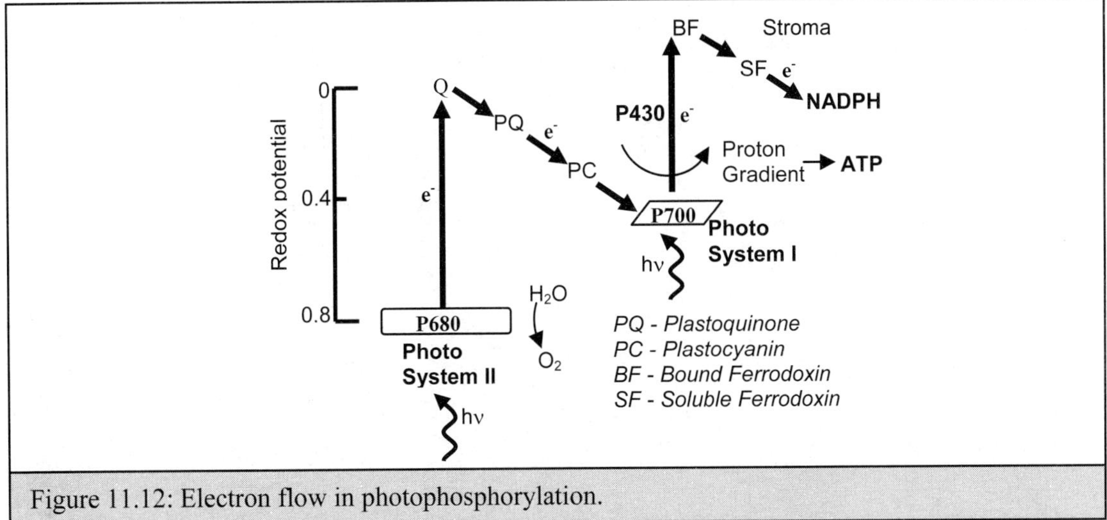

Figure 11.12: Electron flow in photophosphorylation.

The prosthetic groups are arranged in two spirals on either side of the central protein. When light falls on a special pair of chlorophyll molecules situated at the end of the reaction centre near the inner surface, an electron within the pair absorbs the energy and moves to the next prosthetic molecule viz. pheophytin. This step takes 2 picoseconds to complete. During this process the electron passes through another chlorophyll molecule but does not get attached to it. As an electron has left the special pair of chlorophyll molecules, they now have acquired a positive charge. Next the electron moves from pheophytin to a quinone molecule (Q_A) which is at the end of the spiral. The time taken for this process is about 4 picoseconds. Lastly, the electron moves through the central core of the protein to reach the quinone molecule of the other spiral (Q_B) through which electrons do not flow. This step is comparatively very slow and takes about 100 microseconds. In the meantime a cytochrome molecule donates an electron to the reaction centre and neutralises it. Thus the reaction centre is ready to receive another photon and the process repeats itself. When the quinone molecule has acquired two electrons it detaches itself from the reaction centre and participates in the later stages of the photosynthesis, which takes place on the outer surface of the membrane. Thus there will be two cytochrome molecules with positive charges at the inner surface and two electrons near the outer surface. The separation of the two charges represents stored energy. Later, the charge separation drives chemical reactions that are coupled to the bacterial metabolism. In most cases, the transfer of electrons can take place in both the directions and then the energy is lost as fluorescent radiation. But in the photosynthetic reaction centre the forward reaction is more than eight orders of magnitude faster than the backward reaction and hence the electric charges induced by light energy remain separated. The efficiency of the reaction is as high as 50% i.e. the energy stored in the separated charges is about half of the energy of the incident photons.

11.6.1: Photosystem I:

As mentioned earlier, the reaction centre of photosystem I (P700) consists of chlorophyll 'a' molecules. When P700 is excited by the absorption of light by the chlorophyll molecules, it transfers

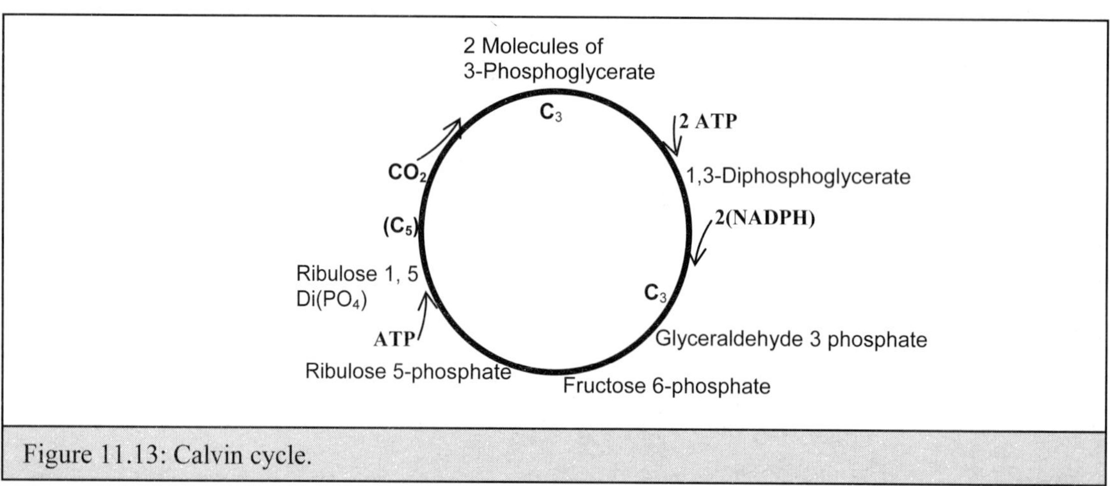

Figure 11.13: Calvin cycle.

an electron to bound ferrodoxin, which is a Fe_4S_4 type iron-sulphur protein. The oxidation-reduction potential of P700 changes from +0.4 V to -0.6 V when it is excited by light. From the bound ferrodoxin the electron moves to soluble ferrodoxin and then to $NADP^+$ reducing it to NADPH. During this process the iron atom of ferrodoxin is alternatively oxidised and reduced. The reduction of $NADP^+$ to NADPH requires two electrons whereas ferrodoxin carries only one electron at a time. Hence electrons from two reduced ferrodoxins come together to convert $NADP^+$ to NADPH (Figure 11.12).

11.6.2: Photosystem II:

The photosystem II (P680) is not very well understood excepting for the fact that the redox potential of this reaction centre is +0.8 V. Photosystem II produces a strong oxidant (A^+) and a weak reductant (B^-) and transfers an electron to P700. A^+ attracts electrons from water and gets neutralised. In this process O is released.

$$4A^+ + 2H_2O \rightarrow 4A + 4H^+ + O_2$$

The electron from photosystem II goes through a bound molecule of plastoquinone, then through mobile plastoquinones, through cytochromes b_{559} and c_{552}, finally to plastocyanin, which transfers it to photosystem I. Thus photosystem I, which was in an oxidised state after absorption of light, is now reduced and is ready for the next cycle of photo reaction. A proton gradient is developed across the thylakoid membrane when electrons flow through the above carriers and, as described earlier, this gradient drives the formation of ATP. The electrons from P700 can also follow a cyclic path, if there is insufficient $NADP^+$ to accept electrons from reduced ferrodoxin. In the cyclic path the electron gets transferred through cytochrome b_{563}, c_{552}, and plastocyanin, back to P700. Thus NADPH and ATP produced during light reactions are now available for subsequent dark reactions in which CO_2 is converted to carbohydrate.

11.6.3: Photophosphorylation and carbon fixation:

In the second phase of photosynthesis, which occurs in the absence of light, CO_2 is fixed into carbohydrate through a series of reactions known as the Calvin cycle. CO_2 condenses with a five-

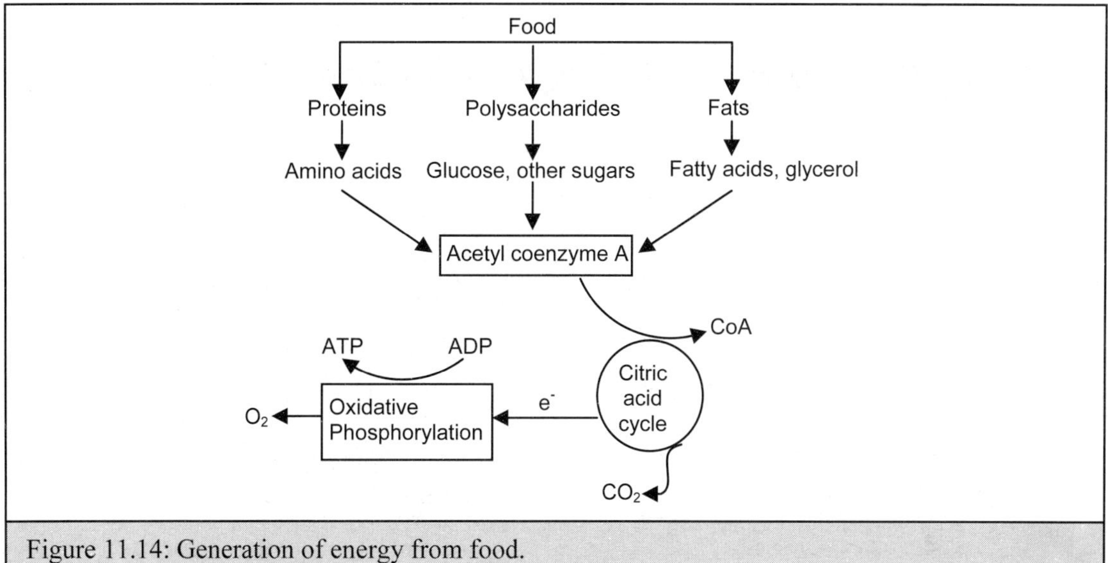

Figure 11.14: Generation of energy from food.

carbon compound (ribulose-1, 5-diphosphate) to form an unstable six-carbon compound. This, in its turn, hydrolyses to two three-carbon compounds (3-phosphoglycerate). This particular reaction can be written as

$$C_5 \rightarrow C_6 \rightarrow C_3 + C_3$$

The cycle is shown in Figure 11.13 on the previous page. This figure illustrates how two molecules of 3-phosphoglycerate, which is a three-cabon compound, combine with three molecules of ATP to produce 1,3-diphosphoglycerate. In the next step, molecules of 1,3-diphosphoglycerate each combine with two molecules of NADPH to yield glyceraldehyde-3-phopsphate, and then fructose and ribulose phosphates, to complete the cycle. Thus, three ATP molecules and two NADPH molecules are consumed in the process of CO_2 fixation as a sugar. As only one carbon is reduced in each cycle, a total of six cycles will be required to form hexose. The complete reaction is given by the equation

$$6CO_2 + 18ATP + 12NADPH + 12H_2O \rightarrow C_6H_{12}O_6 + 18ADP + 18P_i + 12NADP^+ + 6H^+$$

To calculate the efficiency of the system, consider the following. ΔG^0 for the CO_2 reduction is 114 Kcal/mole. $NADP^+$ reduction to NADPH in photosytem I requires two electrons and four photons. Photosystem II replenishes the lost electrons of photosystem I. It is known from a study of the photosystems that, for one electron to be transferred from H_2O to $NADP^+$, two photons must be absorbed by each of the two photosystems. Thus a total of eight photons will have to be absorbed for producing the required number (2) of NADPH molecules, per CO_2 fixed. The energy of the photons may vary from 72 Kcal/mole at 400 nm to 41 Kcal/mole at 700 nm. If the energy of the photon at 600 nm (47.6 Kcal/mole) is taken as the average, the total energy absorbed from the 8 photons is 8 x 47.6 = 381 Kcal/mole. The efficiency of the photosynthetic reaction at 600 nm wavelength is 114/381 = 30% under standard conditions.

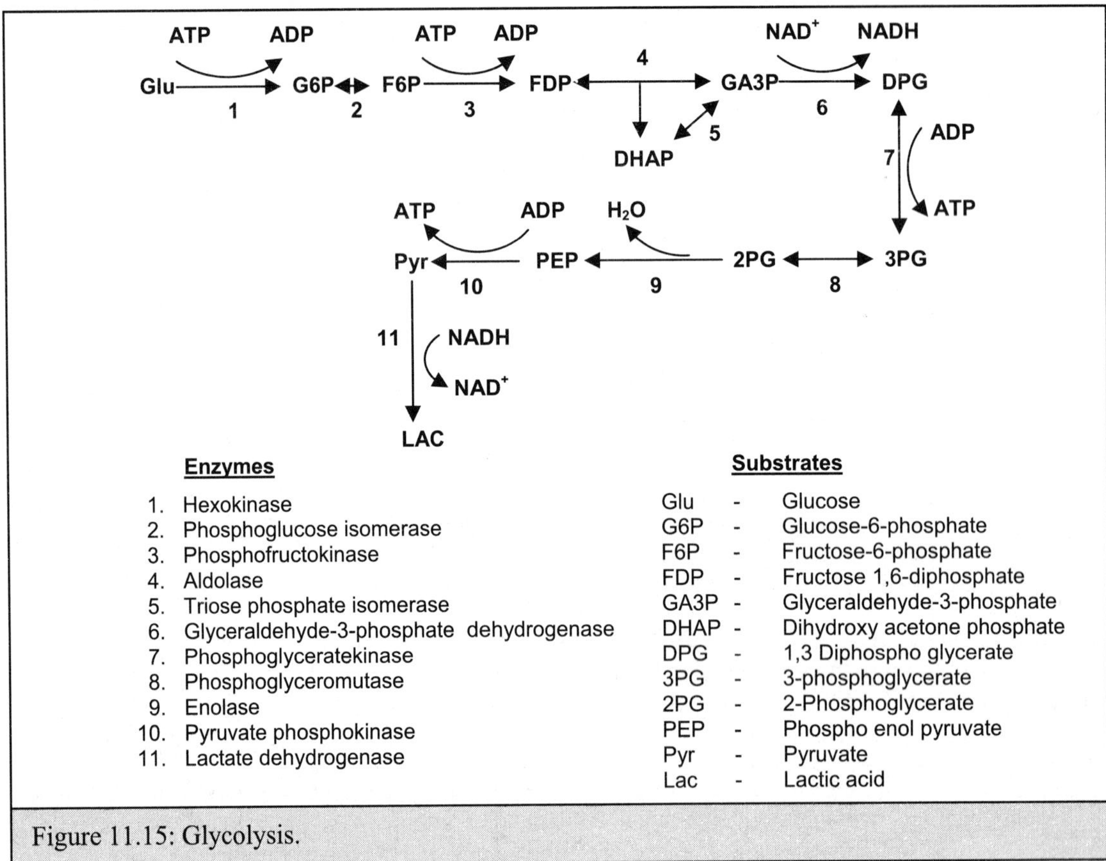

Figure 11.15: Glycolysis.

11.7: Energy conversion pathways

ATP and NADPH are continuously generated and degraded. These processes are regulated and the biosynthesis and the degradation have well-separated distinct pathways. When a high level of ATP is available then anaerobic reactions, which utilise ATP, are stimulated and ATP generating pathways are inhibited. The reverse occurs when the level of ATP is low.

11.7.1: Oxidation:

Oxidation of food is always coupled with the formation of ATP and this process can take place even in the absence of oxygen. In some organisms anaerobic oxidation is the only available pathway for energy conversion whereas in others both anaerobic as well aerobic energy conversion can take place. This is illustrated in Figure 11.14.

11.7.2: Glycolysis:

Glycolysis takes place in the cytosol (Figure 11.15). Glycolysis is a metabolic pathway in which glucose is converted into pyruvate with the concomitant production of ATP. This mechanism precedes

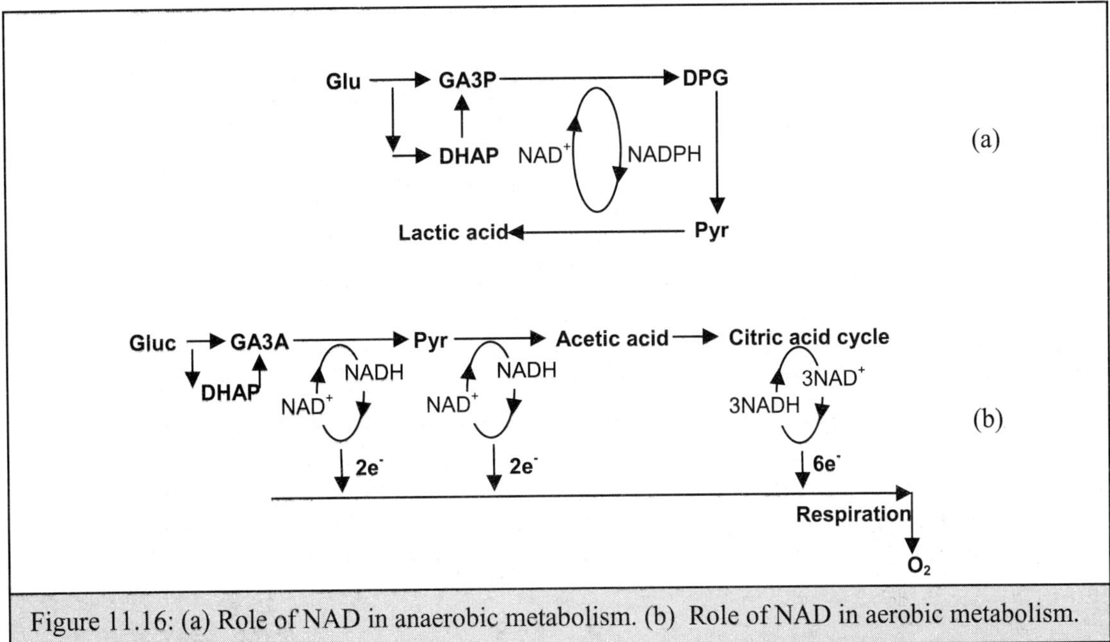

Figure 11.16: (a) Role of NAD in anaerobic metabolism. (b) Role of NAD in aerobic metabolism.

a sequence of other oxidation reactions in higher organisms. Figure 11.16 gives in detail the role of NAD in the glycolytic pathways in anaerobic and aerobic metabolism respectively. When ethanol or lactate is produced from glucose in anaerobic organisms it is known as fermentation. The glycolytic pathway has a dual role in that it degrades glucose to form ATP and also provides building blocks for reactions like formation of long chain fatty acids. Glycolysis is controlled by essentially irreversible reactions, such as those catalysed by hexokinase, phosphofructokinase and pyruvate kinase. In these reactions, two molecules of ATP are generated during the conversion of glucose to pyruvate.

In aerobic organisms, glycolysis is followed by respiration, which takes place in the mitochondria (Figure 11.17). Mitochondria have two membranes. The outer membrane is permeable to small

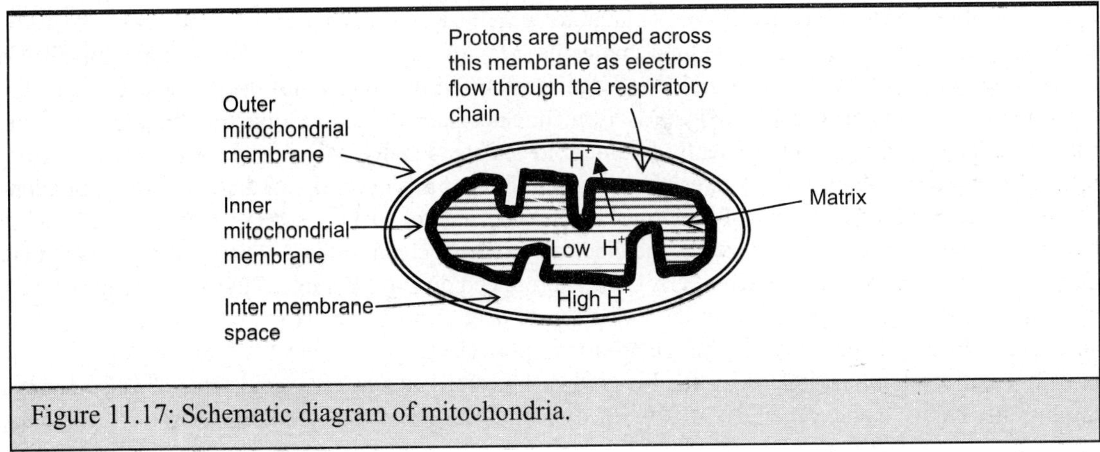

Figure 11.17: Schematic diagram of mitochondria.

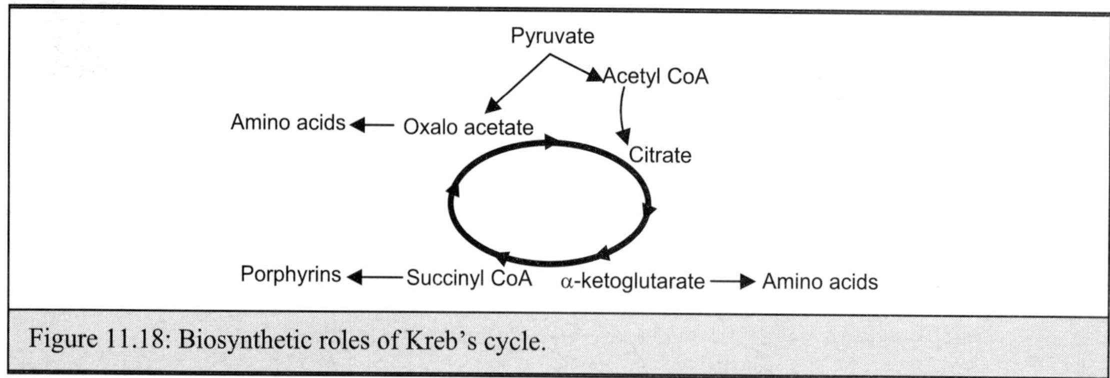

Figure 11.18: Biosynthetic roles of Kreb's cycle.

molecules. The inner membrane forms several inner folds thus increasing the total surface available. These folds are called cristae. They enclose an inner compartment known as the matrix. There are several enzymes in the membrane. The respiratory reactions take place in two steps, the first of which takes place in the matrix. This cycle is named after its discoverer Hans Krebs as Krebs cycle. The second step is part of the respiratory cycle and takes place in the cristae.

11.7.3: The Krebs cycle:

The Krebs cycle is also known as the citric acid cycle, or the tricarboxylic acid cycle (TCA). In addition to the oxidation-reduction reactions that lead to the generation of ATP, the citric acid cycle provides several intermediates, which are branch points for the synthesis of various compounds required by the cell, like amino acids. Most fuel molecules enter this cycle as acetyl coenzyme A (acetyl CoA). The reaction for the formation of acetyl CoA is the following.

$$\text{pyruvate} + \text{CoA} + \text{NAD}^+ \rightarrow \text{acetyl CoA} + \text{CO}_2 + \text{NADH}$$

The Krebs cycle can be written as

$$\text{acetyl CoA} + 3\text{NAD}^+ + \text{FAD} + \text{GDP} + \text{P}_i + 2\text{HO} \rightarrow 2\text{CO}_2 + 3\text{NADH} + \text{FADH}_2 + \text{GTP} + 2\text{H}^+ + \text{CoA}$$

Two carbon atoms from acetyl CoA enter the cycle and condense with oxaloacetate (a four-carbon molecule) to form the six-carbon molecule citrate. Citrate is isomerised to isocitrate and oxidative decarboxylation of isocitrate gives α-ketoglutarate (a five-carbon molecule). One molecule of CO_2 is evolved during this reaction. A second molecule of CO_2 is given out when α-ketogluterate is oxidatively decarboxylated to succinyl CoA (again a four-carbon molecule). In the next stage of the reaction GTP is generated and succinate is formed. Succinate is then oxidized to fumarate (a four-carbon molecule) which is subsequently hydrated to malate. Finally oxaloacetate is regenerated from malate by oxidation. Figure 11.18 gives a schematic sketch of this cycle. In all there are four oxidation-reduction reactions and 3 pairs of electrons are transferred to NAD^+ and one pair to FAD. These reduced electron carriers are oxidized in the electron transport chain to form eleven molecules of ATP. One high-energy phosphate molecule, viz. GTP is produced during the cycle. Thus twelve high-energy phosphate molecules are produced in the oxidation of acetyl group into H_2O and CO_2. When NADH and FADH lose their electrons in the electron transport chain, NAD^+ and FAD are regenerated. As already mentioned, intermediates for the biosynthesis of amino acids, fatty acids etc., are also formed during the Krebs cycle. The loss of intermediates of the cycle due to biosynthesis is constantly

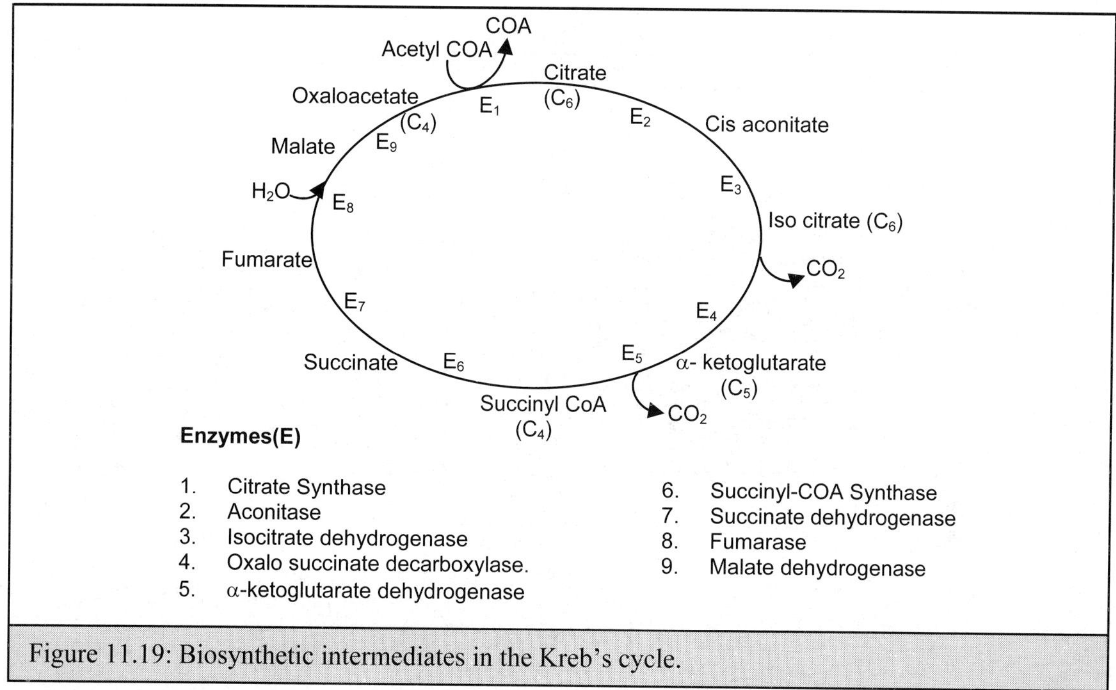

Figure 11.19: Biosynthetic intermediates in the Kreb's cycle.

compensated, since many of the branching off reactions are reversible. In addition to this there are reactions known as anaplerotic (fill up) reactions. For example, if oxaloacetate is converted into amino acids then the supply of oxaloacetate for the Krebs cycle is replenished by the carboxylation of pyruvate.

$$\text{pyruvate} + CO_2 + ATP \rightarrow \text{oxaloacetate} + ADP + P_i + 2H^+$$

The biosynthetic intermediates in the Kerb's cycle are shown in Figure 11.19.

11.7.4: The respiratory chain:

The respiratory chain consists of a series of oxidation-reduction enzymes bound to the cristae membrane of mitochondria. The NADH and $FADH_2$ molecules formed in glycolysis, the citric acid cycle, etc. are energy rich molecules because each is capable of transferring two electrons to molecular oxygen, thus liberating a large amount of energy. This energy can be utilised to generate ATP. This is the major source of ATP molecules in aerobic organisms. The oxidation of NADH, $FADH_2$ and phosphorylation are coupled processes. Figure 11.20 shows the respiratory cycle. The proteins involved in the chain are FMN or NAD dehydrogenases, FAD or succinate dehydrogenases, cytochromes and iron-sulphur protein complexes. Electron transfer from NADH to O_2 takes place through a number of electron carriers like flavins, iron-sulphur complexes, quinones and hemes. It has been observed that when the mitochondrial (cristae) membrane is treated with specific detergents, the components separate into four structured complexes. Complex I contains NADH dehydrogenase and four Fe-S centres. Complex II consists of succinate dehydrogenase and its Fe-S centres. Complex III is made of cytochromes b and c, and a specific Fe-S centre. Cytochrome a and a_3 constitute complex IV.

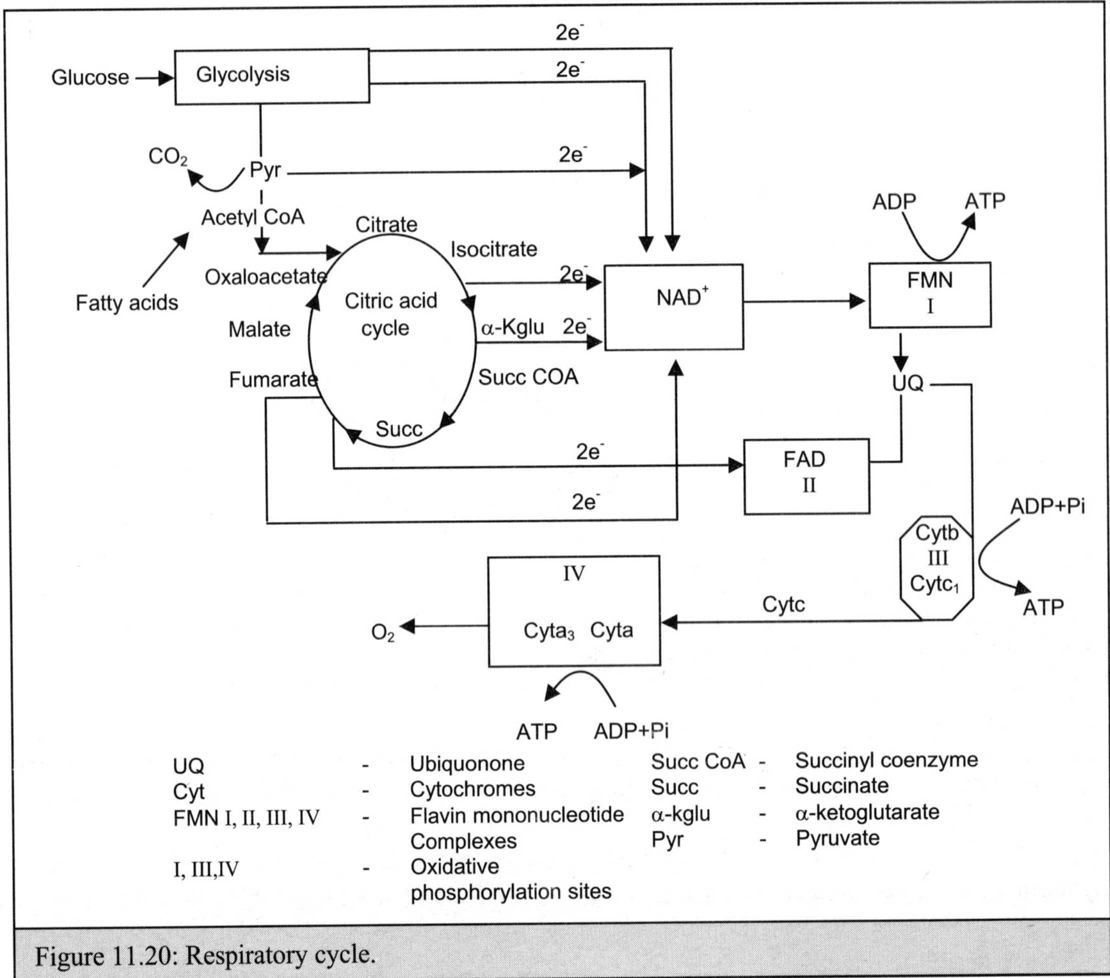

Figure 11.20: Respiratory cycle.

Ubiquinone is the connecting link between complex I, II and III while cytochrome c is the link between complex III and IV.

As discussed in chapter 1, chemical reactions in which electrons are transferred from one molecule to the other are termed as oxidation-reduction reactions. A redox (oxidation-reduction) reaction can be written as:

$$\text{electron donor} \Leftrightarrow e^- + \text{electron acceptor}$$

e.g.
$$Fe^{2+} \Leftrightarrow e^- + Fe^{3+}$$

Fe^{2+} and Fe^{3+} together form a conjugate redox pair. The standard redox potential (E_0') is defined as the electromotive force (emf) in volts measured by an electrode placed in a solution containing both electron donor and its conjugate acceptor in 1.0 M concentration at 25° C and pH = 7.0. Increasingly negative redox potential means an increasing tendency to loose electrons. Similarly increasingly positive potential indicates an increasing tendency to accept electrons. Thus electrons will tend to flow

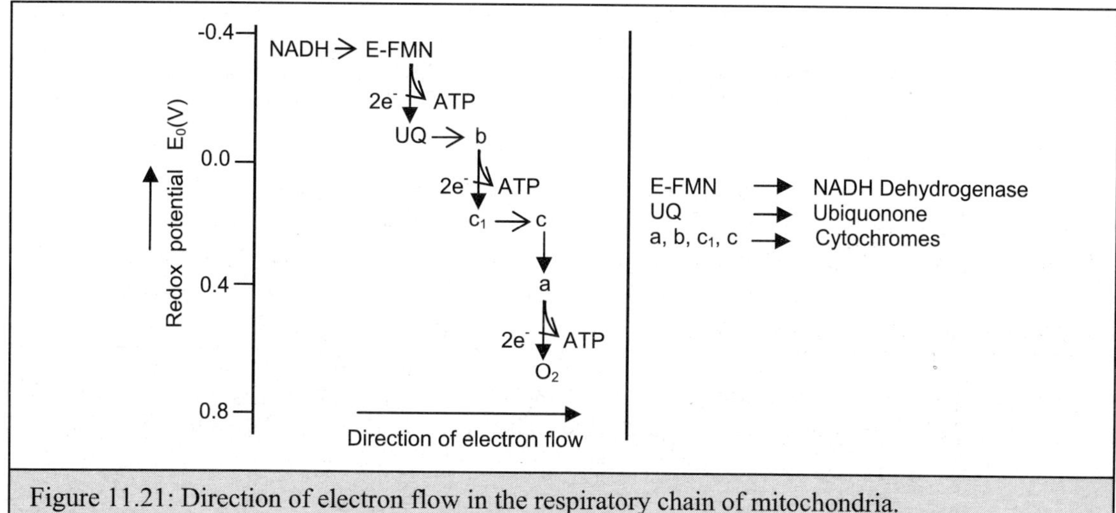

Figure 11.21: Direction of electron flow in the respiratory chain of mitochondria.

from a relatively electronegative conjugate pair like NADH/NAD$^+$ (E$_0$' = -0.32V) to more electropositive redox pair like reduced cytochrome c / oxidized cytochrome c (E$_0$' = 0.23V) (Figure 11.21). This electron flow leads to a loss of free energy, which is utilized for conversion of ADP → ATP. The complete equation for oxidative phosphorylation can be written as

$$NADH + H^+ + 2O_2 + 3Pi + 3ADP \rightarrow NAD^+ + 3ATP + 4H_2O$$

ATP formation by oxidative phosphorylation is regulated by the [ATP]/[ADP][P$_i$] ratio. This ratio is known as the mass action ratio of the ATP system. The square brackets denote molar concentrations. Using absorption, spectrophotometry one can follow oxidation-reduction reactions of the electron carriers. For example, cytochromes, which absorb light in the visible region, undergo spectral changes when the enzyme changes its redox state, and these can be followed. Such coupled reactions can be inhibited by compounds like 2,4-dinitrophenol (DNP), certain classes of antibiotics etc. A protein called coupling factor is required for the coupled reaction to proceed. These coupling factors are located inside the matrix space and connected to the inner membrane by a narrow stalk. (Figure 11.22). The coupling factor, which is referred to as F$_1$, catalyses the synthesis of ATP and is also known as ATP synthetase complex. Electron flow in the electron transport chain results in the pumping of protons across the inner mitochondrial membrane, i.e. from matrix to the cytoplasm. Thus the H$^+$ ion concentration is greater in the cytoplasm and a potential is generated across the membrane, with the cytoplasmic side positive. This proton motive force drives the synthesis of ATP by ATP synthetase

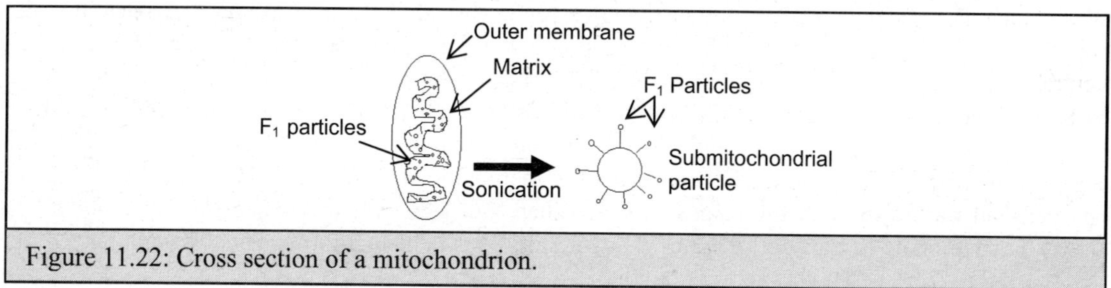

Figure 11.22: Cross section of a mitochondrion.

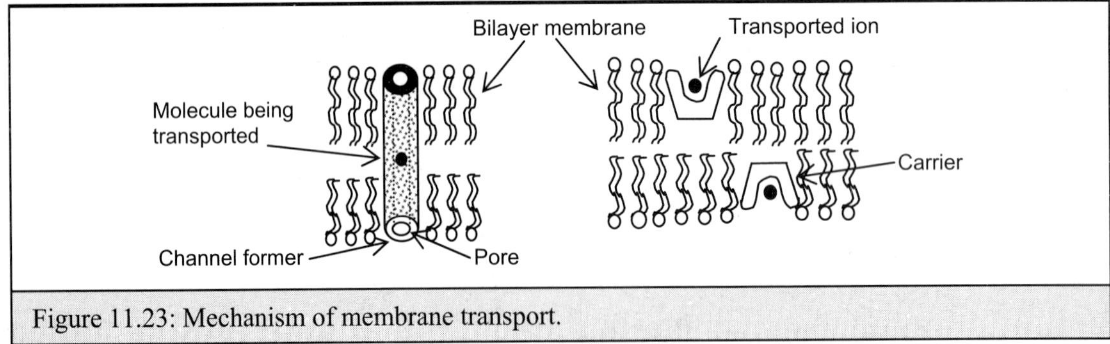

Figure 11.23: Mechanism of membrane transport.

complex. Oxidative phosphorylation generates 32 out of the 36 molecules of ATP produced when glucose is completely oxidized to CO_2 and H_2O. The electron flow from NADH $\rightarrow$ CO_2 is an exergonic process.

11.8: Membrane transport

The inner mitochondrial membrane is impermeable to H^+, OH^-, K^+ and many other ionic solutes. The ATP formed inside the mitochondrial matrix will have to leave the matrix and reach the cytosol to become useful. Similarly any energy converting process requires a continuous supply of substrates and disposal of products and wastes. In several biological functions, products made at one part of the cell will have to be transported to another. Membrane transport, i.e. transport of molecules and ions across the membrane, is achieved in two general ways - (i) active transport, (ii) passive transport. For passive transport no energy input is required and the movement of the solutes is in the direction of the thermodynamic gradient. In active transport the movement is against the gradient and hence requires a source of energy. Membranes are made up of lipid bilayers, which divide the cell into compartments. They have a finite thickness ranging from 60 to 100 Å. They have selective permeability for certain ions and molecules. As the molecules will have to traverse through a hydrophobic lipid bilayer, there is good correlation between the lipid solubility of molecules and the penetration rate across membranes. Thus one would expect the rate of penetration of substances like ions, amino acids, etc. to be poor and this is indeed what is observed. However, this fact alone does not explain membrane transport completely, especially the high degree of selectivity observed.

A carrier is one of several mechanisms proposed for membrane transport (Figure 11.23). There are carrier molecules A in the membrane, which have a high affinity for the species B to be transported. When A and B come together they form a complex A+B. Though B by itself has poor permeability, the complex A+B can traverse the membrane. For simplicity, we assume that the complex is electrically neutral. Valinomycin is an example of a carrier molecule. It is a cyclic peptide antibiotic with a hydrophobic exterior and a hydrophilic interior and is highly specific to K^+ ions. Other antibiotics from micro-organisms, like gramicidin and alamethicin are known as channel formers. As their name suggests, these molecules form channels across the membrane. The specie to be transported binds at one end and travels through the channel to the other end. Channel former molecules do not move across the membrane to transport the ions. The interiors of the channel formers are hydrophilic and the

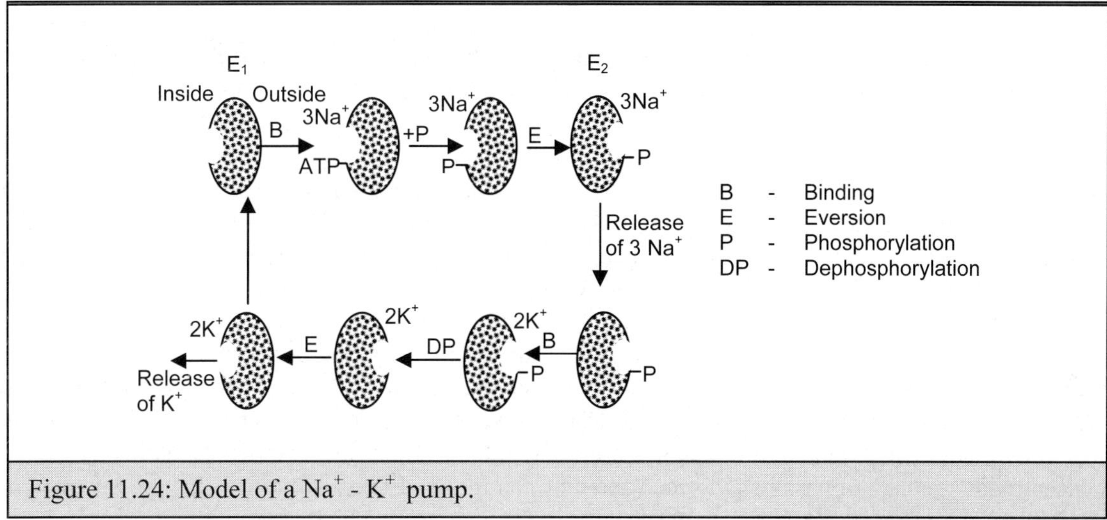

Figure 11.24: Model of a Na^+ - K^+ pump.

exteriors are hydrophobic, just as in carrier molecules. Whether a particular mode of transport is active or passive is determined by the change in the free energy (ΔG) of transported species. If ΔG is positive it is active transport, and the mechanism requires a carrier molecule. If ΔG is negative it is passive transport and can occur spontaneously.

11.8.1: Active transport:

In red blood cells, the cytoplasmic membrane is permeable to both Na^+ and K^+ ions but the K^+ ion concentration inside the membrane is greater than outside. For Na^+ the reverse is true. Thus an ionic gradient is maintained by a specific transport system known as the Na^+–K^+ pump. This ionic gradient is crucial for signal transmission through nerves. This difference in K^+ and Na^+ ion concentration can be maintained only by active transport and in fact the Na^+ - K^+ pump maintains the gradient at the expense of hydrolysis of ATP. Thus active transport is coupled with the process that supplies energy. Under suitable conditions, this pump can be reversed to synthesise ATP from ADP and P_i. A model has been proposed for the Na^+ - K^+ pump. The model takes into account the discovery, by Jens Skou, of an enzyme which hydrolyses ATP and which is present among the molecules that go to make up the pump mechanism. According to this model the hydrolysing enzyme in the Na^+ - K^+ pump must be in a position to adopt two conformations and the affinity for the transported species must be different for the two conformations. These are such that the cavity, or the site at which the species to be transported bind, is on the outside for one conformation and on the inside for the other (Figure 11.24). From the figure we see that if E_1 has a higher affinity for Na^+ while E_2 has a higher affinity for K^+, the pump would function in the required manner. As stated earlier, the action of ATPase is reversible and hence transport of ions across membrane in the direction of thermodynamic gradient must cause the synthesis of ATP from ADP and P_i. In fact the proton gradient across the inner membrane of the mitochondria is linked to the production of ATP. The proof of this comes from the fact that no ATP is synthesised when the membrane is not intact.

11.8.2: Chemi-osmotic theory – passive transport:

As seen in a previous section, during electron transport in the inner membrane of the mitochondria, H^+ ions are ejected, generating a proton gradient across the membrane. According to the model proposed for this process, there are three H^+ conducting loops in the respiratory chain. Each loop translocates two H^+ ions from the inner matrix of the mitochondrion to the exterior through a carrier. The two electrons, which remain after the transport of H^+ ions are carried back to the matrix by a carrier. Thus for every pair of electrons passing from H_2 to oxygen, the loops carry $6H^+$ ions from the matrix to the external medium. Uncouplers like dinitrophenol (DNP) act by carrying protons across the membrane thus dissipating the proton gradient. This is necessary for ATP synthesis. In the case of chloroplasts, light causes water to loose electrons, which flow across the membrane and release protons in the inner medium. The reduced carrier on the other side of the membrane picks up protons from the outer medium and releases electrons to the second light reaction. The alternation of hydrogen and electron carriers on both sides of the membrane causes a net flow of protons across it. The thylakoid membrane and the mitochondrial membrane are permeable to water but not to hydrogen ions. Hence a concentration gradient of protons is formed and a proton motive force is built up. This proton gradient causes an electrochemical potential difference across the membrane which is given by

$$\Delta\mu = \mu_2 - \mu_1 = R\,T\log([H^+]_2/[H^+]_1) + F(\psi_2 - \psi_1)$$

where F is the Faraday constant and $\Delta\psi = (\psi_2 - \psi_1)$ is the contribution from the membrane electric potential. Since $pH = -\log[H^+]$, the above equation can be written as

$$\Delta\mu = 2.303RT(\Delta pH) + F\Delta\psi$$

where $\Delta pH = pH_1 - pH_2$ when hydrogen ions move from position 1 to position 2, and the factor 2.303 is due to conversion from natural logarithms to log base 10. Both ΔpH and $\Delta\psi$ are negative and that is the direction in which ions are pumped in. When the electrochemical potential is divided by the Faraday constant F we get the proton motive force or pmf – analogous to the electron motive force, emf. In chloroplasts almost all the contribution to pmf is from the pH gradient. In mitochondria the contribution from the membrane potential is higher. This is due to the fact that the transfer of ions into the thylakoid space is accompanied by the transfer of either Cl^- in the same direction or Mg^{2+} in the opposite direction leading to neutrality. Hence no membrane potential is developed in chloroplasts.

Summary:

- The ultimate source of energy for all living beings is solar energy. Energy is dissipated when natural processes take place. Living systems retain their internal order at the cost of environment.

- ATP is the universal currency of free energy in biological systems. Free energy is liberated when ATP is hydrolyzed to ADP or AMP.

- Thermodynamically unfavorable reaction can be driven by a favorable reaction if they are coupled through a common intermediate phosphate.

- ATP has a higher group transfer potential as compared to glucose-6-phosphate. Three factors contributing to this are ionization of ATP, repulsive force due to the negative charges in the ATP molecule, and resonance stabilization.

- NAD, NADP and FAD are common cofactors in the oxidation reduction reactions.

- The biological functions of NADP and NAD are different. NADP is used in reductive biosynthesis while NAD is used in the energy metabolism.

- NADPH can be converted to NADH and the enzymes involved in this reaction are known as transhydrogenases.

- Photosynthesis is nature's way of converting solar energy in to chemical energy.

- Green plants, algae and some bacteria are capable of carrying out photosynthesis in the thylakoid membranes of the chloroplast.

- The photosynthetic reaction can be written as follows.

$$nH_2O + nCO_2 \xrightarrow{\text{light}} [CH_2O]_n + nO_2$$

- Two types of reactions viz. photosystem I and II are involved in photosynthesis.

- Photosystem I is maximally excited by the light of wavelength 700 nm and photo system II is excited by 680 nm. Hence they are called as P700 and P680 respectively.

- Photo system I (P700) contains chlorophyll *a* molecules while Photosystem II contains more of chlorophyll *b* molecules.

- NADPH and ATP are produced during light reactions and are utilized in the subsequent dark reactions in which CO_2 is converted into carbohydrate through a series of reactions known as Calvin cycle.

- ATP and NADPH are continuously generated and degraded. The biosynthesis and degradation pathways of these molecules are well separated and are known as energy conversion pathways. These pathways are well regulated.

- Oxidation, glycolysis, Krebs cycle and the respiration chain are the important energy conversion pathways.

- Energy converting processes require continuous supply of substrates and disposal of products and wastes. Hence mechanisms to transport these molecules across the membrane are crucial.

- Transport of molecules and ions across the membrane is membrane transport.

- Active and passive transport are the two main modes of transport across membranes.

- No energy input is required for passive transport while active transport requires energy input.

- Carrier molecules and channel formers play a vital role in membrane transport.

- Channel formers do not move across membranes to transport ions and molecules while carriers move with the molecules to be transported.

- Na^+ - K^+ pump helps in maintaining ionic gradient across the membrane and is crucial for signal transmission through nerves.

- The chemi-osmotic theory describes passive transport.

Chapter 12

Biomechanics

12.1: Introduction

In this chapter we will deal with the conversion of chemical energy into mechanical energy. Mechanical work is done by all living organisms. It gives the organism the ability to move away from a noxious environment or move towards a beneficial region. There are movements inside the cell such as pinocytosis, cell division, etc. However, the mechanism involved in these situations is not well understood. This chapter focuses on larger systems such as the entire organism or the limbs, etc. The best known example in this system of conversion of chemical energy into mechanical work is that of muscle action. The proteins involved in these processes are actin and myosin and the chemical substance providing energy is adenosine triphosphate (ATP). Whenever ATP is converted to ADP (adenosine diphosphate) or AMP (adenosine monophosphate) approximately 8 kcal/mole of energy is released, and is used for mechanical work by the muscle proteins.

Muscles can be classified as smooth muscle, cardiac muscle and striated or skeletal muscle. The walls of some internal organs, like the bladder, the esophagus, intestines, etc. are made of smooth muscles. The contraction of smooth muscles is not under voluntary control but takes place in a slow, generalised manner. The cardiac muscles are fibrous and also slightly striated. They undergo periodic contraction and relaxation automatically. Striated muscles are under voluntary control. They are highly organised and consist of several elongated cells. These elongated cells or muscle fibres in turn contain the myofibrils. In a light microscope the muscle cells show a characteristic repeating structure which gives it a striated appearance.

12.2: Striated muscles

Striated muscle cells are multi-nucleated and are enclosed in an electrically excitable membrane called the sarcolemma. The intracellular fluid surrounding the myofibrils is sarcoplasm. It contains glycogen,

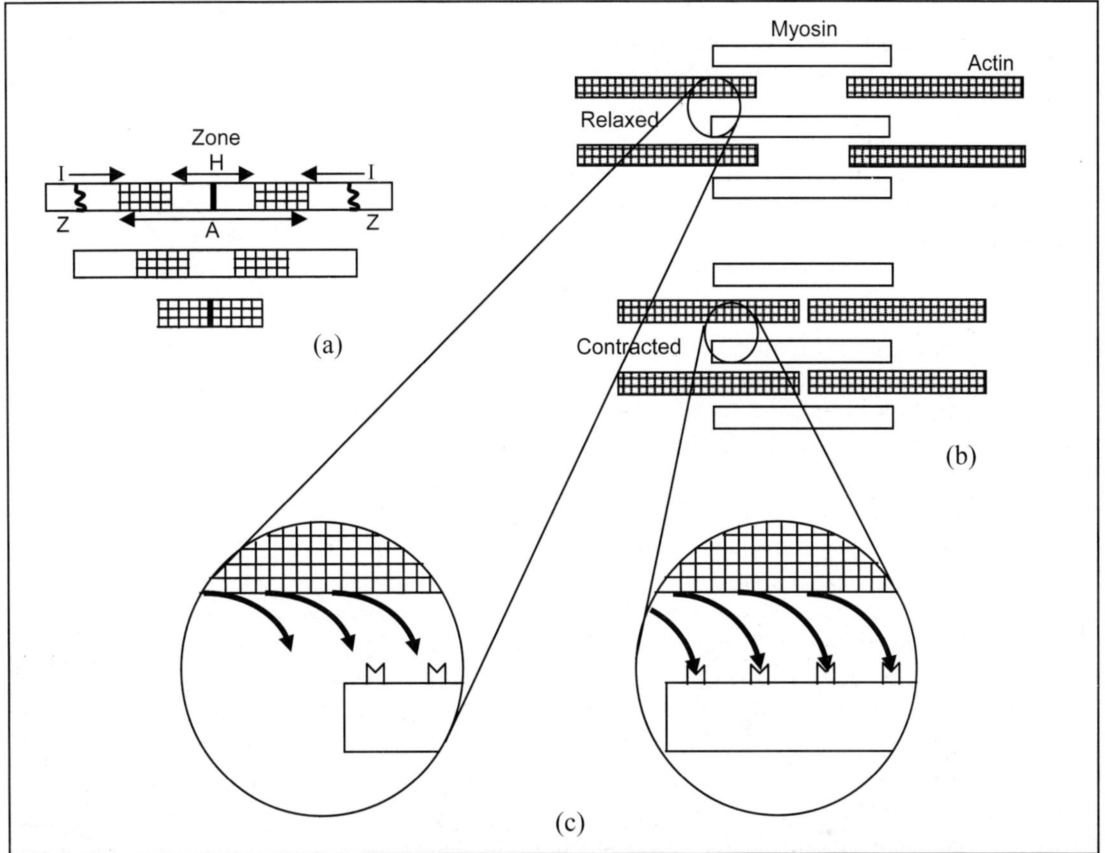

Figure 12.1: A Schematic representation of the sliding model of muscle action. (a) As seen under a microscope. (b) Thick and thin filaments in a contracting muscle.(c) Formation of cross bridges.

ATP, phosphocreatine and glycocyte enzymes. When a longitudinal section of myofibrils is examined under an electron microscope one can see the repeating unit, sarcomere, every 2.3 μm along the fibre axis. A sarcomere has a banded appearance due to regions with different optical properties. A dark 'A' band and a light 'I' band alternate regularly (Figure 12.1). The middle of the A band is known as the 'H' zone and a dark line is found in the middle of the H zone. The 'Z' lines bisect I bands.   There are two kinds of filaments in the sarcomere. The thick filaments make up the H zone of the A band and only thin filaments are found in the I band. These filaments are protein filaments of diameters 150 Å and 70Å respectively. The thick filament consists mainly of the protein myosin while the thin filament consists of actin, tropomyosin and troponin. Each thick filament is surrounded by six thin filaments in the region of overlap, and each thin filament has three neighbouring thick filaments. The filaments interact with each other through what are known as cross bridges. Electron microscopy and optical measurements show that the I zone shrinks when the muscle contracts. Based on this, a 'sliding filament' model of muscle action has been proposed. According to this model, the lengths of the filaments do not change but the length of the sarcomere decreases as the filaments slide past each other during contraction. Thus, though the lengths of the thick and thin filaments do not change, the sizes of the H zone and the I band decrease. The cross bridges change angle during the sliding motion of the filament. When at rest the thin and thick filaments overlap only to an extent of about one third of their length. At maximal contraction, the thick and thin filaments fully overlap, and shortening is about 32 Å (Figure 12.1(b))

12.2.1: Contractile proteins:

The major components of myofibrils are the protein myosin (50 to 55% of the total protein) and actin (20 to 25%). Actin is the major constituent of thin filaments. When extracted from muscle, a globular protein known as G-actin emerges. In the presence of Mg^{2+} and ATP G-actin polymerises to form the fibrous F-actin. F-actin looks like two strings of beads wound round each other when looked under electron microscope. The other two proteins that are closely associated with actin in thin filaments are tropomyosin and troponin. Tropomyosin is a double stranded α-helical rod arranged parallel to the long axis of the thin filament. Each tropomyosin molecule covers about seven G-actin monomers. Troponin is a complex protein with three subunits viz. TnC, TnI and TnT having the molecular

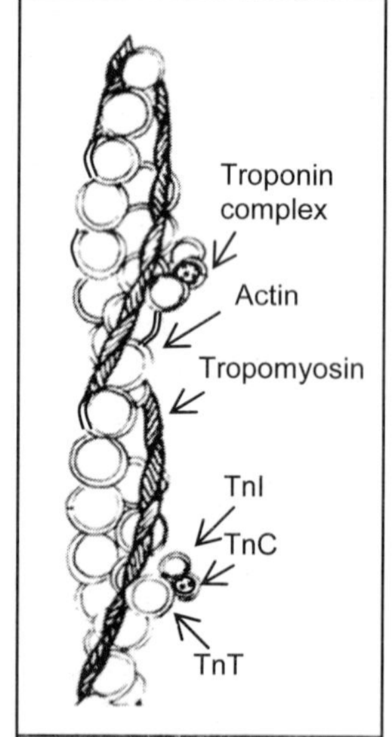

Figure 12.2: Model for a thin filament in the resting stage.

weight of 18, 24 and 37 kDa respectively. Troponin is attached to the tropomyosin filament, one on each molecule. TnC binds calcium ions, TnI binds to actin and TnT binds to tropomyosin. The troponin-tropomyosin complex plays an important role in the regulation of contraction, which is effected by Ca^{2+} ions (Figure 12.2).

The thick filament consists mainly of myosin molecule whose molecular weight is about 500 kDa. Electron microscopy shows that myosin has a double headed globular protein attached to a long rod. The rod is a two-stranded α-helical coiled coil. When myosin molecule is cleaved using the enzyme trypsin, it splits into two fragments, known as light meromyosin (LMM) and heavy meromyosin. The heavy meromyosin molecule, which contains the globular heads, has ATPase activity while the light meromyosin lacks this activity. The heavy meromyosin (HMM) does not form filaments (Figure 12.3). The globular heads also have a strong affinity for actin suggesting that this part of the myosin molecule must be involved in forming the cross bridges with actin in the myofibril. The formation of this complex leads to a large increase in viscosity, but this is reversed on addition of ATP. Figure 12.3(a) is a schematic diagram of the myosin molecule showing the six-polypeptide chains that form this macromolecular assembly. Myosin has a molecular weight of about 500 KDa and is made up of six polypeptide chains, viz. two heavy chains (= 400 KDa) and two pairs of light chains (each with a molecular weight of about 20 KDa). Each one of the myosin heads is globular in nature and has about 850 amino acids contributed by the respective heavy chain and two light chains. The remaining portions of the two heavy chains assemble in an extended coiled coil form, which is about 1500 Å long.

12.3: Mechanical properties of muscles

Muscles are transducers capable of converting chemical energy into mechanical energy. In a typical muscle function a load is moved through a distance due to shortening of the muscle fibres. The contractile unit of a muscle fibre is known as a tension generator. A muscle is elastic and develops tension when it is stretched. When a muscle is greatly stretched, the tension increases to a point beyond which muscle fibres can break. The elastic system of a skeletal muscle basically consists of three components, a contractile component and two elastic components. Sarcoplasm and sliding filament constitute the contractile component and tendons, disc, connective tissues and sarcolemma form the elastic components. Experimental observations of muscle contraction can be done using thermal, mechanical or electrical stimulus. However, electrical stimulus is preferred as the muscle is not damaged in the process and the intensity of the stimulus can be accurately measured using a galvanometer. A muscle twitch, which is a simple response to the threshold stimulus, is useful in the study of properties of muscles. A muscle twitch goes through three stages: (1) Latent period. (2) Contraction phase. (3) Relaxation phase. A muscle fibre can continue to be in tension if a series of stimuli are applied at high frequency without allowing the muscle to relax. This is known as tetanus. A fibre capable of maintaining tetanus for a considerable time can stop contraction due to fatigue if the tetanus is prolonged too long.

12.3.1: Contraction mechanism:

The decrease in viscosity when ATP is added to the troponin-tropomyosin (or actomyosin) complex undergoes slow recovery when ATP is hydrolysed. This is because ATP is required for breaking the cross bridges between actin and myosin rather than for making it. Thus the steps involved in muscle contraction are cyclic and can be designated as the following.

1. Binding of ATP to the myosin head (also known as S_1 head).

2. Activation of myosin ATP complex.

3. Binding of the activated myosin-ATP complex to actin in thin filaments in the presence of Ca^{2+} ions.

4. ATP hydrolysis followed by the change of orientation of S_1 heads with respect to the filament.

5. Binding of ATP to myosin heads leading to the dissociation of myosin-ATP complex from actin filament.

6. The cycle repeats from step 2.

Thus the above model proposes two forms for the actomyosin complex: (1) The activated form. (2) The inactivated form. The binding of myosin-ATP complex to thin filaments in the presence of Ca^{2+} ions is the activated form. The part that is left after hydrolysis of ATP is the inactivated form. Figure 12.3(b) shows the details of the proposed mechanism of muscle contraction. The contractile force is generated by the making and breaking of the complexes between S heads of myosin and actin. In the resting stage, S_1 heads are detached from thin filaments, and myosin contains tightly bound ADP and Pi. When muscle is stimulated, S_1 heads move away from thick filaments, and get attached to the thin filaments in a perpendicular orientation. The tightly bound ADP and Pi are released and the orientation

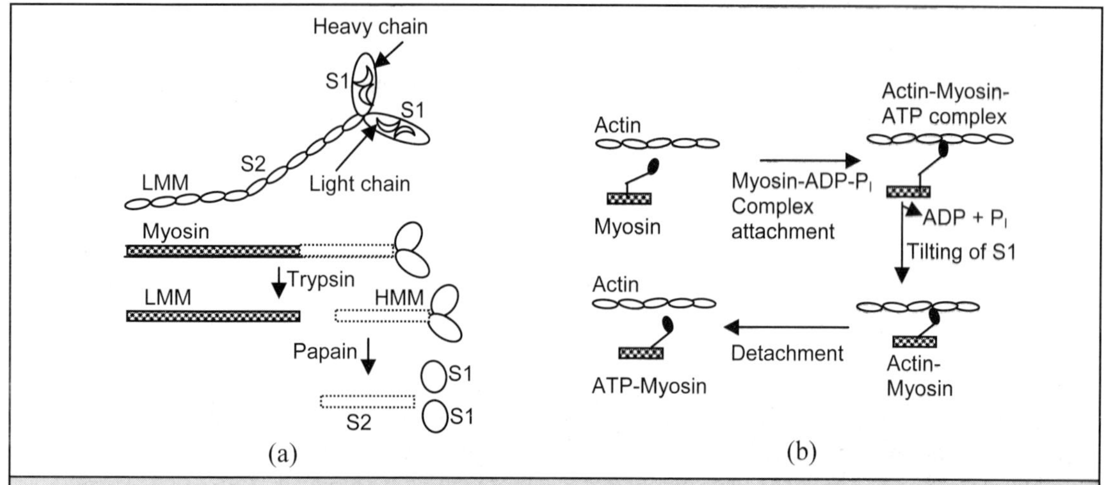

Figure 12.3: (a) Schematic diagram of myosin molecule. (b) Proposed mechanism of muscle contraction.

of S_1 head changes to make an angle of 45° with the thin filament axis. This tilt of the S_1 head is the power stroke of muscle contraction. The tilt is then transmitted to the thick filament through the S_2 unit (rod region) of the myosin molecule. This results in pulling the thick filament by about 75 Å with respect to the thin filament. In the final step, the S_1 head is detached from the thin filament by the binding of ATP and reverts back to its perpendicular orientation. Finally, hydrolysis of ATP by S_1 completes the cycle. Two types of hinges are invoked in the contractile mechanism – one between the S_1 and S_2 units, and the other between S_2 and light meromyosin.

12.3.2: Role of Ca^{2+} ions:

It has been shown that the onset of contraction is mediated by Ca^{2+} ions. The contraction of muscle is initiated by the arrival of a nerve impulse at the end of the muscle fibre and the consequent release of acetylcholine. The muscle membrane (the sarcolemma) has a resting potential of 70mV. The response of the membrane is a change in membrane permeability, with Na^+ ions flowing into the muscle and K^+ ions out of it, and the consequent change in membrane potential. Within a few milliseconds the cell begins to contract as the local response propagates over the membrane. The nerve excitation triggers the release of Ca^{2+} ions by the sarcoplasmic reticulum. The released Ca^{2+} ions bind to TnC component of troponin and causes conformational changes, which are transmitted to tropomyosin and actin. Consequently tropomyosin moves towards the helical groove of the thin filament which enables S_1 heads of myosin molecules to interact with actin unit of the thin filament. Contraction takes place with concomitant hydrolysis of ATP till Ca^{2+} ions are removed and this causes the S_1 head to be blocked by tropomyosin again. Thus the binding of Ca^{2+} to troponin causes the activated myosin-ATP binding site on actin to be exposed. This is otherwise blocked by troponin and tropomyosin (Figure 12.2). The flow of information can be summarised as

$$Ca^{2+} \rightarrow troponin \rightarrow tropomyosin \rightarrow actin \rightarrow myosin$$

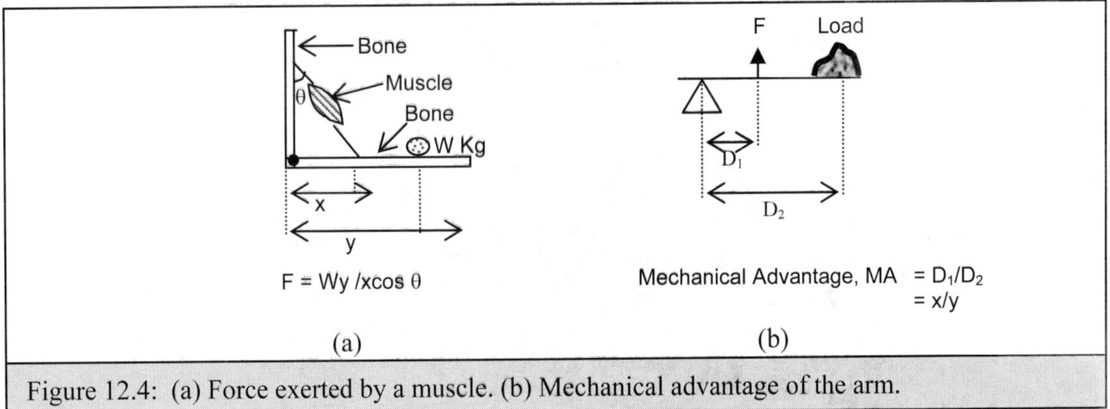

Figure 12.4: (a) Force exerted by a muscle. (b) Mechanical advantage of the arm.

Though it is known that ATP is the source of energy for muscle contraction, the supply of ATP is through a phosphocreatinine which has higher phosphate group transfer potential than ATP.

$$\text{phosphocreatinine} + \text{ADP} \Leftrightarrow \text{ATP} + \text{creatine}$$

However phosphocreatine itself cannot make muscles contract.

The experiments on muscle contraction are performed by applying the potential directly to the tissue. Single fibres respond only when the potential exceeds a threshold and hence are not preferred. Two types of experiments on muscles can be carried out. In one case, both the ends of the muscle are fixed so that contraction occurs without shortening. This condition is known as isometric. In the second case only one end of the fibre is fixed and hence the muscle contracts and shortens and this condition is known as isotonic. A.V. Hill carried out fundamental studies on muscle behaviour and he showed that the maximum tension exerted by a muscle is about 5 Kg-m/cm^2 of the muscle cross section. Figure 12.4 shows how the mechanical advantage and efficiency of muscle contraction may be calculated.

12.4: Biomechanics of the cardiovascular system

The blood circulatory system is a transport system that transports gases like O_2 and CO_2, nutrients, wastes, hormones and immunologically active substances. Blood flow can be considered as a more or less closed circulatory system, which is pumped by the heart. (It is not a completely closed system since some fluid, called lymph, leaves the capillaries and flows into the tissue spaces). Blood is a suspension of different types of single cells in a viscous medium containing proteins, organic salts etc. Kidneys, heart and lungs are the major organs that interact with blood. The vessels into which blood is pumped by heart are known as arteries. They branch into smaller and smaller arteries and finally form capillaries. Most of the exchanges between blood and the surrounding tissues take place at the capillary level. The capillaries then join together to form venules, leading to larger and larger veins which bring back the blood to the heart. The kidney is the most crucial organ. It removes waste products and excess water from the blood, thus maintaining the blood volume, ionic concentrations, arterial pressure etc. Thus the circulatory system is in a dynamic state, losing water to kidneys, lungs, exocrine organs (e.g. salivary glands), etc., and re-absorbing it from the gut and also that formed during metabolic processes.

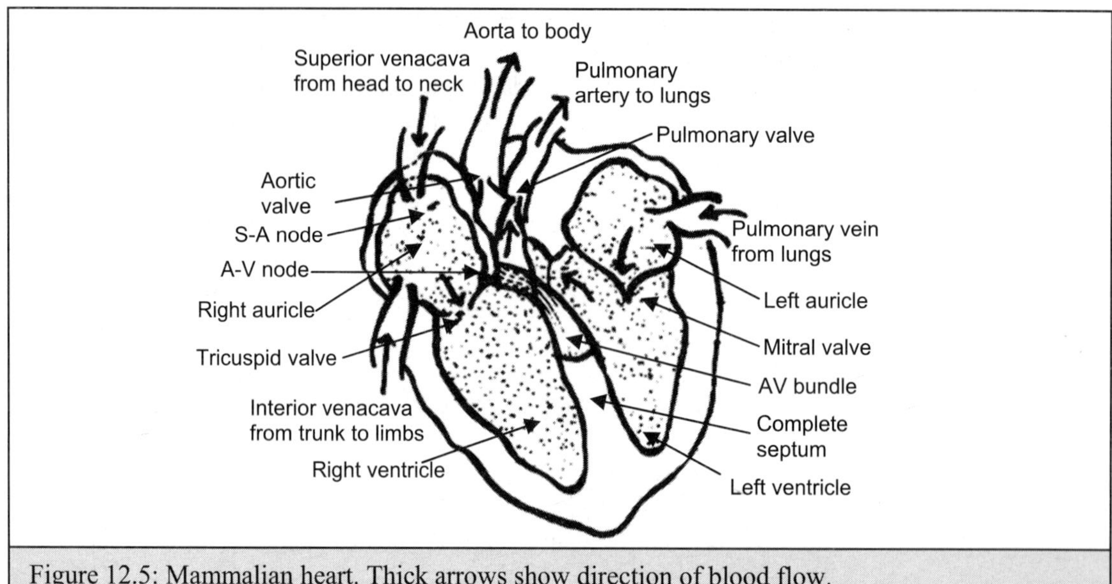

Figure 12.5: Mammalian heart. Thick arrows show direction of blood flow.

12.4.1: Blood pressure:

The fundamental properties of any fluid are pressure, velocity and density. Pressure is force per unit area, and is generally measured in terms of the height of a column of liquid that can be supported. The most frequent system of units used to measure blood pressure is millimetres of mercury, being the height of the column of the liquid metal supported by blood pressure. The maximum arterial pressure is called the systolic pressure and the minimum arterial pressure is called the diastolic pressure. The pressure falls when the blood reaches the capillaries. At the venous system it is still lower. The arteries and veins have very similar inner diameter (0.5 to 12.5 mm) and flow rates, but the pressures are different. This is due to the fact that arteries have thick, elastic walls while veins have thin walls. The larger pressures in arteries prevent any reverse flow of blood from veins. The pulsations in the arteries are due to the stretching of its elastic walls from the force of the heartbeat.

The mammalian heart is shown in Figure 12.5. It consists of four chambers, two auricles and two ventricles. Blood from the entire body, excepting from the lungs, enter the heart at the right auricle. From there it is pumped into the right ventricle, then into the lungs and finally back into the left auricle. From the left auricle the blood goes to the left ventricle and from there through the aorta (which is the largest blood vessel leading away from the heart) to the arteries and is then distributed to the entire body, excepting the lungs. This system of circulating blood is very efficient in supplying oxygen and removing carbon dioxide. Two valves prevent the blood flowing back from the ventricles into the arteries. They are the tricuspid valve and the mitral valve, also called the atrio-ventricular valves. Two other valves prevent the back flow of the blood from the main arteries into the respective ventricles. These are the semilunar valves, viz. the pulmonary valve and the aortic valve (Figure 12.6).

The mechanical action of the heart is due to the heart muscle (myocardium) and is analogous to those of the skeletal muscle. As in other striated muscles the cardiac muscle fibre membranes are also

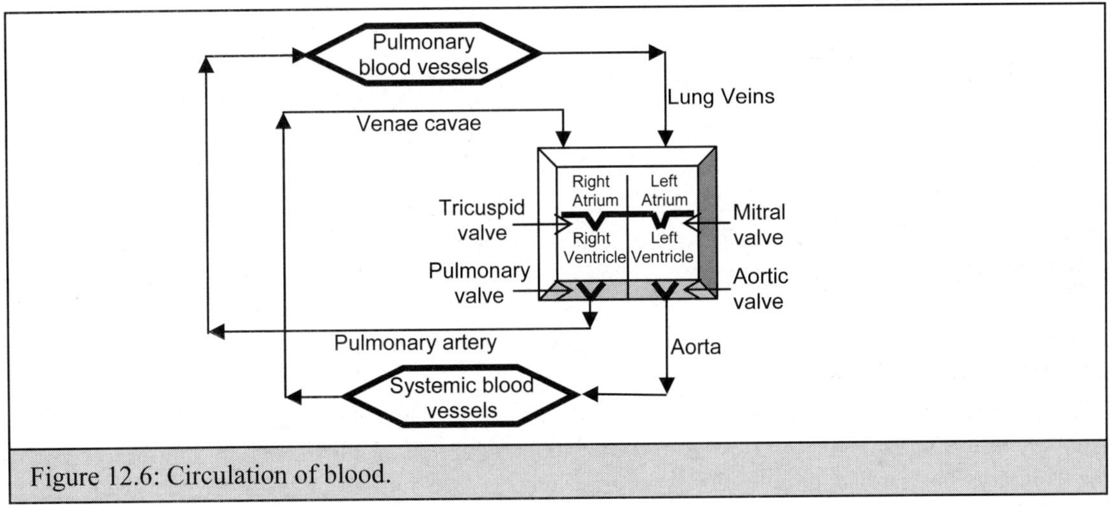

Figure 12.6: Circulation of blood.

polarised, i.e. they have 90mV negative potential between the inside of the membrane and the outside. Just as in nerve fibres, the polarity reverses for a short period of time, when an impulse is received by the membrane, just before contraction occurs. The action potential is around 120mV, i.e. the outside will be 30mV more negative than the inside at the peak of the spike. The large collection of cardiac muscle fibres acts like a group of electric cells and leads to measurable potential changes on the body surface. These potential differences can be measured using an electrocardiogram. The heartbeat is initiated by a sino-auricular (s-a) node and the node acts like an electronic multivibrator letting out one electrical pulse per heart cycle. The heart pulses are periodic and rhythmic. During systole, which is the active part of the heart cycle and when the arterial pressure is at a maximum, the blood from ventricles is forced into aorta and pulmonary artery respectively. In the beginning of this cycle all the four valves are closed. As the pressure increases, and the ventricular pressure just exceeds the pressure at the roots of the aorta or the pulmonary artery, the semilunar valve opens. The ventricular pressure reaches a maximum and gradually decreases and this causes the semilunar valve to close. The second part of the heart cycle known as diastole then begins with the relaxation of the ventricular wall and a drop in the ventricular pressure with respect to the corresponding attrium. Hence the atrio-ventricular valves (a-v) open and the blood enters the ventricles. At the end of the diastolic cycle the atria contract and the pressure inside the ventricles increases. Immediately after this the ventricular valves close and the systol and diastole cycle is repeated.

12.4.2: Electrical activity during the heartbeat:

The heart pulses generated by the s-a node spread over the surface of the auricle and make the muscle fibres to contract. In addition, the electrochemical pulses stimulate the a-v node. After a time delay of 0.1 seconds or less, this node lets out a new electrical pulse that is carried by special fibres known as a-v bundle to the septum. These fibres terminate at different points in the septum and on the outer ventricular wall. Thus the fast conduction of the impulses by a-v bundle synchronises with the ventricular contraction. The (a-v) node can take over the control of heart rate when the (s-v) node has failed. Under these circumstances auricular contraction will not be properly synchronised with

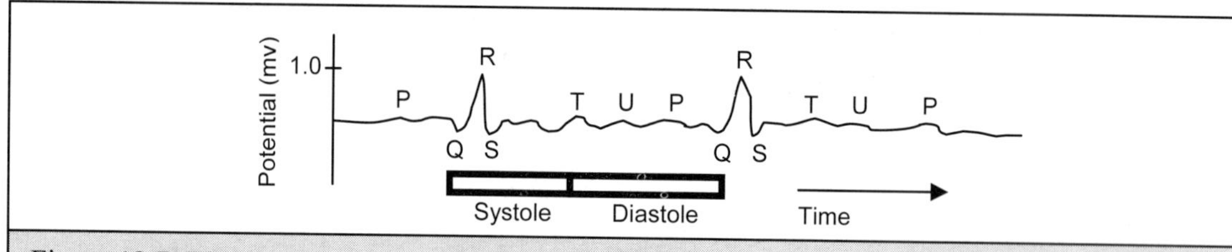

Figure 12.7: Electrocardiogram. P wave precedes auricular contraction. QRS is due to ventricular contraction.

ventricular action. If the a-v node also fails, the auricular and ventricular walls take over the control of heart rate. These conditions are not fatal.

The difference between cardiac muscle fibres and nerve fibres lies in the recovery process. The recovery of resting potential in nerve axons takes about a fraction of millisecond to 5 milliseconds. On the contrary, the cardiac muscle fibres take as long as 200 milliseconds to recover their resting potential. Similarly, there is a subtle difference in the permeability of K^+ ions during and after the action potential in the case of cardiac muscles as compared to nerve axons. This was seen by experiments using squid axons. The permeability of K^+ ions increases suddenly and then drops when the resting potential of a voltage clamped squid axon is suddenly decreased. In the case of the cardiac muscles, the original low or zero permeability to potassium ions is recovered only after the membrane potential returns to its original value.

12.4.3: Electrocardiography:

As mentioned above, electrical potential changes occur during heartbeats and this is spread over the surface of the body. Electrodes at any two points of the surface of the body will be able to measure this potential difference and it can be recorded using an electrocardiogram (ECG). As the heart is not the only organelle capable of generating such potentials at the body surface, adequate precautions will have to be taken while recording an ECG. Even a movement of a skeletal muscle can give rise to body surface potential difference and hence the patient is made to lie down during recording of the ECG. The electrodes are placed generally at the right arm and left leg. Sometimes three wires are attached, one to each arm and the third to left leg. The ECG is then recorded between each pair of electrodes. The maximum potential difference is about 1mV and this is adjusted to show 1cm deflection on the ECG. The electrical signals corresponding to one heart cycle, also known as the waves of the signal, are shown in Figure 12.7. The P wave occurs just before the auricular contraction and the QRS complex occurs at the start of the ventricular contraction. The end of ventricular contraction gives rise to T waves. The ECG cycle is described in Table 12.1. The potential difference between two points can be written as

$$V_1 = V_{LA} - V_{RA}; \quad V_2 = V_F - V_{RA}; \quad V_3 = V_F - V_{LA}$$

(The subscripts LA, RA and F refer to the left arm, the right arm and foot respectively. Instead of the foot, the chest or the back can also be used for potential measurements.) Therefore

$$V_2 = V_3 + V_1$$

Table 12.1. The ECG cycle of signals and the corresponding cardiac functions. The normal QRS interval is about 60 to 100 milliseconds.		
ECG waves	Duration (milliseconds)	Corresponding Heart cycle
P	8	Precedes auricular contraction
P-Q	150-200	a-v delay time
Q	40-80	Precedes ventricular contraction
R		
S		
S-T	100-250	Ventricular ejection
T	100	Follows ventricular ejection
T-P	300	Diastole

This is known as Einthoven's Law. Thus if any two of the three potential differences (also known as leads in ECG terminology) are measured the third can be calculated. Further if V_1, V_2 and V_3 are taken to be vectors, they are chosen such that

$$V_{LA} + V_{RA} + V_F = 0$$

taking into account the signs of the potentials (Figure 12.8). The triangle thus constructed is known as Einthoven's triangle. Einthoven further showed that the three vectors could be considered as projections of a single hypothetical heart vector **H** on this triangle. Though this relation is true for any three points, the **H** vector will be meaningful only when it is related to the electrical axis of the heart. The three points chosen are such that they are equidistant from the heart so that the triangle is an equilateral one.

A realistic model for the electrical events of the heart cycle is based on the assumption that the heart acts as a single dipole, represented by the **H** vector, and the torso is a homogeneous conductor. The recording of the magnitude and spatial orientation of the equivalent heart dipole as a function of time is

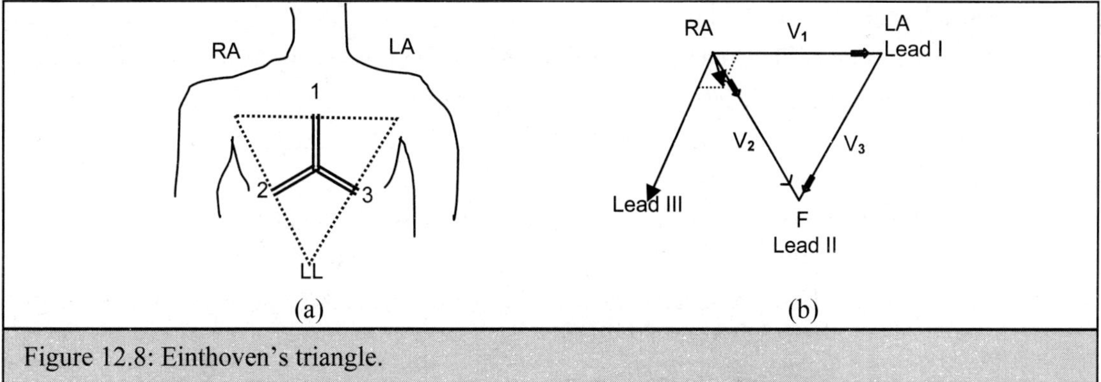

Figure 12.8: Einthoven's triangle.

known as vector cardiography. However, in practice it is not possible to locate a single equivalent dipole and to monitor its magnitude and orientation with respect to time. One of the ways of improving the performance of Vector ECG is to increase the number of electrodes used in order to get a refined value of the dipole. Clinical ECG tests may therefore involve use of 12 or more leads to aid in diagnosis In spite of its great utility, many of the abnormalities in the functioning of the heart may not be obvious in vector cardiography. A more appropriate method would be to represent heart as a collection of dipoles. By the use of such techniques and of computer based automated diagnosis, vector cardiograms have proved clinically very useful.

Summary:

- Muscle action is an example for the conversion of chemical energy into mechanical energy.
- Muscles are classified as smooth muscle, cardiac muscle and striated or skeletal muscle.
- Striated muscles are under voluntary control and consist of elongated cells or muscle fibres.
- These muscle fibres contain the myofibrils.
- Striated muscle cells are multinucleated and are enclosed in an electrically excitable membrane called sarcolemma.
- The intra cellular fluid surrounding the myofibrils is sarcoplasm. There are two kinds of filaments in the sarcomere, the thick filament and thin filament.
- Actin and myosin are the two major contractile proteins.
- Myosin is found in thick filaments while actin is found in thin filaments.
- Actin consists of a globular protein, G actin and a fibrous protein, F actin.
- Tropomyosin and troponin are the other two proteins associated with actin in thin filaments.
- In the sliding filament model of muscle action the lengths of the filaments do not change. However, the length of the sarcomere decreases as the thick and thin filament slide past each other during contraction.
- Muscles are transducers which convert chemical energy to mechanical energy.
- Contractile force is generated by the making and breaking of the complexes between S_1 heads of myosin and actin.
- Onset of contraction is mediated by Ca^{2+} ions and ATP is hydrolysed during muscle contraction.
- The mechanical action of the heart is due to the heart muscle, which is known as myocardium.
- Blood flow helps in transporting gases like O_2 and CO_2, nutrients wastes, harmones and immunologically active substances.
- Like the striated muscles, the cardiac muscle fibre membranes are polarised.
- They have 90 mV negative potential between the inside and outside of the membrane.
- The large numbers of cardiac muscle fibres in the heart enables these potential differences to be measured using an electrocardiogram or ECG.

Chapter 13

Neurobiophysics

13.1: Introduction

The external world is perceived by organisms through stimuli received by them. The stimuli may be mechanical, thermal or chemical. They are transduced by the sensory receptors of the organism into electrical signals, and passed on to the central nervous system. The internal communications within the organism are taken care of by the nervous system. This transmission of information relating to the environment, and the ability of multi-cellular organisms to respond appropriately to them, is essential for survival. This chapter deals with the nerve transmission and other forms of information transport.

As far as the human beings are concerned, vision, hearing and smell play vital roles in communications. Hence, the mechanism of hearing and vision is discussed in some detail here. Chemical signals, like the odours emanating from flowers, are described in the later part of the chapter.

13.2: The nervous system

The nervous system consists of neurons, which transmit information from one part of the organism to the other. This is known as the neuronal response. The nerve cells (neurons and Schwann cells) have a cell body from which small projections known as dendrites and large projections called axons protrude. The dendrites carry impulses towards the central cell body (Figure 13.1). The cell body is the thick region of the neuron, containing the nucleus and most of the cytoplasm. The axons carry impulses away from the cell body. A fatty covering known as the myelin sheath covers most of the larger axons. The myelin sheath is interrupted, somewhat periodically, at sites called the nodes of Ranvier. Several axons are grouped together in each nerve. Similarly, nerve cell bodies form clusters known as ganglia. Some axons are very long and can extent up to 3 or 4 metres. For example, the

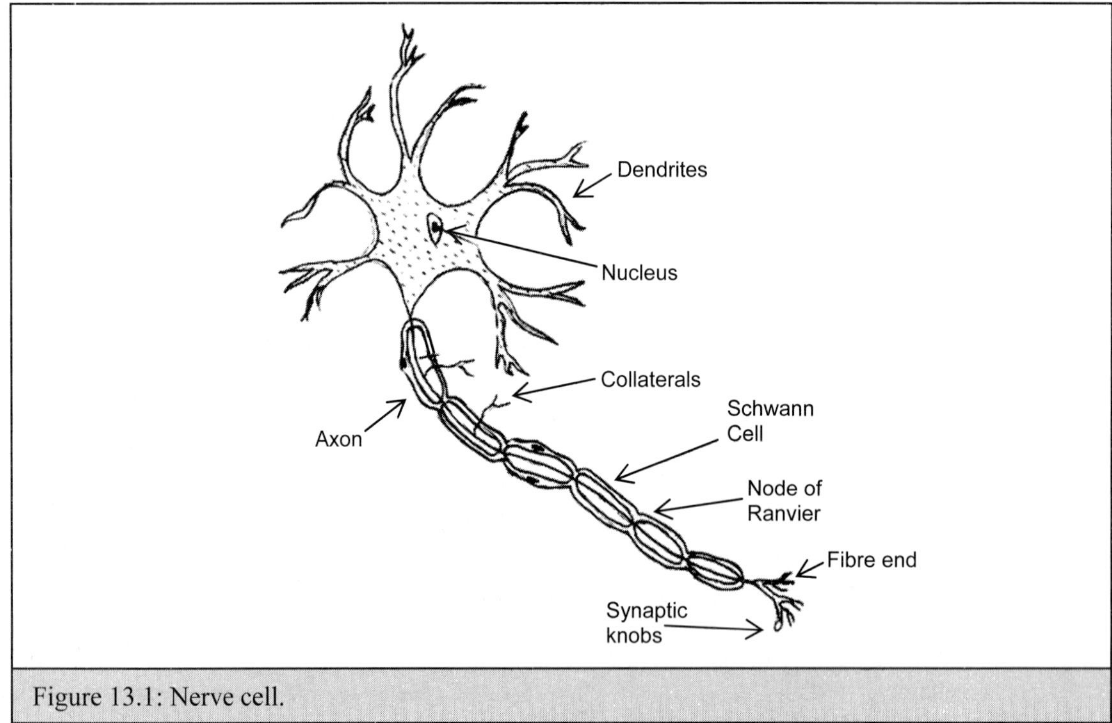

Figure 13.1: Nerve cell.

nerves controlling the muscles in the human finger extend up to the spinal cord, where they have their cell bodies. The diameter of the axons is of the order of 1 to 100 μm. Connections between neurons are called synapses and these occur at the branched ends of axons, dendrites and collateral branches of different neurons. Axons can also end in a synapse at the receptors such as the hair cells of the organ of the corti in the ear.

The nervous system of vertebrates is organised into the central nervous system and the peripheral nervous system. The central nervous system is in the skull and the spinal cord, while the peripheral nervous system consists of nerves and ganglia. Generally the peripheral nervous system contains both sensory axons and motor axons. The sensory axons conduct signals toward the central nervous system while motor axons conduct them away from it. Though the neurons within the central nervous system look similar to those outside it, there are several differences. For example when a nerve fibre within the central nervous system is injured, the entire neuron degenerates. On the contrary, if a peripheral nerve is cut, the fibres will re-grow out of the old nerve trunk.

A single axon can be removed from an organism and be examined in the laboratory. The maximum diameter of a vertebrate axon is about 20 μm while some of the invertebrate axons can be as large as 200 μm in diameter. Squid axons are easy to work with, due to their size, and are used routinely for laboratory studies. It is now possible to insert a microelectrode inside a nerve axon, and to measure the electric potential relative to the surrounding medium (i.e. the interstitial fluid outside the fibre). This potential normally ranges from -50 to 100 mV. Electrical pulses, heat, chemical changes and also mechanical pressures can stimulate axons. When the axon is stimulated, its surface potential changes in

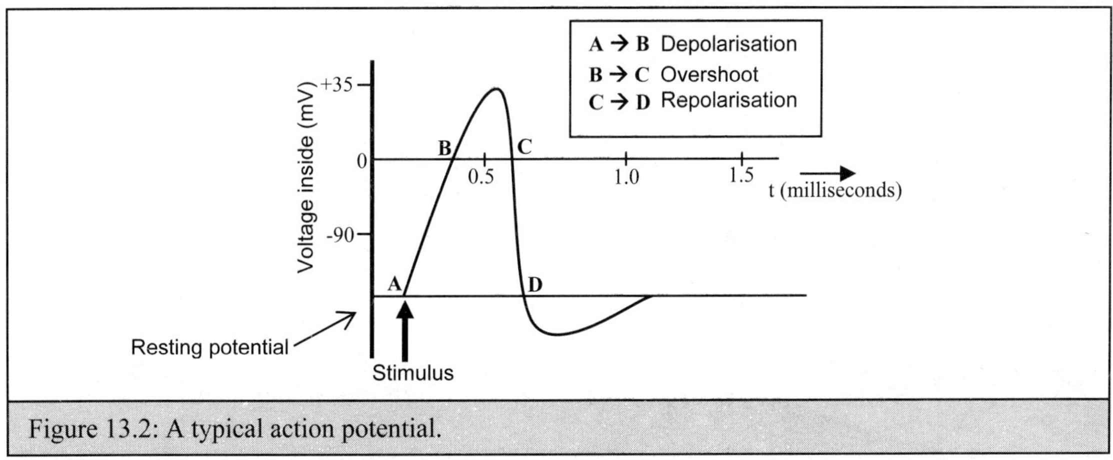

Figure 13.2: A typical action potential.

a specific way, producing a signal called the action or spike potential. The spike potential thus formed is an all-or-none response, and is transmitted down the axon in both the directions. However, due to the nature of the synapse present in an intact nerve, only one of these directions is usually effective. Each spike generally starts with a sudden change from its normal negative resting potential to a positive membrane potential, and then abruptly falls back to the resting potential. The total duration is about one millisecond (Figure 13.2).

13.2.1: Synapse:

In the nervous system the message is transmitted from one point to another. This means that a two-way passage of the message along the axons has to be prevented by some means. This is achieved by the synapse. The point at which an axon and a dendrite associate is called a synapse (Figure 13.3). When an action potential arrives at the presynaptic fibre, transmitter molecules are released from the vesicles. These diffuse through the synaptic gap and reach the sensitive surface of the postsynaptic fibre. The transmitter molecules alter the permeability to Na^+ and other monovalent ions in the membrane of the sensitive

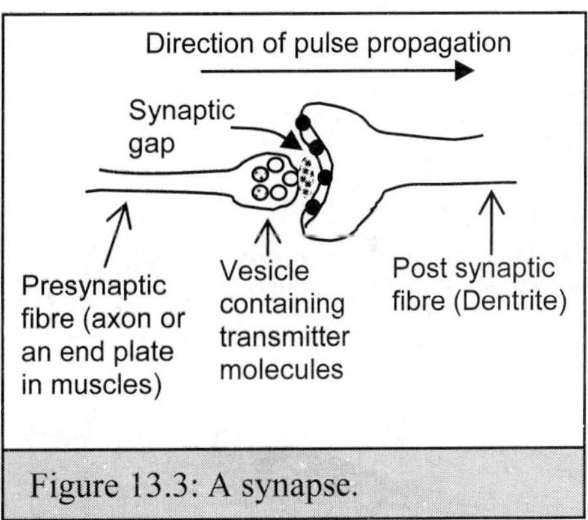

Figure 13.3: A synapse.

surface, thus creating an action potential. The transmitter molecules are destroyed immediately after this so that the synapse could be ready to receive the next signal. Acetylcholine and noradrenaline have been identified as synaptic transmitters and a very small quantity (10^{-18} M) of the substance is enough to evoke an action potential. This amount is known as threshold quantity. The destruction of acetylcholine is achieved by the enzyme acetylcholine esterase, which is situated in the receptor surfaces of the postsynaptic fibres. Synaptic conduction can also occur through charge transfer without any chemical intermediates. In this case the impulses travel from one axon to another electrically with negligible time delay. A single spike potential at the presynaptic fibre can induce a single spike potential at the postsynaptic fibre. In some cases, when the response due to the spike is below the

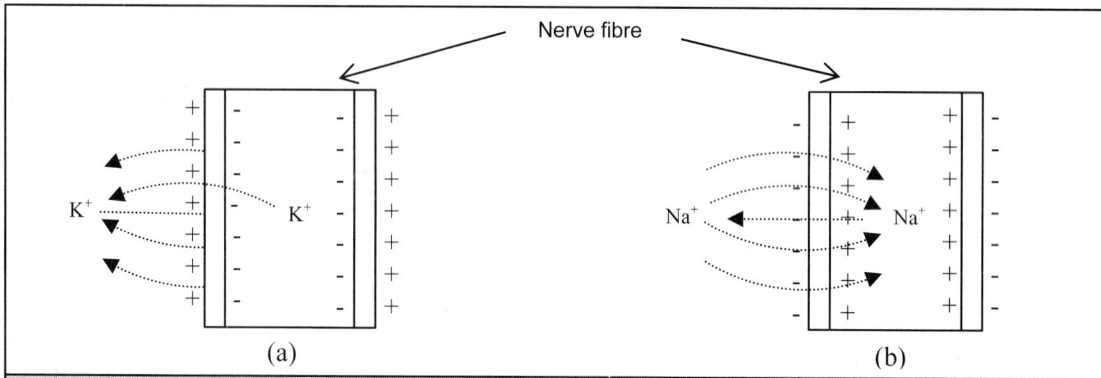

Figure 13.4: Diffusion potential across cell membranes. (a) Nernst potential for K^+ ions. (b) Nernst potential for Na^+ ions.

threshold value, then sub-threshold responses from several synapses can be added together to form a spike potential at the post synaptic fibre. Thus the neurons can act like a digital computer and can add, subtract or do other arithmetical operations.

13.3: Physics of membrane potentials

Membrane potentials can develop in the following two ways. (1) Diffusion of ions through the membrane causes an imbalance of negative and positive charges on the two sides of the membranes (Figure 13.4). (2) Active transport across the membrane again leads to an imbalance of the charges on either side of the membrane. Measuring the membrane potential of a nerve fibre is a difficult task due to the small size of the fibres. A microelectrode with a diameter of 1 micron is inserted through the membrane. The other electrode, called the indifferent electrode, is kept in the interstitial fluid (Fig 13.5). The potential difference is measured using a sophisticated voltmeter, which is capable of

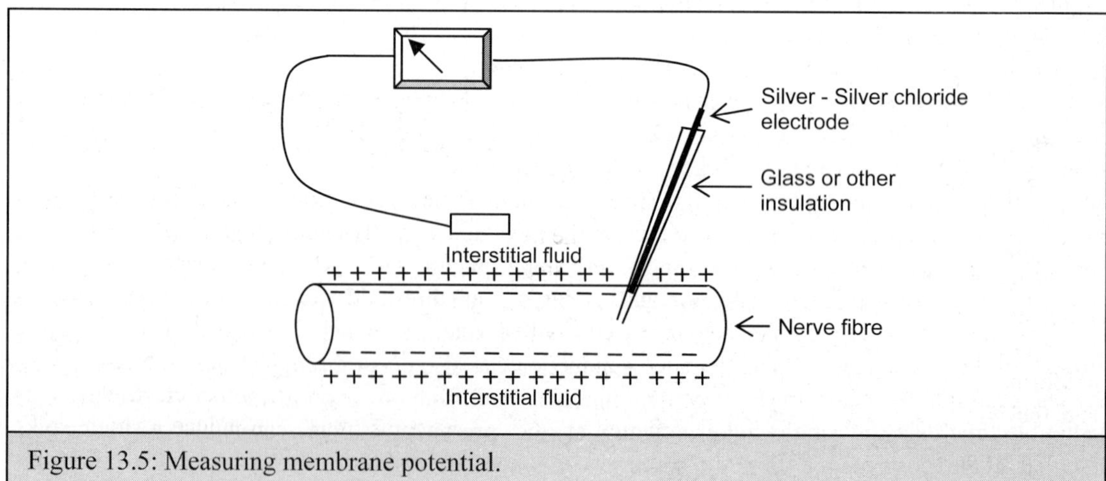

Figure 13.5: Measuring membrane potential.

measuring even very small voltages despite extremely high resistance to electrical flow through the tip of the microelectrode.

13.3.1: Membrane potential due to diffusion:

Generally the K^+ ion concentration is greater on the inner side than on the outer side of the membrane. The reverse is true for Na^+ ion concentration. When there is no active transport, and if we assume that the membrane is permeable only to K^+ ions, then these ions will diffuse from inside to outside the membrane (Figure 13.4(a)), since the K^+ ion concentration inside is greater. The K^+ ions carry the positive charge to the outside and hence a state of electropositivity outside and electronegativity inside is developed. Thus a potential difference across the membrane is developed, and this prevents further K^+ ions moving outward. The potential across the membrane is now known as the Nernst potential for potassium ions. Figure 13.4(b) shows the same effect for Na^+ ions, where we now assume that the membrane is highly permeable to Na^+ ions and impermeable to all other ions. As the Na^+ ion concentration outside is more, sodium ions move inward and a membrane potential with reverse polarity is generated. When the membrane potential rises high enough to prevent further diffusion of Na^+ ions, it is known as the Nernst potential for sodium ions. The magnitude of the potential is given by the following expression.

EMF (in mV) = -61 x log[concentration of the positive ions (inside)/
concentration of negative ions (outside)]

This is the Nernst equation. When the concentration of positive ions inside is ten times that of negative ions outside, then log(10) = 1.0, and the EMF = -61 mV. For sodium ions, the Nernst potential $\cong$ +61 millivolts. For potassium ions, the Nernst potential $\cong$ -94 millivolts. Under resting conditions, the membrane potential averages to -90 millivolts which is near the Nernst potential for potassium. This is because in the resting stage the membrane is more permeable to K^+ and only slightly permeable to Na^+ ions.

We can now see more details of how the action potential, mentioned above, develops across the membrane. There are three stages. (a) Resting potential: This is the stage before the action potential actually occurs. The membrane potential is negative and the membrane is said to be polarised during this stage. (b) Depolarisation stage: At this stage, membrane becomes very permeable to sodium ions, and a tremendous number of sodium ions rush into the interior of the axon. The membrane potential rises in the positive direction and sometimes overshoots the zero value. (c) Re-polarisation stage: Sodium channels close immediately after this and potassium ions flow to the exterior re-establishing the resting stage negative potential (Figure 13.2).

Voltage gated channels and Na^+-K^+ pumps play a vital role in both depolarisation and re-polarisation of the nerve membrane during the action potential. These are membrane proteins, or peptides, or molecular complexes, which open and shut appropriately to allow or prevent the flow of the ions. As the names suggest, the gates play a passive role, while the pumps are proactive. The sodium channel is activated when the membrane potential rises from its resting value of -90 mV, in the positive direction, to somewhere between -70 mV and -50 mV. When the sodium channel opens, it changes its conformation (Figure 13.6). After about 0.00001 seconds, it again changes its conformation and closes, and the Na^+ ions cannot enter the nerve axon. The conformational changes of the sodium channel during depolarisation state are rapid, while those during the closing of the channel are delayed. The

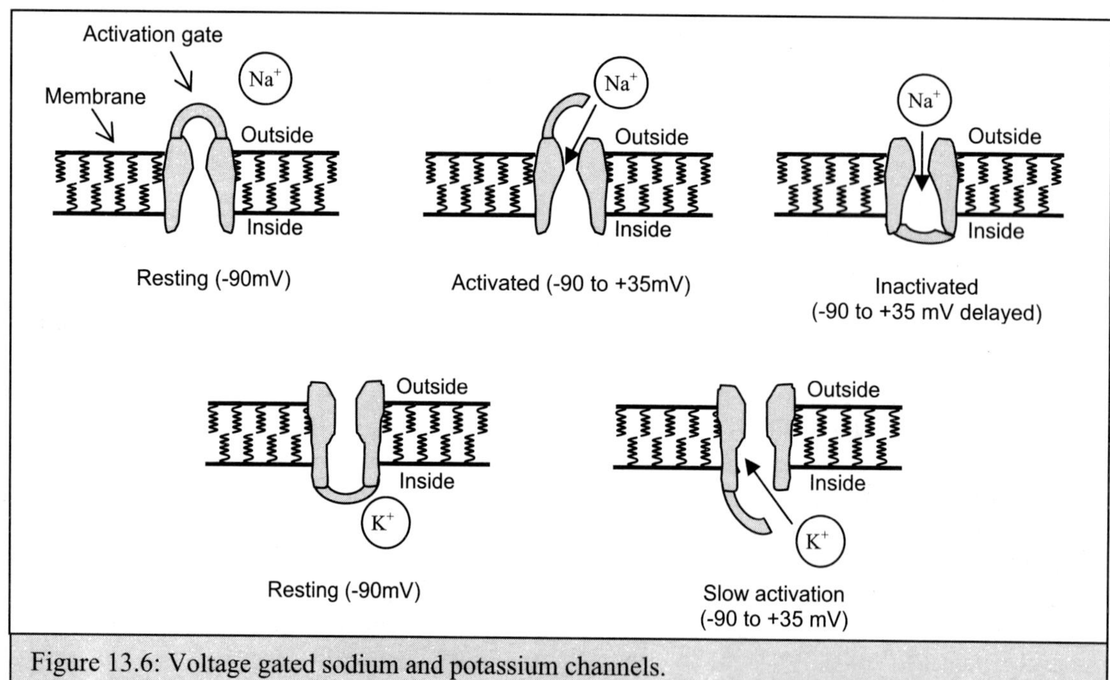

Figure 13.6: Voltage gated sodium and potassium channels.

membrane potential starts recovering back to its resting stage as soon as the sodium channel is closed. The inactivated gate will not reopen till the membrane potential reverses back to the resting negative potential. The potassium channel is closed at the resting stage, preventing potassium ions flowing to the exterior. When the membrane potential moves in the positive direction, i.e. from -90 mV to zero, then the potassium ion channel opens up, and K^+ ions diffuse outward. As this process is relatively slow, it takes place when the sodium channel is inactivated. Thus the closing of the sodium channel and simultaneous opening of the potassium channel speeds up the repolarisation process.

13.3.2: Voltage clamp:

Membrane potentials change very rapidly and it is possible to study the kinetics of these processes only when the membrane potential is held constant at a specified value, and if the measurements are carried out for a specified time. This is achieved by the voltage clamp technique. It was first successfully employed by Hodgkin and Huxley to measure the flow of ions through different channels. The voltage clamp has two electrodes inserted into the nerve fibre. One of these is for measuring the voltage, and the other is to conduct electric current inside or outside the fibre (Figure 13.7). The investigator chooses a voltage to be established inside the nerve fibre, which is greater than the threshold value, and the voltage is clamped at the chosen value by a feedback system. The membrane potential (E_m) is measured by an intracellular electrode, and this is compared with the chosen voltage in a control amplifier. The output of this amplifier is connected to a second electrode, which passes a current into the cell, which is proportional to the deviation of the membrane potential from the desired voltage. Thus the deviation is compensated. It may be noted that the compensating clamp current has the same amplitude as, but opposite polarity to, the membrane current at the chosen potential. The current flow

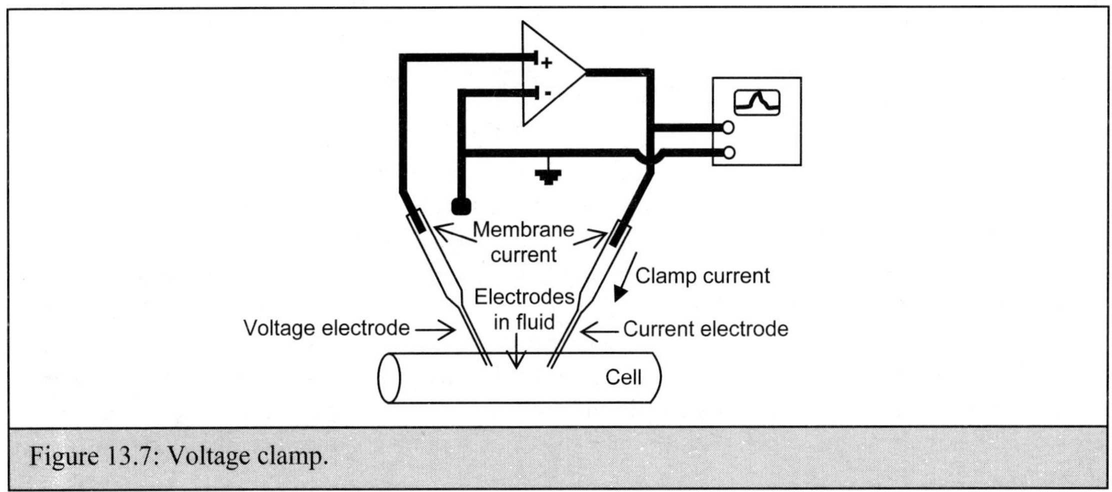

Figure 13.7: Voltage clamp.

is measured by connecting the current electrode to an oscilloscope. For example, when the membrane potential is suddenly increased from -90mV to zero by this method, then sodium and potassium channels open up and the ions begin to flow through the channels. Electric current is injected through the current electrode to compensate for the net current flow through the channels and to maintain the intracellular voltage at zero (or a pre-determined value). The conductance, which varies with time, is measured. By selectively inhibiting one of the channels, the flow of ions through individual channels can be studied. Toxins like tetrodotoxin are used for blocking sodium channels while tetra ethyl ammonium ion is used for blocking potassium channels. Figure 13.8 is an example of the results obtained from such measurements. Changes in conductance when the membrane potential is increased from its normal resting value of -90 mV to +10 mV for 2 milliseconds are shown. The sodium channels are activated and inactivated within this time, while potassium channels are only activated during this time.

The behaviour of the nerve fibre in terms of changes in permeability of the membrane to Na and K

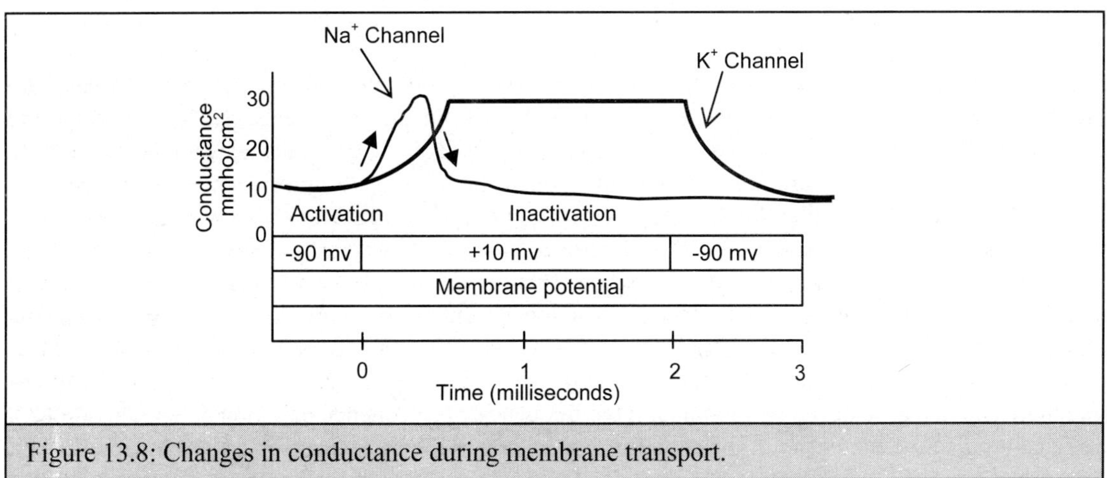

Figure 13.8: Changes in conductance during membrane transport.

ions has been formulated as a set of equations, known as the Hodgkin-Huxley equations. These are given below.

$$g_K = n^4 \overline{g_K}$$

$$dn/dt = \alpha_n(1-n) - \beta_n n$$

for K^+ conductance. $\overline{g_K}$ is the constant maximum conductance. The α-state is that in which the membrane allows K^+ to pass, while β-state is the state in which the membrane is impermeable to K^+ ions. n is the proportion of molecules in α state while (1-n) is the proportion of molecules in the β state. β_n is the reaction rate of the change from state α to state β, and α_n is that of the change from β to α. At the resting stage β_n is large and α_n is small. Consequently most of the regulating molecules are in β state. On depolarisation, α_n increases and β_n decreases, depending on the membrane potential, and this leads to a rise in n and hence in g_K. After about 10 milliseconds, α_n is larger than β_n and an equilibrium is reached where most of the molecules are in α state. n and g_K attain maximum values and this will last as long as the depolarisation state lasts. For sodium conductance, the Hodgkin-Huxley equations are as follows.

$$g_{Na} = m^3 h \overline{g_{Na}}$$

$$dm/dt = \alpha_m(1-m) - \beta_m m$$

$$dx/dt = \alpha_h(1-h) - \beta_h x$$

$\overline{g_{Na}}$ is the constant maximum sodium conductance, m and h denote the state of the two molecules which regulate sodium conductance. α_m, β_m, α_h and β_h are reaction rate constants that depend only on membrane potential. Though the conductance of sodium ions is similar to that of potassium, the equations will have to accommodate complications like inactivation of the sodium channel after about 1 millisecond. The variable 'x' has been introduced to take care of the inactivation process.

13.4: Sensory mechanisms – the eye

An organism is continuously bombarded by different signals from the environment. All the sense organs that are provided to the organism are transducers of energy; i.e. they convert energy in one form into another. Unlike neurons the response in these cases is not an 'all or none' process. The intensity of the response by the sensory organs will vary with the extent of the stimulation. Sensory detectors are classified into teleceptors, chemical receptors, somatic receptors and visceral receptors. Organs like eyes and ears are teleceptors as they receive information from distant objects. Sensors of smell and taste belong to the class of chemical receptors as they are excited by chemical stimulation. Touch, pressure, pain and temperature are experienced through the somatic receptors, while the information regarding body conditions like thirst, hunger, etc. are conveyed to the brain by the visceral receptors. When stimulated, each of these receptors will initiate nerve discharges in the form of action potentials. However, the way in which sensory information, such as the image on the retina, or the pitch of a sound, is converted into the binary code of action potential is unknown. The action potentials generated by sensory nerves are usually evoked by synaptic stimulation and are therefore called generator potentials. Generator potential magnitudes vary with the intensity of the stimulation of the receptors, and this is achieved by modulating the frequency of a sequence of action potentials in such a way that it reflects the intensity of the stimulus. Let us look at the basic question of how a stimulus like

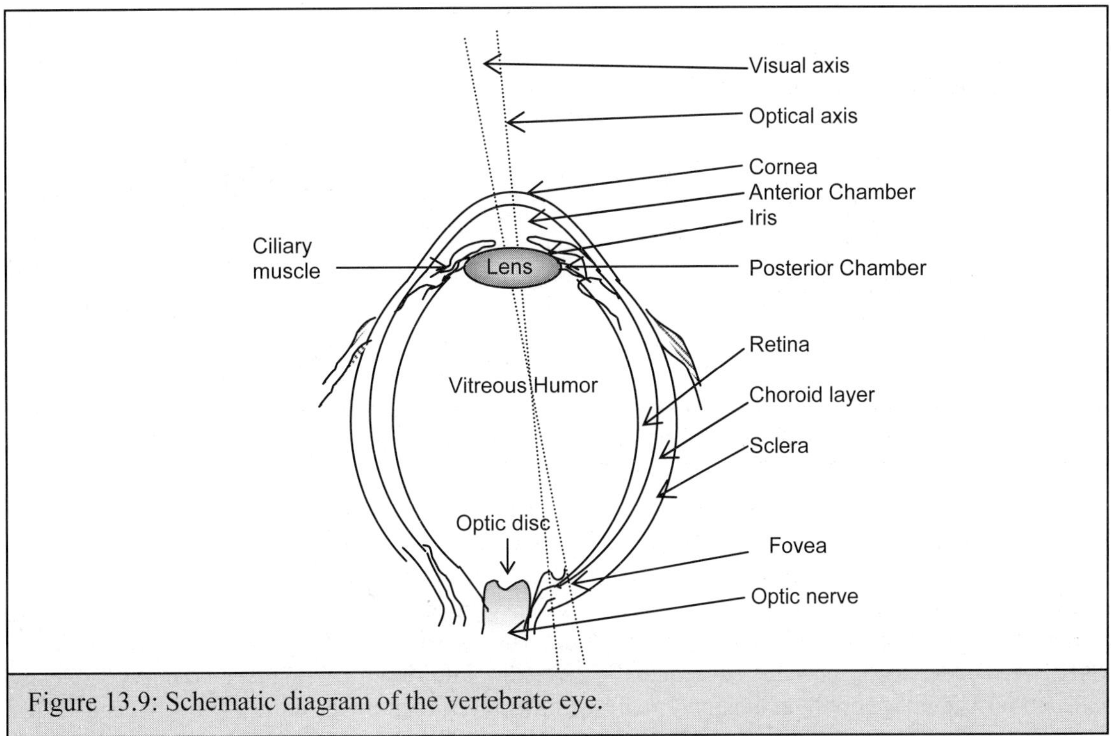

Figure 13.9: Schematic diagram of the vertebrate eye.

light, sound or temperature is converted into electrical polarisation of the membrane in the case of visual receptors and auditory receptors.

13.4.1: The visual receptor:

Light sensitive receptors are of different kinds. They vary from the most primitive system, like photokinesis (light induced motion) in plants, to the most elaborate ones like the vertebrate eye. A schematic diagram of the vertebrate eye is given in Figure 13.9. The lens system of the eye, which consists of cornea, aqueous humour, ciliary muscle, and vitreous humour of the eyeball, helps in focussing the image of the object onto a layer containing visual receptors. The layer on which image is formed is called the retina. The iris is an important structure and it controls the amount of light entering the eye. A small iris opening increases the depth of focus, in addition to restricting the distortions such as spherical aberration, field curvature, etc. For example in the night, sensitivity is more important than depth of focus or acuity. Hence the iris diaphragm will be open to the maximum in low light conditions.

The crystallin lens or eye lens is the most important structure in the eye. The eye lens is a cellular structure, which is more curved at the rear than in front. The focus of the lens is varied by changing the curvature of the front face of this by the action of ciliary muscles. The space between the retina and the lens is filled with a jelly-like substance called vitreous humour. Vitreous humour is optically similar to the aqueous humour filling the cornea and the eye lens. Light enters the eye through the cornea and passes through the aqueous humour and the crystalline lens into the vitreous humour. The light is

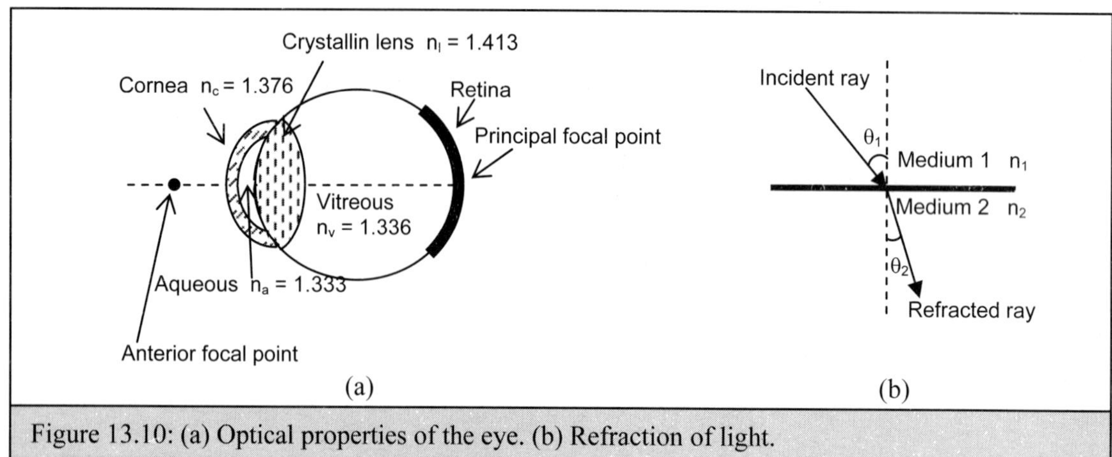

Figure 13.10: (a) Optical properties of the eye. (b) Refraction of light.

received by the retina and an image is formed. Figure 13.10(a) shows the geometrical optics of the eye. Vision itself is made up of several processes. The photons (light) which fall on the retina trigger a series of events in the light sensitive photoreceptor cells. As a result, the energy carried by the photons is transduced into chemical energy, which in turn is transformed into an electrical signal. This signal is carried by the optic nerve to the brain for processing. The transduction of light into chemical energy is fairly well understood, but the subsequent conversion into electrical signals is not completely understood. The image on the retina is formed by refraction of light by the cornea and lens, and can be described by the well-known Snell's law

$$Sin\theta_2 / Sin\theta_1 = n_1/n_2$$

where θ_1 is the angle made by the incident light with the normal and θ_2 is the angle made by the refracted ray with the normal and n_1 and n_2 are the refractive indices of the two media respectively (Figure 13.10(b)). Before reaching the retina, light is refracted at three places: 1) At the surface of the cornea, going from the air into the cornea. 2) At the surface of the lens, going from aqueous humour into the lens. 3) At the posterior surface of the lens, in passing from lens to the vitreous humour, the refractive indices being $n_{air} = 1.0$, $n_{humour} = 1.33$, $n_{lens} = 1.413$. Thus the greatest refraction takes place at the posterior surface of the cornea. In order to calculate the image size we can replace the cornea-lens system by a single equivalent lens. In that case, we know that

$$Object\ size/Image\ size = Object\ distance/Image\ distance$$

This expression can be used to calculate the image size. For example if the object distance is 1 Km and the object size is 10 m then the image size = 0.15 mm, where the average image distance is taken to be 1.5 cm. The process of vision is dynamic, since the eye lens is not a rigid object with fixed focal length. The focal length can be adjusted by changing the curvature of the lens and this process is called accommodation. The standard relation between the object and image distances (p and q) is given by

$$1/p + 1/q = (n-1)(1/R_1 + 1/R_2) = 1/f_L$$

where R_1 and R_2 are the front and rear radii of curvature of the lens, f_L is the focal length of the lens, and n is the refractive index of the lens. If the focal length of the lens is fixed, then objects near the lens will be focussed behind the retina and vice versa. As the lenses have a flexible focal length, the images are always focussed on the retina. The change of focal length is achieved by a change in

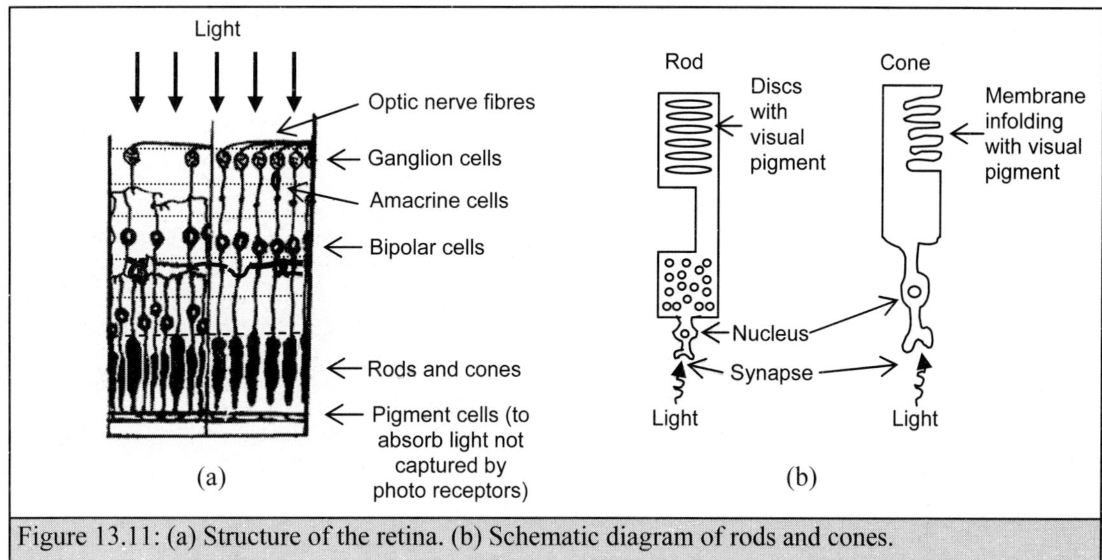

Figure 13.11: (a) Structure of the retina. (b) Schematic diagram of rods and cones.

curvature of the lens, which is made possible by the action of muscles attached to the lens. With aging, the power of accommodation reduces and external lenses are used to correct this defect.

The image formed on the retina is received by photoreceptors, which are called rods and cones (Figure 13.11). Cones are more concentrated at the central part of the retina while rods are more abundant at the periphery. Colour vision is due to cones, while rods are responsible for vision in weak light. As rods are insensitive to colours, three types of cones have been implicated in colour vision. Though Newton's experiments demonstrated that a spectrum of white light consists of a continuum of colours, Maxwell showed that almost all colours could be obtained by mixing three primary colours. This suggests that at least three different receptors are necessary for colour vision. However, the perception of colour is achieved only by a great deal of processing of the signals by the brain. The fovea, which is the central part of the retina, has a high degree of acuity of vision and this is due to the exclusive presence of cones in this region. The optic disk lies very close to the fovea. This disk is also known as the blind spot. Objects focussed on this region cannot be seen as there are no cones or rods in the optic disk. The photoreceptor cells contain visual pigment molecules that are light sensitive. The light sensitive molecule in the vertebrate eye is rhodopsin, and has a molecular weight of about 35 to 40 KDa. This consists of a single-polypeptide protein, opsin, to which other molecules are covalently

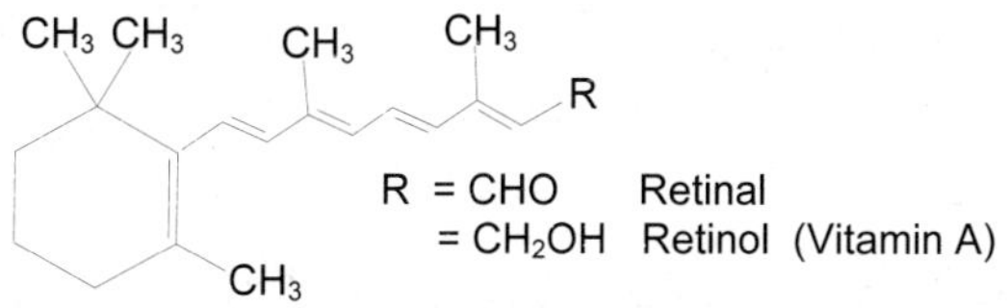

Figure 13.12 : Structure of retinal and retinol.

linked, viz. a chromophore and two oligosaccharide chains. The chromophore is retinal, and is responsible for absorption of light in the visible region. Retinal is a derivative of vitamin A and is obtained by oxidation of vitamin A by NAD . Figure 13.12 shows the two isomeric forms of retinal. Of the two oligosaccharides, at least one is composed of three N-acetyl glucosamine units and three mannose units. Rhodopsin is a membrane protein and is insoluble in water. Figure 13.13 shows the structural arrangement of seven α-helical segments of rhodopsin. The absorption spectrum of rhodopsin is shown in Figure 13.14. The absorption maximum is at approximately 500 nm. This is also the most effective wavelength for human vision in a dark-adapted eye that has been exposed to weak light. In strong light the peak is shifted to 550 nm which is close to the absorption maximum of iodopsin. In strong light when the rod vision changes to cone vision, iodopsin is the visual pigment involved.

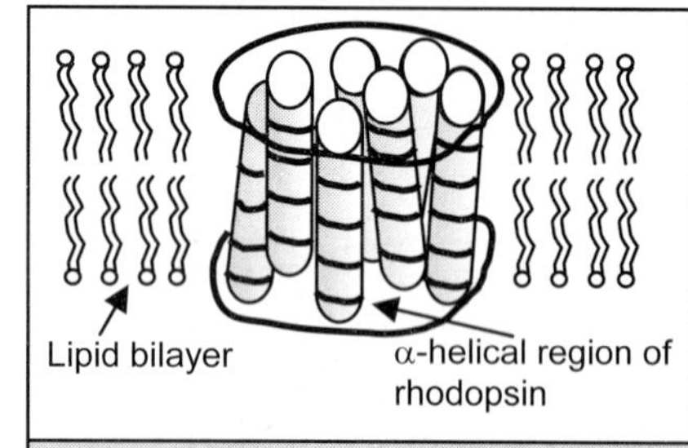

Figure 13.13: Arrangement of the α-helical segments of rhodopsin in the lipid bilayer.

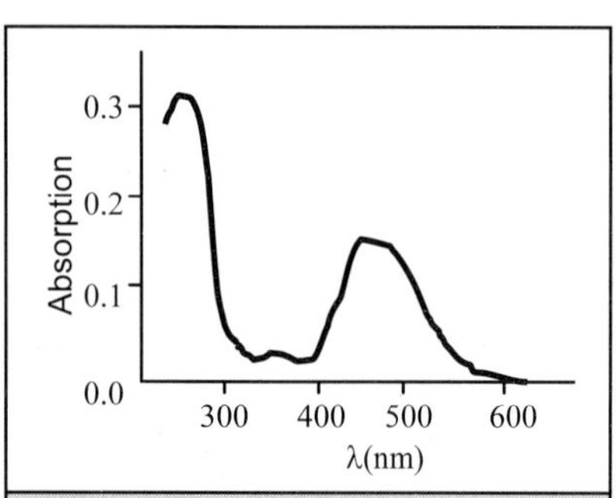

Figure 13.14: Absorption spectrum of rhodopsin.

The first step in the energy conversion procedure is the absorption of photons by chromophores. This causes a change in conformation of retinal from '11-cis' to 'all-trans' leading to detachment of the chromophore from opsin. This is known as the bleaching reaction. The recovery process involves reduction of retinal to 'all-trans' vitamin A, diffusion of vitamin A into the epithelium and re-conversion of vitamin A by light or enzymatically into the '11-cis' form. The '11-cis' vitamin A is then oxidised to '11-cis' retinal and linked to the protein opsin. The protein probably goes through a number of conformational changes before the 'all-trans' retinal comes off. Thus detachment and reattachment of the chromophore is the crucial part of the photochemistry of vision. An enzyme in the retina maintains an equilibrium between retinal and retinol (vitamin A). Generally a higher concentration of the alcohol (retinol) is maintained. However if the aldehyde (retinal) concentration is low, then it is produced from the alcohol by the action of an oxidoreductase enzyme and NADPH. The vitamin produced by oxidoreductase enters the blood through the rod membrane and is maintained at a constant concentration by the liver (Figure 13.15).

13.4.2: Electrical activity and visual generator potentials:

When the rhodopsin molecule absorbs a photon, it undergoes a change in its electrical state, which is passed on as a

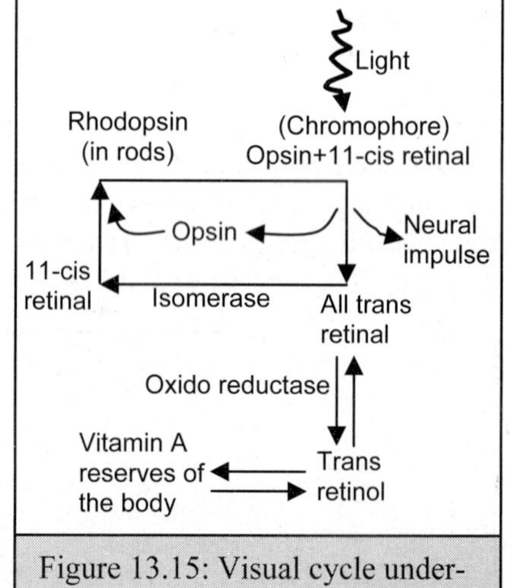

Figure 13.15: Visual cycle undergone by rhodopsin.

membrane potential. This electrical signal, which is known as receptor potential, is proportional to the intensity of the light stimulus. The receptor potential is processed at the synapse and translated as series of nerve impulses, which are carried to the brain for visual perception. Information about the firing of action potentials in optic nerve fibres has been obtained by implanting microelectrodes in selected places. In vertebrates the ionic permeability decreases, while in invertebrates the reverse occurs. The measurements show that the K^+ ion concentration is about 20 times more inside the cell than outside, while the sodium ion concentration is more on the outside than in the inside. This is the same as the situation seen in nerve cells. The association of electrical impulses with the visual process is different from that of transmission of electrical impulse in the axon. The receptor cells behave like photocells. The membrane potential of the vertebrate photoreceptor is also mainly a diffusion potential. The potential difference across the cell membrane in the inner segment is larger than in the outer segment and this causes a permanent ion current in darkness. This current flows from the inner segment to the outer segment through the extra cellular medium. In the dark, the external membrane of the rod outer segment is more permeable to Na^+ ions while the membrane of the inner segment is preferentially permeable to potassium ions. On illumination, the permeability for the sodium ions across the outer segment cell membrane drops, and this drop is greater if the intensity is higher. This reduction of conductance induced by transient light causes a reduction of the membrane current and a transient increase in the membrane potential (i.e. the membrane is more polarised), and this is communicated to the horizontal cells. The signal is actually produced in the amacrine and ganglion cells, and is transmitted to the brain by the optic nerve.

13.4.3: Optical defects of the eye:

A person with normal vision can see objects as close as 250 mm. In order to focus objects from 250 mm to infinity the crystalline lens curvature has to change by about 20%. The strength of the human eye is about 48 diopters. A diopter is a measure of the focussing power of a lens. It is calculated as the reciprocal of the focal length of the lens, measured in metres. Thus a lens of power, say, 3.0 diopters

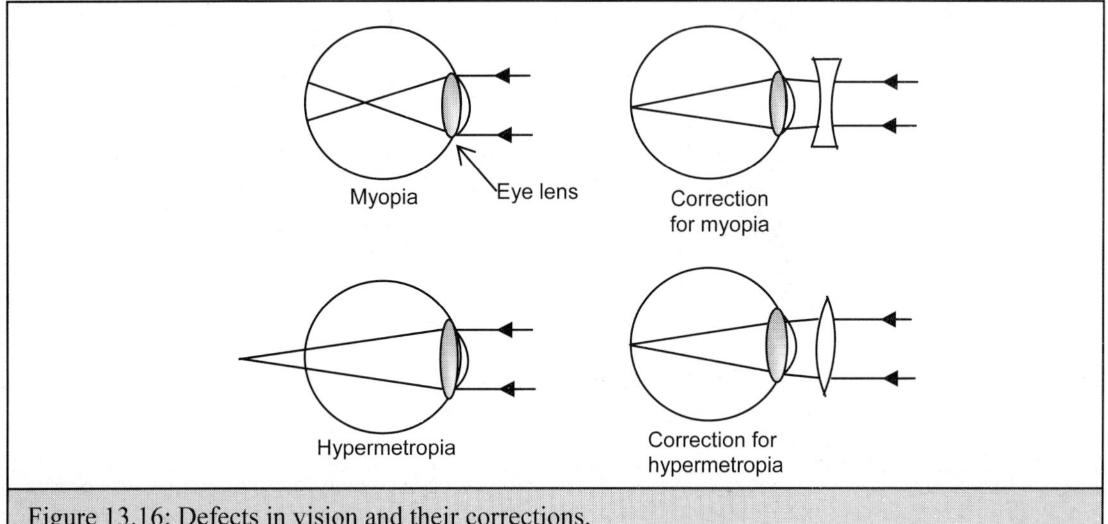

Figure 13.16: Defects in vision and their corrections.

will have a focal length of 0.33 metres. If the lens of an eye has a larger value than about 50 diopters, then images of distant objects will be focussed in front of the retina, and not on the retina. Thus, distant vision will be not be clear as compared to near vision, and the person is said to be myopic or suffering from myopia. This can be corrected by using a diverging lens in front of the eye (Figure 13.16). If the strength of the eye is too weak, then the images of near objects will be formed behind the retina. This condition is known as far-sightedness or hypermetropia, and requires a convergent lens for correction. As a person gets older, the adaptability of the ciliary muscles drastically reduces. This also leads to long sight, and a person who is myopic would require bifocals for correcting the near and far vision independently. Astigmatism is another eye defect. This arises due to the deviation of cornea from sphericity. When astigmatism is present, a point object does not form a point image on the retina. This is because the eye lens has different focal lengths for different directions. This defect can be corrected by using a cylindrical lens.

The response of the eye extends beyond the visible spectrum. For example the cornea absorbs ultra-violet light, and this can damage it. Hence dark glasses are recommended in strong sunlight.

13.4.4: Neural aspects of vision:

The retina acts as a photo-neural transducer, i.e. it converts light energy into a neural response. This response, which is in the form of a spike potential, travels along the optic nerve and reaches the central nervous system (CNS). From the central nervous system, the neural response reaches specific areas in the cerebral cortex.   The information is analysed both in the retina and in the CNS to generate the sensations of shape, colour, brightness, etc. The responses to an object from either eye are projected on to the occipital lobe of the cerebral cortex opposite to the half of the visual field containing the object.

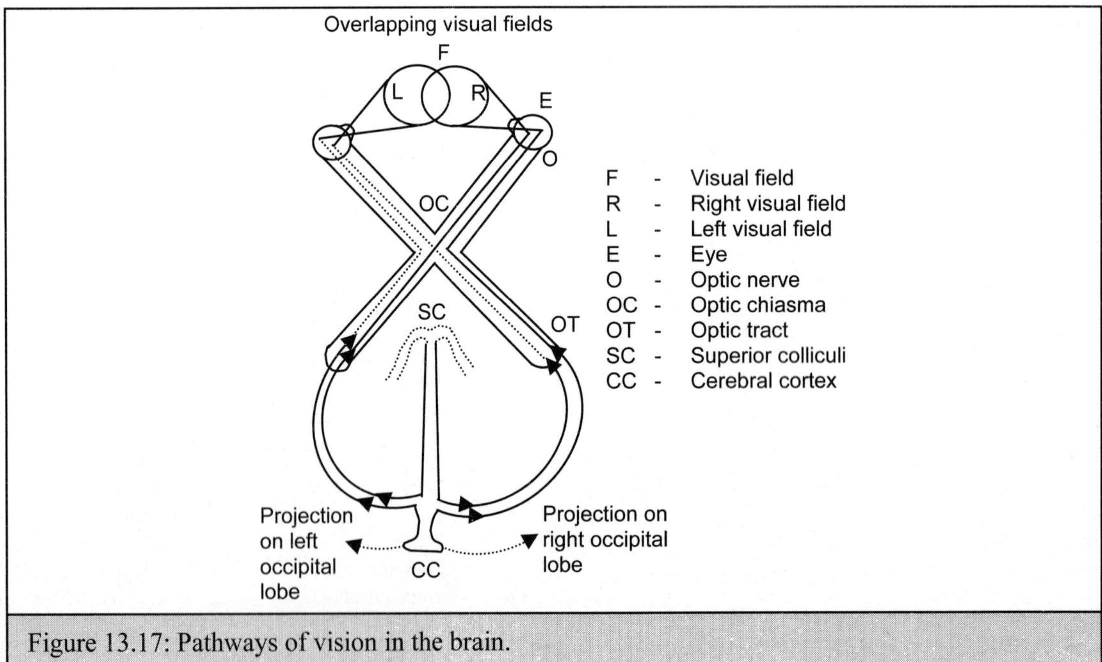

Figure 13.17: Pathways of vision in the brain.

There are three layers of cells which are important in the function, viz. the retinal[10] cells, intermediate cells, and a third layer of cells. These layers are interconnected, and the information flows from retinal cells to the brain through these layers of cells along the optic nerve. The information coming out of the optic nerve is divided into two bundles before reaching the brain. The visual cortex in the brain is responsible for the analysis of these impulses, and for subsequent vision. There are a few fibres that go to the midbrain rather than to visual cortex directly. Information such as the intensity of light, clarity of vision, etc., are processed in the midbrain and a feedback is sent to the appropriate regions of the eye. Figure 13.17 shows the neural connections between the eye and the visual cortex. The left side of the brain receives information from the right side of the visual field, and vice versa. They run through the optic chiasma, and the information from both the eyes are put together to achieve binocular vision. Every point in the retina probably has a corresponding point in the visual cortex. The points that are close together on the surface of the retina are also close together in the visual cortex. However, the central field of vision, which occupies a very small region on the retina, is expanded and is represented by several cells in the visual cortex.

13.4.5: Visual communications, bioluminescence:

Several insect families communicate with others in their species by producing a light, without accompanying heat. The visual signals can vary in intensity, colour and periodicity. The production of this light by living organisms is called bioluminescence. Most bioluminescent signals are not continuous, and are mostly produced in the night or in special habitats like caves, ravines, etc. A diverse variety of organisms like bacteria, fungi, sponges, corals, and fish can produce luminescence. Amphibians, reptiles, birds and mammals do not possess this property of producing light. The bioluminescent reaction in fireflies has been well characterised. The two substances responsible for this reaction, viz. luciferin and luciferase, have been isolated and crystallised, and their structures have been established. Figure 13.18 gives the chemical structure of luciferin. The reaction involving these two chemicals is a luciferase-catalysed oxidation using molecular oxygen and a substrate, viz. D(-)luciferin. A large amount of energy is liberated in this reaction, and the intermediate or product is left in the excited state. When the excited product returns to the ground state, light is emitted. ATP is hydrolysed in this reaction, which is written as follows.

$$\text{enzyme} + \text{D(-)luciferin(LH}_2) + \text{ATP} \Leftrightarrow \text{Enzyme-LH}_2\text{-AMP} + \text{PP}_i$$
$$\text{enzyme-LH}_2\text{-AMP} + \text{O}_2 \rightarrow \text{oxyluciferin} + \text{AMP} + \text{CO}_2 + \text{light}$$

where enzyme-LH$_2$-AMP is the enzyme bound-luciferyl adenylate.

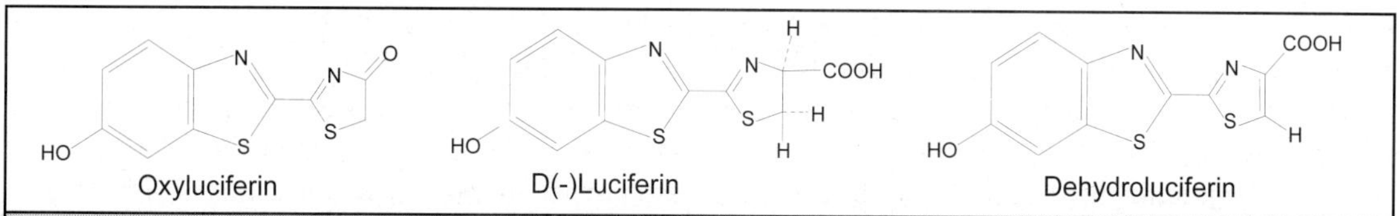

Figure 13.18: Chemical structure of Luciferin.

---

[10] The word 'retinal' in the present context is connected to 'retina' and not to the compound 'retinal' mentioned in the previous chapter.

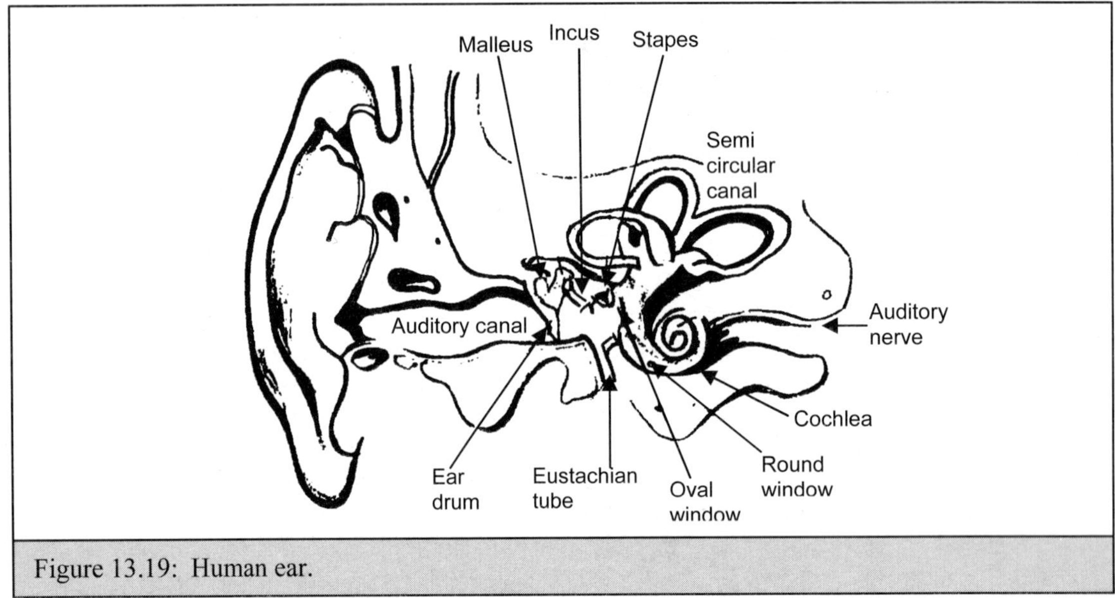

Figure 13.19: Human ear.

13.5: Physical aspects of hearing

Hearing, like vision, is an important sensory mechanism that enables many animals, including humans, to perceive their environment. Though the biophysics of hearing has been one of the earliest to be studied, information on the transduction of sound is meagre. This is due to the fact that the sensory organ, viz. the ear, will have to be intact for it to function and it is very difficult to perform experiments and measurements on the inner ear, where transduction takes place. However the structure of the ear is known in very great detail from anatomical and histological studies, and this has helped to reconstruct the hearing process.

Hearing is actually the perception of pressure waves that are generated by a vibrating medium. The ear, which is the hearing apparatus, is homologous in most species, though their sensitivity varies. The frequency range of the normal human ear is 20 Hz to 20,000 Hz.

13.5.1: The ear:

The ear is divided into three parts – the outer, the middle and the inner ear. Figure 13.19 shows the human ear. The outer and middle ear are filled with air and their primary function is to conduct the sound waves to the inner ear. Here the acoustic energy is converted to neural spikes, which are analysed by the brain. The outer ear consists of an external auricle, a narrow tube called the ear canal or external auditory meatus, and the ear drum or tympanic membrane. The ear canal is similar to a closed end organ pipe with the tympanum serving as the end-closing membrane. The vibration of air molecules produced by an external sound is passed on to the small bones in the middle ear. The middle ear has three small bones or ossicles called malleus, incus and stapes. The ossicles amplify the acoustic pressure of vibrations transmitted via the tympanum. They also restrict the amount of energy that is transmitted at high sound levels. Thus when a high volume of sound reaches the system, the amplification at the middle ear is decreased, and this acts as a volume control. The ossicles also discriminate against the vibrations reaching them through the skull. The outermost ossicle or the malleus presses against the tympanic membrane, while the innermost one, the stapes pushes against a membrane called the oval window. The oval window separates the air-filled middle ear from the liquid-filled channels of the inner ear. There is a second membrane known as the round window, which separates another channel of the inner ear from the middle ear. The maximum pressure amplification achieved by the outer and middle ear together is about 35 dB (or decibels[11]). The middle ear is connected to the pharynx through a tube called the eustachian tube. The eustachian tube prevents distortion in the shape and position of the tympanic membrane when a change in pressure takes place due to atmospheric conditions, or changes in the altitude. The mammalian inner ear consists of several portions, of which only the cochlear portion is involved in hearing. The cochlea has two and half turns, making up three parallel fluid-filled tubes. These tubes are called tympanic, vestibular and cochlear ducts. The individual ducts or canals are separated by membranes. Reisner's membrane separates the cochlear duct and the vestibular duct, while the basilar membrane separates the tympanic duct from cochlear duct. The organ of corti, which is situated in the basilar membrane, is probably a neuro-mechanical transducer. This organ is a complicated system of cells. It includes hair cells. The bending of these hair cells excites the nerve endings located in the organ of corti. These excitations are transmitted to the brain where it is perceived as sound.

13.5.2: Elementary acoustics:

Sound is actually an elastic disturbance in a continuous medium. Suppose a particle in the medium is disturbed from its equilibrium position by a sound wave. If v is the frequency of the displacement, and ψ is the amplitude, then we can write the expression describing this effect as

$$\psi = A\sin(2\pi vt) + B\cos(2\pi vt)$$

or

$$\psi = Ce^{2\pi ivt}$$

where the amplitude varies as a function of time, t. Most sound waves are mixtures of frequencies. If the velocity of sound waves is c, then λ = wavelength = c/v. In a gas the pressure variation produced by the passage of sound wave is very rapid, and hence the process is assumed to be adiabatic, i.e. occurring at a constant temperature. The process of hearing can be considered as the conversion of mechanical energy in the form of

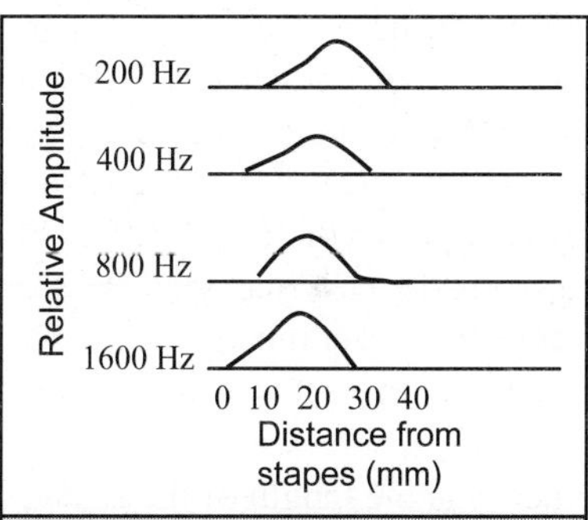

Figure 13.20: Amplitude of the basilar membrane displacements.

pressure variations in the air to electrical energy in the form of nerve pulses, which are transmitted to the brain for processing by auditory nerves. The conversion starts at the tympanic membrane in response to the pressure variations in the atmosphere, and causes the three bones in middle ear to deflect. The three small bones in the middle ear have a mechanical linkage, and cause compressions in the cochlear fluid. The inward motion of the stapes causes downward motion of the basilar membrane, which triggers the nerve signal. The movement of the basilar membrane is a travelling wave, with its

---

[11] dB represents 'decibel', which is a unit of measurement of sound intensity on a logrithmic scale, i.e. a sound that is 20 dB is ten times as loud as one that is 10 dB. 0 dB is the threshold of human hearing.

maximum amplitude at some small distance from the oval window. The responses of basilar membrane to sound have been measured using microelectrodes placed in the cochlear space and in the fibres of auditory nerves. Some typical responses are shown in Figure 13.20. The figure shows that the displacements are inversely proportional to the frequency of the exciting sound wave. The magnitude of amplification is estimated as follows. The force on the stapes is

$$F_{stapes} = 1.3 \times 52 \times P_{tympanum}$$

where 1.3 is a constant signifying the mechanical advantage of ossicles, 52 mm^2 is the area of contact between the membrane and the hammer, and P is the pressure. The area of contact between stapes and round window is 3.2 mm^2. Therefore the pressure at the window is

$$P_{window} = F_{stapes}/3.2 \quad \text{(since pressure = force/unit area)}$$

Therefore

$$P_{window}/P_{tympanum} = 68/3.2 \cong 21$$

This is roughly the maximum amplification observed. In humans, the average value of magnification is about 17, and is less than the calculated value because of friction, which has not been taken into account in the above calculations.

13.5.3: Theories of hearing:

There are two different theories that have been formulated to explain hearing. They are known as the spatial and the temporal theories. The spatial theory proposed by Helmholtz suggests that a particular point on the basilar membrane responds to a particular audible frequency. Anatomical investigations have shown the basilar membrane to have a fibrous structure, and accoring to this theory, each individual fibre resonates at a particular frequency. The Helmholtz theory met with a few difficulties. Experiments conducted with resonators showed that the fibres respond to a narrow band of frequencies rather than a single frequency. Secondly, any resonating system produces an 'echo' which continues even after the excitation of resonance has stopped. However, no such effect is observed in the hearing process. Nevertheless, hearing loss due to damage to particular regions of the membrane clearly supports the view that the loss is due to the lack of response by the fibres expected to resonate at those frequencies. Helmholtz compared the resonance of the fibres to that of a stretched string. The resonating frequency of such a stretched string is given by

$$\nu_{res} = \tfrac{1}{2}\, l\sqrt{(T_0/\rho)}$$

where l is the length of the string, T_0 is the tension on the string, and ρ is its density. Using the above equation we can calculate the frequencies over which the ear is sensitive. When this is done we find that the audible frequency range is reduced by five octaves from the experimentally known values. This shows that Helmholtz's theory does not completely represent the hearing process.

Bekesey showed that, to explain the process of hearing, in addition to the variable compressions, one should take into account the hydrodynamic waves that are produced in the cochlear fluid. He could successfully predict the point of maximum response in the basilar membrane as a function of frequency. The maximum displacement position is dependent on the initial frequency. The mathematical theory dealing with hydrodynamic waves is quite complex, and a simplified explanation is given here. We have seen earlier that the vibration of the stapes produces a wave in the cochlear fluid, which spreads from the oval window throughout the cochlea in about 25 microseconds. This leads to vibrations in the basilar membrane. However, since the membrane is less elastic at the base, the displacement moves towards the apex. Thus, the fluid and the membrane form a coupled system, and individual parts cannot resonate separately. As mentioned earlier, the position of the maximum displacement depends on the initial frequency; for example when the frequency is low the entire membrane moves at the tip but if the frequency is high then the oscillations are produced near the oval window. This explains why damage to certain regions of the membrane leads to hearing loss at certain frequencies. Thus, the high frequency response depends on the functional response at the base, while low frequency depends on the condition of the apex. The organ of corti acts as a transducer and the motion of the basilar membrane is converted into an electrical signal and propagated along the nerve fibres to the auditory cortex of the brain. The threshold potential required to fire a cell depends on the frequency of stimulation. The processing at auditory cortex of the brain is still not fully known.

13.6: Signal Transduction

Biologically significant information is generally transmitted through chemical signals. The message carriers are special chemicals. Messages are also transmitted by intracellular substances, like mRNA, cAMP (cyclic adenosine monophosphate), etc. Communications between cells and organs are carried out through substances like hormones. As we have seen earlier, neural communication is through neuro-transmitters. Chemical signals like odours emanating from bugs, skunks, flowers, etc., are useful in communication between organisms of the same species or different species. The antigens that stimulate the production of antibodies can also be considered as biological signals.

In any signal mechanism there is a sender, a receiver, and a mode of transport of the signal. In biological systems the receivers of chemical signals are the chemosensitive organs (such as the nose or the tongue), chemoreceptor cells, and other receptor molecules. For the recognition of the signal, the receiver molecule must have chemical specificity. The nature of the response depends on the concentration and nature of the signal substance. In most cases, the chemical specificity is due to macromolecules which are the receptors. The molecules on the cell surface that enable cell-cell recognition (e.g. glycoproteins, glycolipids) can also be considered as chemical signals.

There are three stages in molecular recognition. a) The capturing of the signal substance by the receptor molecules. b) The causing of the physiological effects. c) The removal of the signal substance by decomposition or by removal from the receptor.

13.6.1: Mode of transport:

The signal molecules may be transported by diffusion or convection. The diffusion process is effective only when the distance between receiver and sender is small (of the order of a few tens of nanometres), and the concentration of signal substance is high. For example, in nerve transmission the neurotransmitters travel a distance of only about 20 nm. In the case of hormones, which may travel from one organ to the other, the signal substance is first adsorbed on the cell membrane, and then diffused through the membrane for action. If the affinity for the signal substance to the receptor molecules on the membrane is high, then these molecules could accumulate specifically on this cell, and help in transporting the signal. Thus the strong affinity of the lipophilic surface of the olfactory organ for odour-producing substances helps insects to respond to even low concentrations of the signal substances in the air.

Convection is another method of transport of signal molecules and is especially useful when the distance between the sender and the receiver is large. The time taken for transport by diffusion is given by the following expression.

$$t = x^2/D$$

where x is the distance to be travelled, and D is diffusion coefficient. Hence the time taken for transport by diffusion over large distances is large. Convection plays an important role under these circumstances. For example, odorous substances are known to move by convection across distances of even several kilometres, borne by winds or by hot air, and be recognised at the destination.

13.6.2: Signal transduction in the cell:

The chemical signals are recognised by receptor proteins. Due to its non-covalent nature, the binding of the signal molecules to the active site of the receptor is reversible. The strength of the binding is also dependent on the extent of complementarity between the signal molecule and the active site of the receptor. The receptor active site could be, for example, the opening of an ion channel in the membrane of the muscle end plate, when the neurotransmitter acetylcholine binds to the receptor molecule. Another such example is the binding of antigen molecule to the hypervariable region of antibodies. In the case of insect olfactory cells, even a single odour molecule is sufficient to trigger a nerve impulse. This is made possible by opening of a single ion channel in the sensory cell membrane.

If the signal substance is continuously produced, then accumulation of the signal substance in the transporting medium can take place. This has to be avoided by removal of the signal substance by decomposition, inactivation or by dilution due to diffusion. For example, in the antigen-antibody complex, irreversible chemical binding inactivates the signal substance. In the case of neurotransmitters, the acetylcholine molecule is broken down very rapidly by acetylcholine esterase. The receptor surface of the olfactory cells in animal noses is cleaned by being continually rinsed by convection, and by the flow of fresh mucosa over the cells.

Summary

- The external world is perceived as mechanical, thermal or chemical stimuli by organisms.

- These stimuli are received by sensory receptors and converted into electrical signals, which in turn are passed on to the central nervous system.

- The nervous system consists of neurons, which transmit information from one part of the organism to the other.

- The nervous system of vertebrates is divided into the central nervous system and the peripheral nervous system.

- The central nervous system is in the skull and the spinal cord, while peripheral nervous system consists of nerves and ganglia.

- Nerve cells have a cell body from which small projections known as dendrites and large projections called axons protrude.

- Connections between neurons are called synapses and these prevent two-way passage of the messages along axons.

- The peripheral nervous system contains both sensory axons and motor axons. The sensory axons conduct signals toward the central nervous system while motor axons conduct them away from it.

- Membrane potentials develop due to diffusion of ions or due to active transport of ions.

- K^+ ion concentration is more on the inner (cytoplasmic) side of the membrane than on the outer (extracellular) side. In the case of Na^+ ion the reverse is true.

- Resting potential, depolarization and repolarization are the main steps involved in membrane transport.

- A voltage clamp is a useful device to maintain the membrane potential at a specified value. The flow of ions through individual channels can be studied using this method.

- The behavior of a nerve fibre in terms of changes in permeability of the membrane to Na^+ and K^+ ions is given by Hodgkin-Huxley equations

- All the sense organs are the transducers of energy. The eye, the ear, and the skin are sensory organs. Unlike in the case of neurons, the response by sense organs is not an 'all or none' process.

- The sensory receptors are classified as telereceptors, chemical receptors, somatic receptors and visual receptors.

- The eye is a visual receptor. Rod cells and cone cells are the photoreceptors in the retina. Photoreceptor cells contain visual pigment molecules like rhodopsin

- When the rhodopsin molecule absorbs a photon it undergoes a change in the electrical state, which is transmitted as membrane potential, also known as 'receptor potential'.

- This electrical signal is processed at the synapse and translated as a series of nerve impulses.

- The response travels along optic nerve and reaches the central nervous system (CNS).

- The information is analysed both in the retina and CNS to retrieve sensations like shape, colour, brightness, etc.

- Hearing is the perception of pressure waves that are generated by a vibrating medium.

- The ear is divided into three regions, viz. outer ear, middle ear and inner ear. The outer ear and middle ear conduct the sound waves to the inner ear where the sound energy is converted into neural spikes.

- The middle ear helps in amplifying the acoustic pressure of vibrations transmitted by the tympanum (ear drum), and also in the volume control.

- Biological information is also transmitted through chemical signals, e.g. odors emanating from bugs, flowers etc. Chemical signals are recognized by receptor proteins.

- Molecules on cell surface involved in cell-cell recognition, such as glycoproteins, may also be considered as chemical signals.

- There are three major steps involved in molecular recognition. a) Capturing of the signal substance by the receptors. b) Causing the physiological effects. c) The removal of the signal substance by decomposition, etc.

prevalent at that time, or by looking for evidence in fossils. Fossils are the remains of living matter after their slow mineralisation into stone. These fossils can only give information about the gross structure of the cell, as most of the organic matter would have been destroyed by repeated heating and cooling of the rocks due to change in the environment. However, molecular fossils, which is the name given to those structures or functions in the cell shared by diverse present-day organisms, may yield clues to build a coherent picture of primordial events.

14.3: Theories of origin and evolution of life

The evolution of life after the formation of the first cell can be considered to have been controlled by fairly well established Darwinian principles. However, prebiotic evolution is an area of intense debate as it depends on the interpretation of fossil remains or on extrapolations from existing life forms. There are no fossils of the early period of evolution i.e. 4 billion years ago. Whenever experimental evidence is lacking, extrapolations have been made or theories have been proposed. Not surprisingly, these contain several contradictions.

Eigen proposed a theory for the self-organisation of macromolecules and the emergence from chaos of ordered structure, which in turn created information. He said that this is possible only when we have an autocatalytic open system, which is far away from equilibrium. Accidental ordering from chaos has a very low probability. For example, given that a polynucleotide chain of 100 nucleotides is to be formed from four nucleotides, the number of possible sequences is equal to 4^{100} or 10^{60}. For any particular primary structure to appear out of this large number of possibilities, the probability is close to zero. Eigen proposed a model in which he assumed that macromolecules do not make an exact copy of themselves, (i.e. do not self-reproduce) but replicate by synthesising complementary strands. In a more complex system of reactions, there could be a cycle of molecules and reactions, each supporting the next in line. Such a cycle of reactions is called a hypercycle. Eigen's hypercycles have the following properties: (1) All species (molecules) linked through a hypercycle have a stable and controlled coexistence. For example, a hypercycle can be formed by a sequence of reactions in which the product of one reaction may become the reactant of the next reaction and so on. The citric acid cycle is one such set of reactions. (2) A hypercycle can increase or decrease in size if these changes have a selective advantage. (3) Once established, a hypercycle has a distinct advantage over a new, yet-to-be-established species (of hypercycle). To explain how these hypercycles come into existence in the first place, Eigen proposed that the first self-reproducing system with enough information content should have been a short tRNA-type molecule. This lead to the proposal of a so-called 'RNA world', in which the catalytic molecules as well as the information storage and transfer molecules are RNA. Eigen and Schuster have extended this theory with models for the origin of the genetic code and the translation system.

Kuhn's model proposes the formation of short polynucleotide chains like pre-tRNA molecules that have secondary and tertiary structure. The periodic changes in the environment act as a selection factor. Molecules are assumed to have been synthesised in the pores of clay minerals. This theory also proposes steps related to self-assembly of tRNA molecules, synthesis of proteins, formation of membranes, the genetic code, etc.

Ebeling and Feistel's theory assumes that compartmentation of the reactants and products started in the early stages of chemical evolution. This compartmentation, also known as coacervate formation, gave rise to micro-reactors, which played the role of primitive cells during early evolution.

14.4: Laboratory experiments on the formation of small molecules

At the molecular level all forms of life appear to be similar and hence it is appropriate to assume that life on earth descended from single ancestral life form. Chemists have shown that essential building blocks of life can be synthesised from elements that were present on the primitive earth. This is also known as prebiotic synthesis. Miller's experiment showed for the first time that simple amino acids could be assembled spontaneously from the constituents of primitive earth.

14.4.1: Prebiotic synthesis of amino acids – Miller and Urey's experiment:

This experiment attempted to simulate the pre-biotic atmosphere of the early earth. Air was pumped out of a small flask and then a mixture of ammonia, methane and hydrogen was added. Water vapour was introduced by adding water and boiling it. This also helped to mix and circulate the gases into a larger flask that was connected to the smaller one. To simulate lightning, an electric spark was generated across a spark gap set into the arrangement (Figure 14.1). The product of the discharge were condensed by a condenser and washed through a U-tube into the small flask. The non-volatile products remained in the small flask, but the volatile products re-circulated past the spark. The reduced forms of

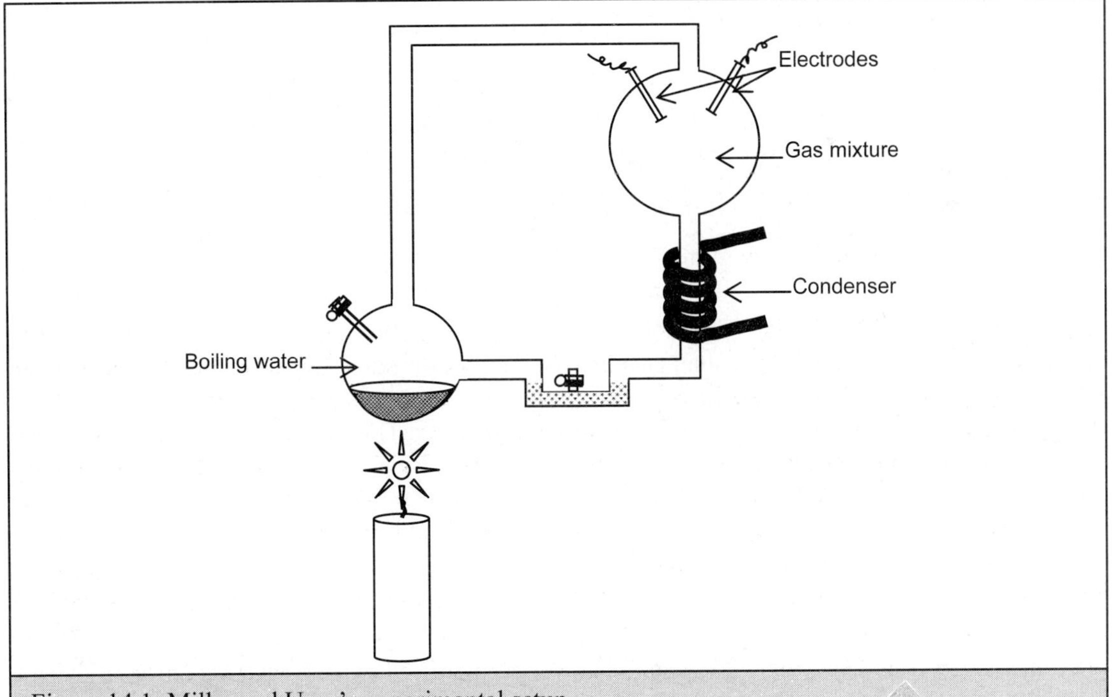

Figure 14.1: Miller and Urey's experimental setup.

carbon, nitrogen and oxygen in the gas phase represented the prebiotic atmosphere; the liquid phase represented the ocean. When the products were analysed, it was seen that many different organic molecules had been synthesised. The results of Miller and Urey's experiment were unexpected in two respects. A small number of relatively simple compounds accounted for a large proportion of the products. Further the major products themselves were not a random selection of organic compounds but included a surprising number of substances that occur in living organisms (Table 14.1).

Table 14.1. The compounds that were synthesised in the Miller-Urey experiment.

Compound	Yield (%)
Glycine	2.1
Sarcosine	0.25
α-Alanine	1.7
α-amino isobutyric acid	0.007
Aspartic acid	0.024
Formic acid	4.0
Urea	0.034
β-Alanine	0.76

α-Alanine was produced in greater quantity than β-alanine. When the experiment was performed over a longer period, large amounts of cyanide and aldehydes were formed during the first 125 hours of sparking and then the rate of their synthesis fell off, whereas the amino acids were formed at a more or less constant rate throughout the run. The mechanism proposed for the formation of glycine is the following.

$$CH_4 + H_2O \rightarrow H_2CO + 2H_2$$
$$CH_4 + NH_3 \rightarrow HCN + 3H_2$$
$$H_2CO + HCN + NH_3 \rightarrow CH_2(NH_2)\text{-}C\equiv N + H_2O$$
$$CH_2(NH_2)\text{-}C\equiv N + 2H_2O \rightarrow CH_2(NH_2)\text{-}COOH + NH_3$$

The general reaction can be written as

$$RCHO + HCN + NH_3 \rightarrow \underset{\underset{NH_2}{|}}{RCH}\text{-}C\equiv N + H_2O$$

$$\underset{\underset{NH_2}{|}}{RCH}\text{-}C\equiv N + 2H_2O \rightarrow \underset{\underset{NH_2}{|}}{RCH}\text{-}COOH + NH_3$$

The above reactions depend on the formation of hydrogen cyanide.

14.4.2: Prebiotic synthesis of purines, pyrimidines and nucleosides:

The most important purines are adenine, guanine and, to a lesser extent hypoxanthine. If concentrated ammonia solution containing 1-10 M cyanide is refluxed, adenine is formed in yields up to 0.5 % along with large amounts of the usual hydrogen cyanide polymer. This reaction does not easily explain the synthesis of purines under prebiotic conditions since useful yields of adenine cannot be obtained except in the presence of 1.0 M or stronger ammonia. The highest reasonable concentration of ammonia or ammonium ion that can be postulated in oceans and lakes on the primitive earth is about 0.01 M. In order to accommodate this fact, an alternative mechanism has been proposed. As shown in Figure 14.2, this mechanism makes use of an unusual photochemical rearrangement, which does not involve additional reagents.

Synthesis of pyrimidines, viz. C, U and T has not been studied extensively. Action of an electric discharge on a mixture of methane and nitrogen led to the formation of cyanoacetylene.

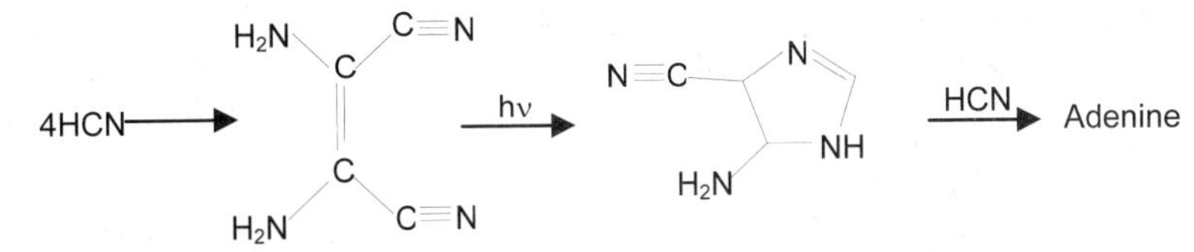

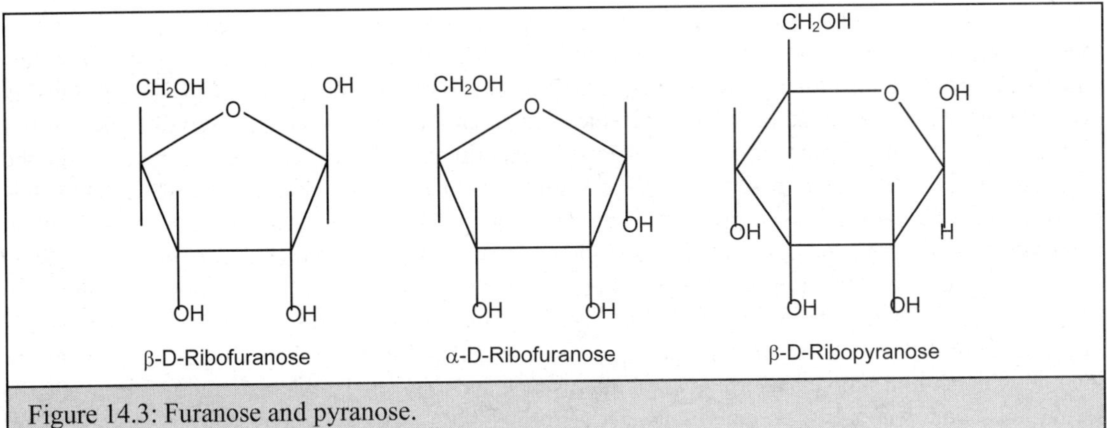

Figure 14.2: Photochemical rearrangement.

Cyanoacetylene reacts with aqueous cyanate to give cytosine in quite good yield (20%). Uracil may be obtained by hydrolysis of cytosine.

14.4.3: Sugars:

Formaldehyde is one of the simplest of organic molecules that is readily formed under pre-biotic conditions, as attested by its presence in interstellar clouds. Sugars are formed by treating formaldehyde with certain alkaline catalysts like sodium hydroxide.

$$nH_2C=O \longleftrightarrow \quad triose, tetrose, pentose, hexose$$

with the inorganic ashes acting as catalysts. While the suggested mode of synthesis is attractive, it is not without difficulties, such as the following. a) Sugars are unstable in aqueous solution. b) The mixtures of sugars obtained usually contain only a very small percentage of ribose, which is the constituent of nucleic acids.   c) The reaction requires a minimum concentration of 0.01 M formaldehyde.

   Groth and Suess produced organic compounds by irradiating a gaseous mixture of H_2O and CO_2 during their work on the origin of life. Similarly Garissian and coworkers bombarded an aqueous solution of carbon dioxide with some Fe^{2+} and formed formaldehyde, formic acid and succinic acid.

14.4.4: Nucleotide synthesis:

Ribose exists in two different ring forms, viz. furanose and pyranose (Figure 14.3). The natural

Figure 14.3: Furanose and pyranose.

nucleosides are all furanosides, but most organic syntheses give a mixture of isomers. Further, heterocyclic bases can form bonds to a sugar at more than one position. Thus, when purines are heated along with sugars in the presence of certain inorganic salts, a mixture of nucleosides is obtained in reasonable yield. A good percentage of the yield consists of natural nucleosides. The mixture of salts obtained by evaporating seawater is one of the most effective catalysts tested so far. Two alternative approaches to the synthesis of nucleosides are possible. Either a sugar can be attached to the preformed base or a base can be constructed on the preformed sugar. The later model has been accepted as the correct one in the case of prebiotic synthesis. Phosphorylation of nucleosides can be achieved by heating them with acid phospates such as NaH_2PO_4 and $Ca(H_2PO4)_2$. The reaction occurs quite rapidly at $130°$ C. This yields the required nucleotides. Thus, there is substantial evidence to prove that a variety of small organic molecules relevant to living systems could be synthesised under prebiotic conditions.

14.4.5: Optical activity:

Most biological molecules exhibit optical activity i.e. they rotate the plane of polarisation of polarised light. In an organic synthesis reaction, optically inactive starting materials will give rise to optically inactive products. In this case, the product molecules may either be individually optically inactive or the product may contain L and D molecules in equal quantities so that the product solution is optically inactive. There is no obvious reason why molecules of one chirality should be produced more than the other in a symmetric environment. In biomolecules only L-amino acids and D-ribose (Figure 14.4) occur, even though molecules with opposite chirality have an equal chance of forming a polymer. This asymmetry needs to be explained. Among the many plausible explanations that have been offered are the following. There are several asymmetric phenomena in nature that could contribute to an asymmetric environment. For example sky light is circularly polarised to a small extent. Radioactive β-decay exhibits handedness or chirality. Similarly the rotation of earth, the magnetic field of earth, etc. may also contribute to the asymmetric environment and this could lead to the preference for one handedness of the synthesised molecules over the other.

14.4.6: Polymerisation:

Once the relevant monomers have been synthesised, the second step in prebiotic chemical evolution is the aggregation, or self-organisation of molecules. The following experiment demonstrates how this could have occurred for proteins. When a pure amino acid solution is heated to about $180°$ C, varying quantities of diketopiperazine (a cyclic dipeptide) and small amounts of polypeptides are formed. If a mixture of 20 natural amino acids is used in this experiment, all of them are incorporated in the polypeptide-like materials. These materials are given the name "proteinoids". Most of the bonds between the different residues in proteinoids are the normal HN-CO peptide bonds. But those formed from mixtures which include aspartic acid and glutamic acid also contain α-β and α-γ amide linkages. A surprising feature of such proteinoids is that many of the polypeptides thus synthesised begin with glutamic acid. If the proteinoids are boiled in water, a dispersion of microspheres of size 2 microns is obtained. Some microspheres are surrounded by structures that look like bilayer membranes. Some of them look much like simple bacteria under a microscope and they can be made to divide into two, or to form buds, under appropriate conditions. These proteinoids, also known as thermal polypeptides,

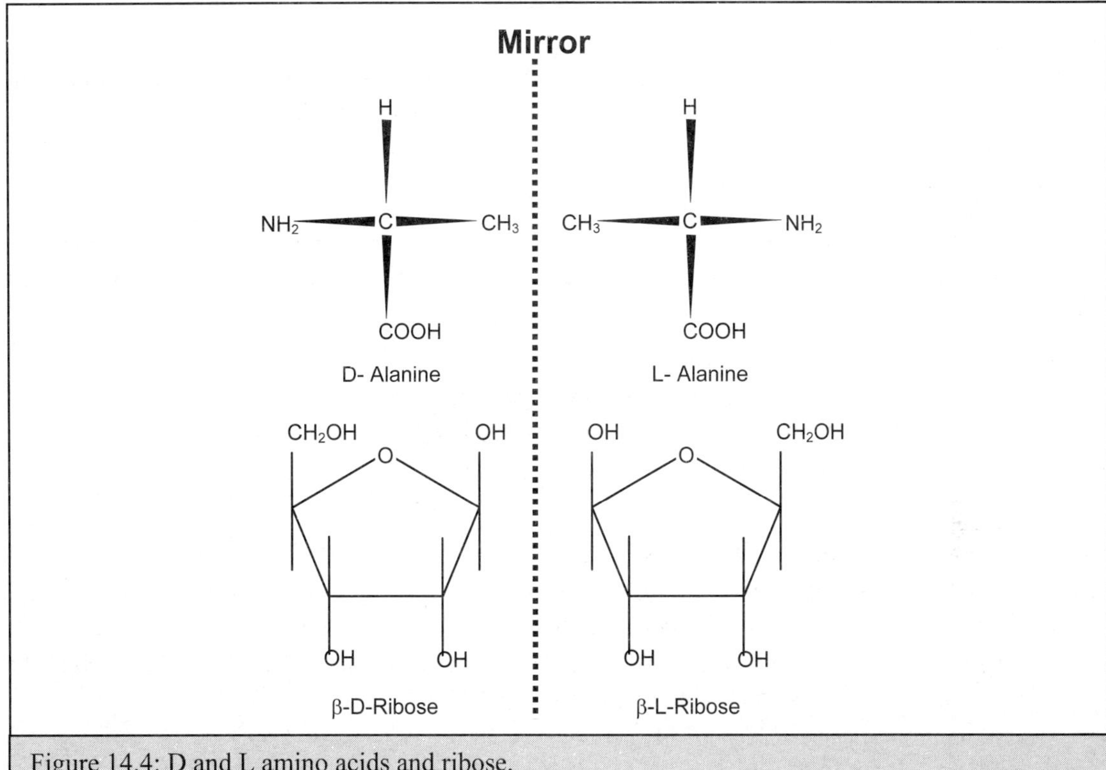

Figure 14.4: D and L amino acids and ribose.

display catalytic activity in some reactions. However, there are some facts that argue against thermal synthesis of polypeptides as a mechanism for prebiotic polypeptide synthesis. One such fact is that temperatures of the order of 150 to 180° C are required. In the prebiotic atmosphere at the time when the monomers are available, such temperature would be found only several kilometres below the surface of the earth, or in some hot springs or volcanoes. However, other mechanisms that do not require high temperatures have also been proposed for condensation of amino acids into polypeptides.

The formation of short oligonucleotides from the monomers could also have occurred purely by thermal processes. However, the specificity that exists in the polymerisation of nucleotides in the present day world, viz. the 5' → 3' direction of the polynucleotide chain, may not have been present under prebiotic conditions.

The prebiotic synthesis of fundamental molecules described above requires special conditions, and all of them may not have existed at the same time at the same place. For example the reactions that produce the monomers require very high concentrations of the reactants and the subsequent yield of biologically important monomers is low. It is believed that repeated evaporation and rehydration, freezing and thawing might have concentrated both the reactants and the products. The building blocks of life must have been condensed into still larger molecules by heating or chemical treatment. Thus, when the prebiotic solution of molecules (the prebiotic 'soup') cooled, it could have contained oligonucleotides, peptides and oligosaccharides. However, a mere collection of these molecules will not make a living system as we know it today.

To summarise, though the above experiments and arguments do not unequivocally establish the mechanisms for origin of life on earth, and for its subsequent evolution, they definitely show that the conditions on primitive earth were suitable for the emergence of some primitive form of life.

Summary

- Evolution is the change or adaptation by the organisms in response to environmental factors.

- Biological evolution can be classified into three phases: chemical, molecular and Darwinian evolution.

- Though several theories of evolution exist, the most widely accepted concept is that life originated on earth in a primitive form, and then gradually evolved to the present form.

- The various stages of evolution have been reconstructed by conducting experiments in the laboratory under conditions known to be prevalent at that time.

- Such reconstruction has also been carried out based on fossil evidence.

- Miller and Urey showed that essential building blocks of life can be synthesised from elements like C, H, O, etc., which were present even as the earth was beginning to form.

- These and other experiments showed that simple amino acids, nucleotides and sugars could be synthesised from elements present under prebiotic conditions.

- Most biological molecules are optically active.

- L amino acids and D sugars are preferred over their optical isomers.

- An asymmetric environment generated by the rotation of earth, the magnetic field of earth, or sky light might be the reason for the preference of one chiral form over the other.

- The thermal synthesis of oligo-peptides and oligo-nucleotides has also been demonstrated.

- These experiments show that the conditions on primitive earth were suitable for the emergence of some primitive form of life.

Further Reading

General

1. Introductory Biophysics. P. Narayanan (1999). New Age Publishing Co., Mumbai, India.
2. Biochemistry. D. Voet and G.J. Voet (1990). John Wiley & Sons, Inc., New York, USA.
3. Biophysical Chemistry, Vol. I, II and III. C.R. Cantor and P. Schimmel (1985). W.H. Freeman and Company, New York, USA.
4. Biophysics. W. Hoppe, W. Lohman, H. Markl and H Ziegler (eds) (1983). Springer Verlag, New York, USA.
5. Physical Biochemistry. D. Freifelder (1982). W.H. Freeman and Company, New York, USA.
6. Aspects of Biophysics. W. Hughes (1979) John Wiley and Sons, Inc., New York, USA.
7. Biophysical Science. E. Ackerman, L.B.M. Ellis and L.E. Williams (1979). Prentice-Hall Inc., New Jersey, USA.
8. An Introduction to Biophysics. C. Sybesma (1977). Academic Press, New York, USA.
9. Molecular Biophysics. M.V. Volkenstein (1977). Academic Press, New York, USA.
10. Organic Spectroscopy: Principles and Applications. Jag Mohan (2000). Narosa Publishing House, New Delhi, India.

Chapter 1

1. Fundamentals of Physics. D. Halliday, R. Resnick and J. Merill (1988). John Wiley & Sons, New York, USA.
2. College Physics. F.W. Sears, M.W. Zemansky and H.D. Young (1985). Addison Wesley Publishing Company, Massachusetts, USA.
3. Chemical Principles. R.E. Dickerson, H.B. Gray, M.Y. Darensbourg and D.J. Darensbourg (1984). The Benjamin Cummings Publishing Company, California, USA.
4. Organic Chemistry. T.L. Finar (1969). ELBS and Longmans, Green and Co., Ltd., London, UK.
5. Stereochemistry of Carbon Compounds. E.L. Eliel (1962). McGraw-Hill Book Company, New York, USA.

Chapter 2

1. Principles and Techniques of Practical Biochemistry. K. Wilson and K.H. Goulding (1986). Edward Arnold (Publishers) Ltd., London, UK.
2. Biophysical Chemistry, Vol. II. C.R. Cantor and P. Schimmel (1985). W.H. Freeman and Company, New York, USA.
3. Physical Biochemistry. K.E. van Holde (1971). Prentice-Hall Inc, New Jersey, USA.
4. Physical Chemistry of Macromolecules. C. Tanford (1966). John Wiley & Sons, New York, USA.

Chapter 3

1. Principles and Techniques of Practical Biochemistry. K. Wilson and K.H. Goulding (1986). Edward Arnold (Publishers) Ltd., London, UK.
2. Biophysical Chemistry, Vol. II. C.R. Cantor and P. Schimmel (1985). W.H. Freeman and Company, New York, USA.
3. Physical Biochemistry. D. Freifelder (1982). W.H. Freeman and Company, New York, USA.
4. Physical Chemistry. P.W. Atkins (1976). W.H. Freeman and Company, San Francisco, USA.
5. Physical Chemistry of Macromolecules. C. Tanford (1966). John Wiley & Sons, New York, USA.
6. Mass Spectrometry, E. de Hoffman and V. Strobant (1999), John Wiley & Sons, New York.

Chapter 4

1. Organic Spectroscopy. W. Kemp (1987). ELBS/Macmillan, Basingstoke, UK.
2. Biophysical Chemistry, Vol. II. C.R. Cantor and P. Schimmel (1985). W.H. Freeman and Company, New York, USA.
3. Fundamentals of Molecular Spectroscopy. C.N. Banwell (1983). Tata McGraw-Hill Publishing Company Ltd., New Delhi, India.
4. Molecular Spectroscopy. G.M. Barrow (1962). McGraw-Hill Book Company Inc., New York, USA.

Chapter 5

1. Light Microscopy in Biology. A.J. Lacy (ed.) (1989). IRL Press at Oxford University Press, Oxford, UK.
2. Optical Techniques in Biological Research. D.L. Rousseau (ed.) (1984). Academic Press, New York, USA.
3. Fundamentals of Optics. F.A. Jenkins and H.E. White (1975). McGraw-Hill Book Company, New Delhi, India.
4. Optics. F.W. Sears (1964). Addison-Wesley Publishing Company Inc., Massachusetts, USA.

Chapter 6

1. Electron Microscopy in Biology. J.R. Harris (ed.) (1991). IRL Press at Oxford University Press, Oxford, UK.
2. The Principles and Practice of Electron Microscopy. I.M. Watt (1985). Cambridge University Press, Cambridge, UK.
3. Optical Techniques in Biological Research. D.L. Rousseau (ed.) (1984). Academic Press, New York, USA.
4. The Electron Microscope. M.E. Haine and V.E. Cosslett (1961). E. and F.N. Spon Ltd., London, UK.

Chapter 7

1. Fundamentals of Crystallography. C. Giacovazzo, H.L. Monaco, D. Viterbo, F. Scordari, G. Gilli, G. Zanotti and M. Cati (1998). International Union of Crystallography and Oxford University Press, Oxford, UK.
2. Principles of Protein X-ray Crystallography. J. Drenth (1994). Springer-Verlag, New York, USA.
3. X-ray Structure Determination—A Practical Guide. G.H. Stout and L.H. Jensen (1989). John Wiley & Sons, New York, USA.
4. Crystals, X-rays and Proteins. D. Sherwood (1976). Longman Group Ltd., London, UK.
5. Crystallography and Crystal Chemistry. F.D. Bloss (1971). Holt, Rinehart and Winston Inc., New York, USA.

Chapter 8

1. Magnetic Resonance. C.L. Khetrapal and G. Govil (1991). Narosa Publishing House. New Delhi, India.
2. Nuclear Magnetic Resonance. Atta-ur-Rahman (1986). Springer Verlag, New York, USA.
3. Biophysical Chemistry, Vol. II. C.R. Cantor and P. Schimmel (1985). W.H. Freeman and Company, New York, USA.
4. Magnetic Resonance. K.A. Mclauchlan (1972). Clarendon Press, Oxford, UK.
5. Interpretation of NMR Spectra. R.H. Bible (1965). Plenum Press, New York, USA.

Chapter 9

1. Molecular Modeling. A.R. Leach (1996). Addison-Wesley Longman Ltd. Essex, England, UK.
2. Molecular Dynamics Simulation. J.M. Haile (1992). John Wiley & Sons, New York, USA.
3. Computer Graphics. D. Hearns and M.P. Baker (1990), Prentice-Hall Inc., New York, USA.

4. Computing for Biologists. A. Fielding (1985). The Benjamin Cummings Publishing Company, California, USA.

Chapter 10

1. Protein Structure. M. Perutz (1992). W.H. Freeman and Company, New York, USA.
2. Introduction to Protein Structure. C. Branden and J. Tooze (1991). Garland Publishing Company, New York, USA.
3. Nucleic Acids in Chemistry and Biology. G.M. Blackburn and M.J. Gait (eds.) (1990). IRL Press at Oxford University Press, Oxford, UK.
4. Biophysical Chemistry, Vol. I. C.R. Cantor and P. Schimmel (1985), W.H. Freeman and Company, New York, USA.
5. Principles of Protein Structure. G. Schultz and R.H. Schirmer (1984). Springer-Verlag, New York, USA.
6. Principles of Nucleic Acid Structure. W. Saenger (1984). Springer-Verlag, New York, USA.

Chapter 11

1. Biochemistry. L. Stryer (1995). W.H. Freeman and Company, New York, USA.
2. Biochemistry. G. Zubay (1993). Wm. C. Brown Publishers, Iowa, USA.
3. Biochemistry. D. Voet and J.G. Voet (1990). John Wiley & Sons, New York, USA.
4. An Introduction to Biophysics. C. Sybesma (1977). Academic Press, New York, USA.

Chapter 12

1. Biochemistry. L. Stryer (1995). W.H. Freeman and Company, New York, USA.
2. Biophysics. W. Hoppe, W. Lohman, H. Markl and H. Ziegler (eds) (1983). Springer-Verlag, New York, USA.
3. Aspects of Biophysics. W. Hughes (1979). John Wiley & Sons, New York, USA.

Chapter 13

1. Textbook of Medical Physiology. A.C.Guyton and J.E. Hall (1998).W.B. Saunders Company, New York, USA.
2. Cell and Molecular Biology, Indian Edition. E.D.P. Robertis and E.M.F. Robertis (1995). B.I. Waverly Pvt. Ltd., New Delhi, India.
3. Biophysical Science. E. Ackerman, L.B.M. Ellis and L.E. Williams (1979). Prentice-Hall Inc., New Jersey, USA.

4. Bioelectric Phenomena. R. Plonsey and D.J. Fleming (1969). McGraw-Hill Book Company, New York, USA.

Chapter 14

1. Molecular Biology of the Gene. J.D. Watson, N.H. Hoplans, J.W. Roberts, J.A. Steitz and A.M. Weines (1987). The Benjamin Cummings Publishing Company, California, USA.

2. Seven Clues to the Origin of Life. A.G. Cairns-Smith (1985). Cambridge University Press, Cambridge, UK.

3. Exobiology. C. Ponnamperuma (ed.) (1982). North-Holland Publishing Company, Amsterdam, Holland.

4. The Origins of Life on Earth. S.L. Miller and L.E. Orgel (1974). Prentice-Hall Inc. New Jersey, USA.